Contributions to Management Science

The series *Contributions to Management Science* contains research publications in all fields of business and management science. These publications are primarily monographs and multiple author works containing new research results, and also feature selected conference-based publications are also considered. The focus of the series lies in presenting the development of latest theoretical and empirical research across different viewpoints.

This book series is indexed in Scopus.

Kushal Anjaria

Sustainable Enterprise Resource Planning (S-ERP) for Industry 4.0

A Secure and Ethical Deployment Approach

Springer

Kushal Anjaria
"Tribhuvan" Sahkari University
Anand, Gujarat, India

ISSN 1431-1941 ISSN 2197-716X (electronic)
Contributions to Management Science
ISBN 978-981-96-7733-7 ISBN 978-981-96-7734-4 (eBook)
https://doi.org/10.1007/978-981-96-7734-4

This Springer imprint is published by the registered company Springer Nature Singapore Pte Ltd.
The registered company address is: 152 Beach Road, #21-01/04 Gateway East, Singapore 189721, Singapore

To my daughter, Meshva.

Preface

This book provides a comprehensive exploration of Sustainable Enterprise Resource Planning (S-ERP) within the framework of Industry 4.0. It delves into the integration of sustainability principles into traditional Enterprise Resource Planning systems, addressing the challenges and opportunities posed by the fourth industrial revolution. The scope extends to examining the role of advanced technologies such as IoT, AI, and big data analytics in enhancing business operations, ensuring environmental stewardship, and promoting social responsibility. The book also covers the implementation strategies, ethical considerations, and security aspects associated with S-ERP, offering practical insights for organizations seeking to navigate the complexities of digital transformation while maintaining a commitment to sustainability.

The primary purpose of this book is to serve as a guide for organizations and professionals looking to implement Sustainable Enterprise Resource Planning (S-ERP) systems within the context of Industry 4.0. It aims to bridge the gap between traditional ERP systems and the need for sustainability in modern business practices, providing readers with the knowledge and tools necessary to integrate environmental, social, and economic considerations into their operations. The book also seeks to address the ethical and security challenges that arise with the adoption of advanced technologies, ensuring that organizations can achieve long-term resilience and competitiveness while adhering to sustainability principles.

The intended audience for this book includes business leaders, IT professionals, and sustainability practitioners who are involved in the digital transformation of their organizations. It is also targeted at academics and researchers interested in the intersection of technology and sustainability, particularly in the context of Industry 4.0. Additionally, the book is suitable for policymakers and consultants who are looking to understand the implications of S-ERP systems and their role in promoting sustainable development in various industries. Through its detailed exploration of S-ERP implementation and challenges, the book aims to equip readers with the insights needed to drive sustainable innovation in their respective fields.

I would like to express my deepest gratitude to Dr. K. V. Yadav for his invaluable input and discussions throughout the development of this book. His expertise and insights have significantly enriched the content, and his support has been instrumental in shaping the direction. I am truly grateful for his guidance and the time he dedicated to helping me bring this book to fruition.

I would like to express my sincere gratitude to the Institute of Rural Management Anand (IRMA) for their unwavering support and encouragement throughout the creation of this book. The resources, intellectual environment, and community at IRMA have been invaluable in enabling me to conduct the research and develop the ideas presented here. I am deeply appreciative of the institute's commitment to fostering academic excellence and innovation.

August 2024

Kushal Anjaria
Institute of Rural Management Anand
(IRMA)
Anand, Gujarat, India

Contents

Chapter 1
Introduction to Sustainable Enterprise Resource Planning (S-ERP) in Industry 4.0

1.1 Introduction

Enterprise Resource Planning (ERP) software is an all-encompassing information management system that unifies all aspects of a business's operations onto a single, integrated platform. This robust software has been designed to automate a broad spectrum of business processes, ultimately aiming to enhance the organization's overall operational efficiency. The versatility of ERP systems is reflected in their widespread adoption across various industries. From manufacturing, retail, and distribution to service industries, organizations in these fields have found immense value in employing ERP systems to manage their various processes seamlessly. At its core, an ERP system is built to automate many business processes, thereby driving improvements in operational efficiency. Nevertheless, its benefits extend beyond just automation. One of its key advantages is its capacity to provide real-time data, acting as a powerful tool that can be leveraged for informed decision-making. This data-driven insight can be a game-changer for businesses, helping them navigate their path forward with clarity and confidence.

A significant feature of ERP systems is their ability to serve as a centralized database [3]. They consolidate all essential organizational information in one cohesive and comprehensive location. This centralization of data offers companies easy and immediate access to crucial data. With this data at their fingertips, businesses can make well-informed decisions, optimizing their operations. Furthermore, ERP systems play a critical role in streamlining operations by obliterating the need for laborious manual tasks traditionally associated with business management [23]. This digitization of processes represents a significant shift toward greater efficiency, leading to notable time savings and increased productivity. As such, the digital transformation enabled by ERP systems can be instrumental in propelling businesses toward achieving their strategic goals while bolstering their competitive standing in the market.

K. Anjaria, *Sustainable Enterprise Resource Planning (S-ERP) for Industry 4.0*,
Contributions to Management Science, https://doi.org/10.1007/978-981-96-7734-4_1

Several ERP systems are available on the market, each tailored to cater to the unique needs of different organizations. Oracle, for instance, provides various enterprise-grade software products specifically designed to accommodate businesses of different sizes and with varying budgetary constraints. Other key players like SAP and Microsoft Dynamics also offer many solutions [13], each catering to the nuanced requirements of various industries. Regardless of the nature of the business, rest assured that an ERP system is designed to meet one's specific needs. Selecting the right ERP system can be an involved process, but by using a systematic approach and keeping specific critical points in mind, one can choose the system best suited for your business. Here are some vital factors to consider when selecting an ERP system for your organization:

1. Determine which features one needs: The available features will vary from system to system, so choose one that meets your needs. Some systems may suit larger enterprises, while others will be better suited to smaller businesses.
2. Choose a solution that can grow with your business: There is no point in investing much money in an ERP system if one is unsure, it will handle the needs in the long run. Ideally, a scalable solution to accommodate expanding business would be best.
3. Choose a user-friendly and intuitive system: An ERP system can be difficult and time-consuming. Fortunately, there are some options available that are designed to make the process easier and less confusing. Look for a software package that is simple to use and easy to navigate so your employees can get up to speed quickly.

The judicious selection of an appropriate ERP system can confer numerous benefits upon a company. For instance, ERP systems can enhance the coordination of a company's operations across its supply chain. This improved coordination can lead to waste reduction and enhance the level of customer service provided. Another advantage of ERP systems is the improved customer service they facilitate. The system can create a unified view of each customer, enabling companies to tailor their services to meet the unique needs of every individual client. ERP systems can also support precise planning. They enable companies to forecast future sales more accurately, aiding in managing cash flow and preventing product shortages. In the realm of supply chain management, ERP systems can considerably increase efficiency. This not only reduces costs but also bolsters customer service by ensuring timely delivery of products. The convenience offered by ERP systems is another notable benefit. Since they can be accessed from anywhere, these systems make it easier for companies to manage their operations irrespective of geographical location. Improved Business Intelligence is yet another advantage conferred by ERP systems. They provide companies with easy access to critical business intelligence data, enhancing decision-making capabilities in areas like supply chain management and customer service. Lastly, the strategic use of ERP systems can enhance a company's competitiveness in the market. By improving operations and providing more straightforward access to management systems, ERP systems enable companies to stay a step ahead in the competitive business landscape.

The present issue with ERP is that its benefits do not lead to sustainable business processes and development [15]. World Commission on Environment and Development of the United Nations Brundtland defined sustainable development as "development that meets the needs of the present without compromising the ability of future generations to meet their own needs." Sustainable development is a continuous improvement and innovation process in which the present generation ensures that future generations have the same opportunities for sustainable development. The definition of sustainable development is related to the triple bottom line (TBL) [24]. Corporations, governments, and international organizations such as the United Nations have widely adopted the TBL approach to sustainability. The TBL approach is a valuable way of thinking about sustainable development because it highlights the importance of economic growth, environmental protection, and social equity.

Nowadays, consumers demand that companies do more to tackle sustainability issues [22]. The trend toward sustainability is evident in most industries. For example, the consumer demand for healthier, more sustainable products has increased sales of organic foods and other environmentally friendly products. External resources, such as carbon dioxide emissions and water consumption, are considered essential to a firm's performance by investors, who expect companies to share information about them. Investors want to know how the business's operations may impact its long-term value. Investors also want to understand how companies are managing these risks and how they will respond if there are any changes in trends or regulations. Companies have responded to this trend by integrating sustainability into their business strategies and operations. From a financial perspective, companies that adopt sustainable practices can reduce costs, improve brand image, and safeguard against environmental risks. Most managers believe a sustainability strategy is essential for their company's success. However, most companies struggle to define sustainability in a way that is relevant to their businesses.

In the present work, we argue that sustainability requires sustainable business practices. For doing so, specially designed ERP systems are essential in the organization's transition toward sustainability initiatives. S-ERP allows organizations to improve productivity, reduce costs, and increase profitability with sustainable business practices. S-ERP can be a possible solution to solve sustainability issues in organizations. The system provides the organization with information about its resources, including energy and water use, waste management, pollution, and carbon emissions. The system also includes information on how to improve the sustainability of these resources by providing recommendations for more sustainable practices. Figure 1.1 shows the concept of S-ERP.

The forthcoming section delves into S-ERP and its pivotal role in driving sustainability initiatives within modern organizations. We lay out an in-depth discussion that elucidates the significance of S-ERP systems as powerful tools in orchestrating and managing sustainable business operations. In the same vein, we explore some of the salient challenges in this domain. These obstacles, ranging from integrating sustainable practices into business models to deploying advanced Industry 4.0 technologies in ERP systems, can often stifle the potential benefits of S-ERP systems. By shedding light on these hurdles, we aim to provide a more rounded perspective on

Fig. 1.1 The concept of S-ERP

implementing S-ERP systems. Moving forward, we will revisit traditional ERP and sustainability research. Drawing on a wealth of academic and industry experience, we aim to glean valuable insights and learnings from these well-established fields. We intend to harness these insights to inform and enhance the design and operation of S-ERP systems.

In synthesizing the knowledge from these research areas, we provide a comprehensive approach to S-ERP systems. Our goal is to equip readers with a deeper understanding and practical guidelines that will enable them to navigate the complexities of sustainable ERP systems effectively in the context of Industry 4.0.

1.2 Basics of the S-ERP System and Its Related Challenges

Sustainability in a firm requires an integrated approach that encompasses not just the product and its manufacture but also the entire supply chain. It means that the firm has to take responsibility for its impact on the environment, not just its operations but those of all suppliers and sub-contractors. In addition, a sustainability strategy should ensure that economic efficiency and social justice considerations are balanced with the need to protect the natural environment. Although the scale and scope of the challenges associated with achieving sustainable development are significant, there is considerable evidence to suggest that organizations may reduce the impact on the environment by controlling the use of resources more effectively and increasing energy efficiency [26]. It also means that firms must be open to feedback from their stakeholders and respond accordingly. The firms perform sustainability modeling and implement specific environmental or social management systems without integrating these efforts with the firm's core business processes. The specialized ERP system should be designed to capture sustainability-related data and support the management of all the dimensions associated with sustainability.

Despite the benefits of an Enterprise Resource Planning (ERP) system, numerous challenges can arise when companies attempt to leverage a traditional ERP system to support their sustainability goals. These challenges can generally be segmented into three broad categories: The first pertains to the scarcity of accurate and reliable data on sustainability. Many companies grapple with collecting and preserving precise, dependable, and consistent data. The quality of this data is often compromised due to an industry-wide lack of standardization, varying interpretations of sustainability, and a limited understanding of how to collect the correct data. The second category is the absence of a fully integrated approach toward sustainable business functions and processes. Implementing a Sustainability-oriented ERP (S-ERP) entails integrating multiple systems across different organizational departments. Without a consolidated view, it becomes problematic for users to comprehend and utilize sustainability-related data in their decision-making processes.

The third challenge is the difficulty associated with integrating sustainability-related data with third-party tools. Many companies find it arduous to merge their data with external data sources. These sources could include field-level data from sensors or weather data feeds from meteorological forecasting agencies. Moreover, they might struggle when combining internal data with external sources. For instance, using carbon footprint data from an environmental service provider or a third-party logistics provider may not always yield optimal results due to disparities in their data collection methodologies or the specific data required for computations.

The S-ERP system is designed and proposed to solve the above challenges and integrate sustainability. It captures sustainability-related data and supports the management of all the dimensions associated with sustainability. A sustainability-focused S-ERP is a good foundation for a company's sustainability management system [4]. S-ERP manages and optimizes an organization's resources to minimize adverse environmental and societal impacts. The organization can undertake this activity by minimizing waste, reducing emissions, and conserving natural resources to help organizations manage their resources sustainably. S-ERPs also help organizations plan for future needs by anticipating resource requirements and predicting future environmental impacts. S-ERP is a "holistic, systematic approach that integrates social, financial, and environmental considerations into decision-making [11]." The S-ERP system is capable of implementing core sustainability elements. The core sustainability elements are as follows:

1. **Social sustainability:** S-ERP incorporates social equity and environmental responsibility into business decisions and processes using social data to inform decision-making. S-ERP is designed to help companies achieve long-term social sustainability by addressing risks associated with volatile economic conditions. S-ERP is a system for managing the triple bottom line of people, planet, and profit.
2. **Financial sustainability:** S-ERP uses financial performance metrics to guide decision-making. S-ERP is designed to help companies achieve long-term financial stability by addressing risks associated with volatile economic conditions.
3. **Environmental sustainability:** S-ERP uses ecological data to inform decision-making. S-ERP is a system for managing environmental risks and opportunities to create value for society and the business enterprise.
4. **Institutional sustainability:** S-ERP uses institutional data to inform decision-making. S-ERP is a system for managing institutional risks and opportunities to create value for society and the business enterprise.

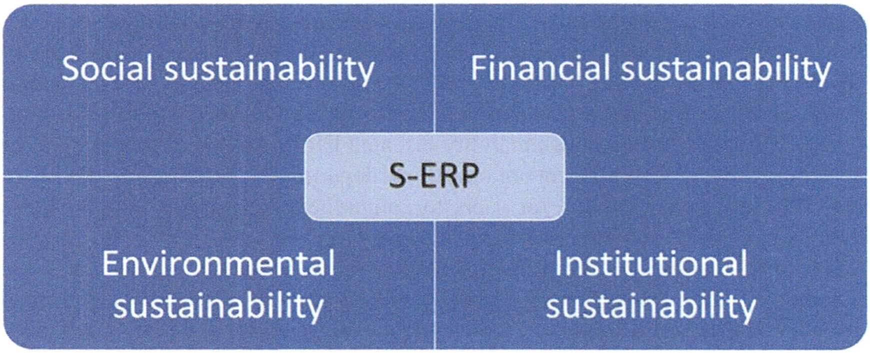

Fig. 1.2 The concept of S-ERP

Figure 1.2 shows the core sustainability elements handled by S-ERP. The primary consideration in an S-ERP is developing an environmental impact statement. An impact statement is a document that describes the potential environmental impacts of a new product or service produced by an organization. The second step of an S-ERP is implementing the plan. One can define the S-ERP system as "an information system driven by sustainability considerations that cover all aspects of the value chain." S-ERP systems would be holistic, integrative, and comprehensive rather than fragmented and individualized. Several vendors, such as SAP, Oracle, and Microsoft, have started their efforts to design S-ERP software packages to solve the segregation problem. However, the issue is that one must develop a robust S-ERP system that can cater to the dynamic manufacturing environment [16]. Presently, novel and evolving manufacturing environments that challenge conventional S-ERP are digital twins and Industry 4.0. Three key areas need attention to build such a responsive S-ERP system. First is the customization of software in light of changing requirements; second is developing software to respond efficiently and effectively; third is the creation of training strategies so people can learn how to use it. In the present work, we focus on S-ERP for Industry 4.0. In the following section, we study the fundamentals of Industry 4.0.

1.3 Fundamental of Industry 4.0: The Fourth Industrial Revolution

Industry 4.0 is the next industrial revolution, and it is already happening! The first phase of this revolution was the digital age, which involved technological changes that led to new ways for people to communicate and share information. The second phase was the information age, which led to using computers to store and access data.

The third phase was the age of automation; the development of automated machinery characterized it to perform physical tasks faster and more efficiently than humans could. The fourth phase is the age of cyber-physical systems. It will combine the power of digital information systems with physical devices to create more intelligent and connected systems and products. Industry 4.0 has the potential to be as revolutionary as the first three industrial revolutions because it will transform how we make and use products and services in virtually every industry.

Many technologies at the heart of this revolution are available today or will be available in the next few years. For example, 3D printers are making it possible to manufacture products on demand instead of producing them in batches and storing them in warehouses until customers can purchase them. The Internet of Things (IoT) connects devices throughout the supply chain and between industries to communicate with each other and improve the efficiency of business processes. Advanced analytics technologies are helping businesses make sense of all this new data to identify patterns, trends, and anomalies in real time. The result is a new era of innovation in which companies can use technology to solve problems, enhance customer experiences, and increase profits. The fourth revolution is the future that many companies are looking forward to, but it can only come about if they take the necessary steps now.

Let us discuss some of the advantages that Industry 4.0 brings to manufacturing by providing greater agility without sacrificing quality, speed, or cost [20]. Industry 4.0 connects machines, devices, and people to create a seamless data flow and processes. It also focuses on automation and using artificial intelligence and machine learning to maximize efficiency and productivity while decreasing costs. With Industry 4.0, companies can monitor their manufacturing processes in real time to detect defects before they become a problem and make immediate adjustments. They can simultaneously reduce their overall waste and maintenance costs and improve workforce safety, facilitating smarter, faster, and better decisions, leading to increased profitability and greater customer loyalty.

Furthermore, with Industry 4.0, companies can automate their operations, so they do not need as much human intervention, which enables them to free up skilled personnel to focus on more important things. As a result, they can better respond to customer requests and create personalized products to meet their specific needs. The flexibility and scalability that Industry 4.0 offers allow manufacturers to respond quickly to changing market demands by shortening the time between development and production [7]. The more efficient production process also allows companies to respond to changing market demands, which can help them stay ahead of their competitors. For example, suppose a customer requests a new product or feature previously unavailable. In that case, manufacturers can use Industry 4.0 technology to build it quickly and efficiently rather than wait for a team of designers and engineers to develop the solution.

In a nutshell, we list the prominent advantages of Industry 4.0 as follows:

1. Improved efficiency, productivity, and quality of products
2. Better knowledge management through data collection and analysis
3. Better decision-making with the help of real-time data
4. Reduction in costs and time for production
5. Enhanced product innovation
6. Improved supply chain management
7. Increased customer satisfaction
8. Improved employee morale through enhanced job satisfaction
9. Reduced training costs
10. Increased customer loyalty
11. Enhanced competitive edge
12. Improved risk management
13. Better employee retention
14. Enhanced data security
15. Improved regulatory compliance
16. Reduced operational costs
17. Reduced fraud risk.

It is evident from the above list that Industry 4.0 will generate a lot of real-time data. The ERP system or S-ERP system has to deal with many data. It must be able to store and process this data in real time. The S-ERP system has to connect with machines and devices that generate data, such as sensors and robots. It has to collect data from these devices through protocols such as MQTT, OPC UA, or Web Sockets and process and analyze the data in real time [21]. It has to send this data back to machines and devices so that they can respond accordingly. S-ERP system must allow humans to access real-time data as well. Access to real-time data means S-ERP should have an exemplary user interface with dashboards, reports, and alerts. The S-ERP system must have a high-performance and scalable architecture to handle this data in real time. The S-ERP system must also be able to scale horizontally by adding additional nodes with just a few clicks. The S-ERP system has an integral role in Industry 4.0 since it enables visibility into the entire manufacturing process across all business units by providing data at the point of production. The S-ERP system is also responsible for integrating the line of the segment with custom functionality and data visualization.

The issue with the present literature is that most authors have addressed Industry 4.0 and ERP from a technological perspective, overlooking sustainability challenges regarding this recent paradigm. Primarily, Industry 4.0 is a movement that encourages companies to use technology to improve efficiency and speed up processes. The global industry is estimated to be worth $11 trillion by 2025 [14], so businesses must stay on top of emerging technology and innovations to remain competitive. However, many companies fail to focus enough on sustainability and eco-friendliness during their digital transformation process. Implementing sustainable practices throughout your business will save you money, make your customers happier, and increase your brand awareness, among other important factors. One of the key themes in this research is how companies must be thoughtful about integrating sustainable practices into their digital transformation process.

The book's main objectives are to provide a sustainability perspective on Industry 4.0 and discuss the implications for an S-ERP system. We believe that one can achieve these objectives with the S-ERP system. The S-ERP system can bridge sustainability and Industry 4.0. S-ERP in the Industry 4.0 context contributes to circular economic goals by achieving social, economic, and environmental benefits as S-ERP's triple bottom line theory focuses on Industry 4.0 adoption and implementation, sustainable supply chains, innovative and sustainable cities, and smart factories. Collectively, S-ERP and Industry 4.0 ensure ecosystems remain healthy. The S-ERP system uses lifecycle thinking to identify and prioritize resources throughout the product lifecycle. It is implemented using Industry 4.0 to identify opportunities for sustainability improvements across value chains. The following section discusses S-ERP's relationship and implementation challenges in the Industry 4.0 setup.

1.4 S-ERP in Industry 4.0: Relationship and Implementation Challenges

The fourth industrial revolution is upon us, and business leaders are starting to understand the potential implications. However, before businesses can reap the rewards of this new era, they must first navigate its sustainability-related challenges. To do so, they will need to consider several key issues. For example, how does technology affect environmental sustainability? What impact will this have on society? Moreover, how can businesses ensure they play their part in reducing pollution and waste? How will advances in artificial intelligence (AI), robotics, the Internet of Things (IoT), and other technologies transform our lives over the next decade? The implementation of S-ERP in Industry 4.0 answers the questions mentioned above. There is a strong bond between sustainability, enterprise resource planning, and Industry 4.0. The three concepts are closely related and aim to create a more sustainable future. S-ERP can achieve a circular economy and net zero goals in Industry 4.0.

The present book aims to provide steps to implement S-ERP in Industry 4.0 and solve the challenges around implementation. Understanding the challenges will allow for more successful deployments. Implementing the S-ERP will go beyond automating the manufacturing process; it will also address and overcome the challenges that arise during the implementation process. An S-ERP system implementation typically involves changing business processes and investment in hardware and software infrastructure. The S-ERP will have to align seamlessly with your existing business processes and adapt and change to suit your business requirements as it grows. S-ERP is an organization's information backbone. According to Gartner, the ERP implementation failure rate was 65% in 2019, exceeding 75% if businesses make business processes lengthy by including sustainability-related aspects [19]. Meanwhile, global consultancy McKinsey estimated in 2018 that more than 70% of all digital transformations fail. In such a scenario, the present book will be a helpful guide for practitioners.

The present book explains the S-ERP implementation strategy, business process modeling in S-ERP, change management, and open innovation with S-ERP, S-ERP Road-mapping, ethical aspects of S-ERP in the context of Industry 4.0, and security aspects of S-ERP, Digital transformation, circular economy, and technology integration. The following subsection discusses the S-ERP implementation strategy for organizations adapting to the fourth industrial revolution.

1.4.1 Foundation of S-ERP in Industry 4.0

With Industry 4.0 becoming a reality, many businesses realize the benefits of implementing an S-ERP solution to support their digital transformation journey. However, organizations must consider a few challenges before an S-ERP implementation starts, as with any complex IT project. Onboarding a new S-ERP system can be a complicated process for any organization. Implementing a new S-ERP system can be overwhelming for everyone involved. From managers to employees, everyone will need to understand how the S-ERP system will affect their roles within the organization, especially when organizations re-evaluate each business process for sustainability consideration. We recommend that organizations start employee training efforts early to help mitigate some of the confusion surrounding the S-ERP implementation project. For employee training, strategy building, and change management activities, it is crucial to break down all of the features of the S-ERP into individual lessons. This break-up will help organizations equip employees with the knowledge to perform their jobs effectively once the S-ERP system goes live. The organization must also take the time to explain each system module to the employees and how it will benefit their business. It is also essential to provide employees with a way to learn about the new S-ERP system by giving them access to training videos, instructional guides, and other educational tools. The organization should consider hiring outside consultants, training specialists, and sustainability experts to help develop these materials.

The next step is to decide on specific modules to implement first. The organization should train employees on those modules which they are implementing first. This module selection activity will allow employees to become familiar with new features and applications while still in a controlled environment. The selection of modules and sequential implementation requires decision-making and a strategic framework. The third step is to analyze the current state of your organization. This analysis will help you determine how well-suited your company is for an S-ERP system and identify what changes must be made before implementation.

This analysis and evaluation should include an assessment of sustainable business processes, information technology infrastructure, and human resources [6]. The organization must consider data format, existing systems, and business processes in this step. The fourth step is to develop an implementation blueprint. The blueprint consists of creating a project plan with timelines and milestones and identifying which modules will be implemented first. The fourth step is to develop a business case for implementing S-ERP. This business case document will help you understand the benefits and costs of implementing an S-ERP system and provide the information needed to justify the project in your organization. The fifth step is to select a vendor and perform due diligence on their product offering. This process can be lengthy, so it is vital to research possible vendors before choosing one.

The present work discusses how practitioners can follow these five steps around S-ERP implementation in the organization that follows Industry 4.0. Chapter 2 elaborates on the five steps mentioned above. Chapter 2 guides the organization on how businesses can design the strategy for S-ERP implementation. It will discuss integrating sustainability principles and metrics into the S-ERP system. We also discuss the business process reengineering for sustainability that can build robust business processes in S-ERP. Building on the strategic foundations outlined in Chap. 2, we must now focus on the dynamic aspects of the S-ERP lifecycle within Industry 4.0 environments. Change management and open innovation are pivotal to ensuring that S-ERP systems meet current organizational needs and adapt to future challenges and opportunities. In the forthcoming discussion, we delve into how change management practices can be applied to facilitate a smooth transition to S-ERP systems, aligning people, processes, and technology with the evolving demands of Industry 4.0. Moreover, we will explore the role of open innovation in enhancing the capabilities of S-ERP systems, making them more agile and responsive to the fast-paced changes characteristic of the fourth industrial revolution. This conversation is crucial for practitioners who seek to leverage S-ERP systems as a platform for continuous improvement and competitive advantage. The following subsection explains change management and open innovation with S-ERP in organizations adapting to the fourth industrial revolution.

1.4.2 Change Management and Open Innovation with S-ERP in Industry 4.0

Change management refers to planning, implementing, and controlling changes to an organization's products, processes, or systems. In the context of Industry 4.0, change management is often used to adopt and integrate new technologies, such as S-ERP systems, that enable organizations to operate more efficiently and effectively [1]. Open innovation refers to actively seeking and incorporating ideas and inputs from external sources, such as customers, partners, and experts, intending to drive innovation and improve the organization's performance. S-ERP systems can support open innovation by providing tools and platforms for collaboration, communication, and idea generation that facilitate the exchange of information and ideas between different stakeholders [2]. In addition, S-ERP systems can provide the technology and tools needed to facilitate data sharing and collaboration among stakeholders, which is essential for open innovation initiatives. However, organizations must understand their objectives and goals to maximize these tools. They must also develop an appropriate strategy for implementing S-ERP systems and monitor their progress toward those goals over time.

In the quest to effectively implement change management and embrace open innovation with Sustainability-oriented ERP (S-ERP) in the context of Industry 4.0, organizations are advised to heed several best practices: Clear identification of goals and objectives is the first crucial step. Gaining a comprehensive understanding of what the organization aims to accomplish through adopting S-ERP and open innovation paves the way for guiding the planning and implementation process, ensuring that the organization remains aligned with its strategic path.

Involving key stakeholders in the planning process is another pivotal practice. Organizations can foster buy-in by including essential players such as employees, customers, and partners in the planning stage and ensure that the S-ERP system caters to the requirements of all relevant parties. Effective communication is an indispensable component of successful change management and open innovation. Keeping all stakeholders abreast of the impending changes and the advantages they offer is imperative to facilitate a smooth transition. Providing training and support to employees and other stakeholders is equally essential. Such support can help equip them with the skills and knowledge necessary to effectively utilize and benefit from the new S-ERP system. Finally, organizations should actively monitor the progress of the S-ERP implementation and open innovation initiatives, making necessary adjustments along the way. Regular check-ins ensure the organization remains on track to achieve its goals and objectives.

Further, the S-ERP system can promote open innovation in several ways. It provides a single, unified platform for all business operations, which makes it easier for employees to access and share information and ideas across departments and teams. S-ERP systems often have built-in collaboration and communication tools, such as forums, chatrooms, and document sharing, which can facilitate open innovation by making it easier for employees to connect and share ideas. The system can also help organizations capture and track ideas and innovations, making it easier to manage and measure the success of open innovation initiatives. S-ERP systems even have built-in innovation management modules, which provide a structured process for capturing, evaluating, and implementing new ideas and innovations. However, some issues persist with the organizations that follow Industry 4.0. These issues are related to managing Intellectual Property (IP), ensuring data security, identifying the right external partners to collaborate with, managing cultural differences, ensuring alignment with business goals, and managing complexity.

The present work discusses how practitioners can solve the challenges around S-ERP open innovation and implement best practices using S-ERP in organizations that follow Industry 4.0. The third chapter will discuss change management and open innovation using an S-ERP for Industry 4.0. Building upon the foundation laid by embracing change management and nurturing open innovation, the fourth chapter will introduce the concept of S-ERP roadmapping as a strategic tool. Roadmapping is a critical process that provides a structured visual outline detailing the trajectory of S-ERP initiatives from conception to full-scale implementation. This chapter will guide practitioners through the intricacies of S-ERP roadmapping, highlighting how to align S-ERP initiatives with long-term business objectives and Industry 4.0's evolving technological landscape. We will dissect the phases of roadmapping, from the preliminary assessment of current systems to the deployment of S-ERP solutions that are sustainable, scalable, and synergistic with Industry 4.0 principles. The discussion will underscore the importance of a well-articulated road map in navigating the complexities of digital transformation, ensuring that S-ERP systems are implemented and continuously refined to drive innovation and value creation. The following subsection discusses S-ERP roadmapping in the context of Industry 4.0.

1.4.3 S-ERP Roadmapping in the Context of Industry 4.0

An S-ERP roadmap should focus on helping organizations leverage the technologies and practices of Industry 4.0 to improve business operations' sustainability, environmental aspects, and competitiveness [10]. It must integrate Industry 4.0 technologies—such as the Internet of Things (IoT), artificial intelligence (AI), and advanced analytics into the S-ERP system to improve sustainability, decision-making, automation, triple bottom line, and efficiency [9]. The business must also consider adopting agile, flexible, sustainable, and scalable S-ERP roadmapping that can adapt to the rapidly changing needs and requirements of Industry 4.0. The S-ERP roadmapping should implement sustainability-related data governance and security measures to ensure safe and secure data handling in Industry 4.0. In a nutshell, developing a clear, strategic plan for how the S-ERP system will support the organization's long-term goals and objectives along with sustainability aspects in Industry 4.0 should be the ultimate aim of the S-ERP read mapping. Beyond these technological integrations, a successful S-ERP roadmapping requires a well-articulated Information Systems Strategy and Sustainability Strategy. These strategies are crucial to ensuring that the implementation of S-ERP systems aligns with the broader organizational goals of digital transformation while promoting sustainable business practices.

The fourth chapter of the present book aims to help organizations leverage the technologies and practices of Industry 4.0 to transform their business operations and achieve a competitive advantage in their market using S-ERP roadmapping. There are several challenges that companies may face when it comes to developing a roadmap for implementing an S-ERP in the context of Industry 4.0, which one can characterize by the use of advanced technologies such as artificial intelligence (AI), the Internet of Things (IoT), and automation. The first challenge of roadmapping is integration with the existing systems. S-ERP systems often must be integrated with various other systems and technologies within an organization. This integration can be a complex process that requires significant time and resources to get right. The positive side, however, is that once the integration is complete, an S-ERP system will provide a single source of truth for all accounting and business management data. The integration improves efficiency across the board by providing a single place for users to look for information about sales, costs, and revenue with the triple bottom line.

Data management is the second challenge that needs to be addressed in S-ERP roadmapping. In Industry 4.0, companies generate and collect vast amounts of data from various sources. Ensuring this data is appropriately managed and integrated into the S-ERP system to manage sustainability aspects can be challenging. In addition to the data collected by IoT devices, companies must also manage environmental impact data related to their supply chain. This data can include information on greenhouse gas emissions, waste generated, water usage, etc. For example, a company may be able to monitor the amount of electricity its facilities use in real time. However, suppose it does not know how much energy is used by its suppliers (or their suppliers), manufacturers or even customers using its products. In that case, it cannot truly understand its impact on sustainability.

The third challenge one should address in S-ERP roadmapping is building and maintaining an S-ERP system. Building and maintaining an S-ERP system requires specialized skills and expertise, which may be in short supply. It is not uncommon for organizations to hire consultants or IT professionals to help them build and maintain their S-ERP systems. If you plan to develop an S-ERP system, it is essential to remember that this may not be a task that one person alone can accomplish. The present work discusses how practitioners can develop S-ERP roadmapping in organizations that follow Industry 4.0. The fourth chapter plans to discuss S-ERP roadmapping for Industry 4.0 in detail.

The present work guides practitioners in developing S-ERP roadmapping for organizations that follow Industry 4.0 and emphasizes the intricacy of such a process. With the roadmap outlined in the fourth chapter, we will explain the steps, milestones, and checkpoints critical to S-ERP implementation. As we pave the way for a comprehensive understanding of S-ERP roadmapping, addressing the ethical and security implications of deploying such systems becomes imperative. The fifth chapter will, therefore, pivot to a crucial yet often underappreciated facet of S-ERP systems within Industry 4.0: the ethical use of data and the robustness of cybersecurity measures. We will explore the responsibilities of organizations to safeguard data privacy, ensure transparency in AI decision-making, and protect against cyber threats that could undermine the integrity of S-ERP systems. This discussion will present the challenges faced and the best practices and proactive strategies that must be in place to navigate the ethical and security landscape of S-ERP systems in the digital age of Industry 4.0. The following subsection discusses S-ERP's ethical and security aspects in the context of Industry 4.0.

1.4.4 *Ethical and Security Aspects of S-ERP in Industry 4.0*

We address organizations' ethical and social responsibilities in the context of Industry 4.0 with S-ERP. As Industry 4.0 is characterized by integrating advanced technologies, such as artificial intelligence and the Internet of Things, into manufacturing and other industries, it is essential to consider the ethical implications of these technologies.

One critical ethical issue in the context of Industry 4.0 is the potential for automation to lead to job displacement [8]. As machines and algorithms take on more tasks, there is a risk that some jobs will become obsolete, leading to economic disruption. It is essential for organizations to consider the social and ethical implications of automation and to work to minimize the negative impacts on workers. Consequently, a crucial ethical obligation is to ensure a just transition for the workforce, which involves providing displaced workers with income support and opportunities for retraining. With these aspects, one more critical ethical consideration is the potential for new types of work to emerge due to Industry 4.0 technologies. As individuals lose their jobs due to automation, they must find new ways to provide value to society and earn an income.

Another ethical issue related to Industry 4.0 is the potential for data misuse. Integrating advanced technologies generates vast amounts of data, which can be used to improve efficiency and decision-making. However, this data is vulnerable to misuse by those with access to it. The European Commission has recognized the need for safeguards against data breaches and is working on guidelines and regulations that will help protect citizens' privacy. It is vital to ensure that this data is collected, stored, and used ethically and to protect the privacy of individuals.

Algorithm ethics is becoming increasingly important as AI and machine learning become part of standard business operations in Industry 4.0 [25]. Whether or not a company has an ethical algorithm in S-ERP is becoming increasingly important. As the use of artificial intelligence and machine learning continues to increase, businesses must build ethical algorithms that meet regulatory standards and protect consumer privacy rights. There are many ethical considerations regarding algorithms, mainly as they are used in decision-making processes. The first ethical issue related to algorithm ethics in S-ERP is that algorithms can perpetuate and amplify biases in the data used to train them, leading to unfair treatment of certain groups. The decision-making processes of algorithms must be transparent so that one can identify and address any biases or errors. When algorithms are used to make crucial decisions, such as in hiring or lending, it is essential to consider who is responsible for the consequences of those decisions.

In S-ERP for Industry 4.0, businesses must consider ethical considerations related to digital information, like privacy, ownership, and freedom of expression [18]. The ownership of digital information can be complex and may not be clearly defined. This ownership of digital information can lead to issues such as plagiarism and unauthorized use of content. In addition, businesses should be aware of the ethical considerations of data privacy and security. The General Data Protection Regulation

(GDPR) states that personal data can only be used for the purpose for which it was collected. Digital information can be collected, shared, and used in ways that impact an individual's privacy. It is essential to consider who has access to personal information and how it is used. S-ERP provides a platform for individuals to express their thoughts and ideas, but it is vital to consider the impact this expression may have on others.

Overall, S-ERP in the context of Industry 4.0 involves considering the ethical implications of advanced technologies and minimizing any negative impacts on society and individuals. These steps may include developing ethical guidelines for using these technologies, working to reduce job displacement, and protecting individuals' privacy and data. Most businesses of different sizes widely use S-ERP to manage their operations in Industry 4.0 setups. However, in Industry 4.0, the proliferation of mobile devices, web applications, cloud computing services, the Internet of Things (IoT), and other technologies have fundamentally changed how organizations manage their data. The introduction of new technologies in Industry 4.0 has raised security concerns that businesses must address to ensure the integrity of an organization's data and the security of its information systems. Companies will use S-ERP with other applications and technologies in this environment. For example, most business processes in Industry 4.0 use mobile devices to access data and perform tasks related to their operations. Businesses must secure confidential information on these mobile devices by implementing a Secure Mobile Application Management (SAM) solution [17] that provides strong encryption capabilities.

S-ERP systems often contain sensitive data, such as financial information and personal data. Implementing strong access controls and authentication measures to prevent unauthorized users from accessing this information is essential [5]. Employees may have access to sensitive data and systems, and they may accidentally or intentionally compromise the system's security. It is necessary to have strict controls in place to prevent unauthorized access and monitor suspicious activity. S-ERP systems regularly need security patches and updates to reduce the risk of exploited vulnerabilities. The security of the S-ERP system is critical to its success. Configuring and managing S-ERP systems securely to protect your business from cyberattacks is vital. A firm security policy outlining the rules and regulations for protecting sensitive data and systems is crucial. The policy should also include procedures for reporting suspicious activity and enforcing access control measures.

Along with unauthorized access, data breaches, and insider threats, the other challenges in S-ERP of Industry 4.0 are unsecured IoT devices and lack of visibility. Unsecured IoT devices mean many do not have strong security measures, making them vulnerable to hacking and other cyberattacks. Lack of visibility indicates that with the increasing complexity of Industry 4.0 systems, it can be challenging to identify and track all connected devices and systems, making it harder to detect and respond to security threats [12]. To address the challenges of S-ERP in Industry 4.0, companies must ensure that their systems are secure and adhere to stringent regulations and standards. They must also leverage technologies that provide visibility into connected devices and systems. Doing so can increase the chances of identifying threats before they cause substantial damage and provide a timely response to cyberattacks.

To address the risk, it is essential to have a robust incident response plan in place to detect and respond to security threats quickly. An incident response plan is a set of procedures, rules, and guidelines to help you during a security incident. It should be updated regularly to ensure it is current with your organization's needs and any changes in your environment. The plan should include the following:

- Regular security assessments of your organization's systems and networks;
- A strategy for protecting against ransomware attacks, including backups (which can help you regain access to files) and regular updates (to prevent vulnerabilities in software from being exploited);
- A strategy for protecting against ransomware attacks, including backups (which can help you regain access to files) and regular updates (to prevent vulnerabilities in software from being exploited); and
- A strategy for containing the spread of ransomware by isolating infected devices.

The fifth chapter of the present book plans to discuss ethical and security aspects related to S-ERP for Industry 4.0 in detail. Having delved into S-ERP systems' ethical and security aspects, we recognize that these principles are crucial for compliance and protection and serve as the cornerstone for sustainable and responsible business practices within Industry 4.0. The integrity of data and the assurance of ethical operations are indispensable when it comes to transforming business models to align with the principles of the circular economy.

As we progress to the subsequent theme, we focus on incorporating the circular economy within the S-ERP framework. The circular economy represents a systemic shift to sustainability, emphasizing the reuse and recycling of materials and energy, which aligns seamlessly with the ethos of Industry 4.0. The sixth chapter will examine how S-ERP can orchestrate and optimize business functions to adhere to circular economy principles. It will discuss strategies for integrating circular practices across various business domains, such as procurement, production, and distribution, and how S-ERP can enable the tracking, analysis, and redesign of products and processes to minimize waste and maximize resource efficiency. This integration is not only a strategic imperative but also an ethical one, reflecting the evolving values of industry and society in pursuing a more sustainable future. The following subsection discusses the circular economy in S-ERP and its integration with business functions in Industry 4.0.

1.4.5 Circular Economy in S-ERP and Its Integration with Business Functions in Industry 4.0

We begin our discussion by diving deep into the integration of circular economy concepts in various business functions through the S-ERP system. Implementing a circular economy through S-ERP is a multifaceted approach that encompasses a wide array of strategies. At the forefront of these strategies is the focus on the entire product lifecycle, where S-ERP systems facilitate the design and development of products for prolonged use, ease of repair, and possibilities for recycling or remanufacturing. This is achieved by providing comprehensive data and insights into materials, design choices, and manufacturing processes. Accurate Bill of Materials (BOM) and efficient engineering change control within S-ERP systems are pivotal for ensuring material efficiency and waste reduction. These systems manage the hierarchy of resources, items, and parts of a product and can trace and report the percentage of recycled materials in the final product. Moreover, during the product development phase, they aid in the planning and integrating of pre-used parts or recycled materials.

Material Requirements Planning (MRP) and kitting functionalities within these systems further augment this approach by ensuring the just-in-time delivery of necessary items and quantities for production, thus minimizing stockpiling and material wastage. Additionally, S-ERP systems enhance inventory management by meticulously tracking expiry dates and the conditions required for temperature-sensitive stock, ensuring optimal use of items and preventing waste. The management of co-products and by-products through S-ERP systems also plays a crucial role, leading to less waste, better machine utilization, and enabling the reuse of materials in the production process. This not only aids in sustainable manufacturing but also opens avenues for generating additional revenue from these by-products.

Supply Chain Management (SCM) within S-ERP systems is another critical element in improving supplier collaboration. This includes efficient communication regarding changes in material requirements, like the inclusion of recycled materials, and the facilitation of automated requests for quotes that encourage suppliers to consider reused or recycled items. Furthermore, ERP systems streamline the processing of sales and purchase returns, aligning with the principles of the circular economy by supporting the recycling and repurposing of products. They also assist in disassembling returned products into their parts, ready for remanufacturing, thereby reducing manufacturing overheads and promoting recycling.

In the context of Industry 4.0, S-ERP systems are instrumental in achieving a circular economy, primarily due to their integration with advanced technologies and digital innovations. Industry 4.0 is distinguished by its digitization and interconnection of industrial processes, embracing technologies such as the IoT, big data analytics, AI, and cloud computing. These advancements significantly bolster the capabilities of S-ERP systems in promoting the principles of a circular economy. A key feature of S-ERP systems in this era is the enhanced management of a product's lifecycle. These systems can leverage IoT and AI to gather and analyze extensive data from the design phase to the product's end-of-life. This enables companies to create products with longer lifespans, ease of repair, and recyclability, aligning with the circular economy's objective to minimize waste and maximize product use. Integrating big data analytics into S-ERP systems further optimizes material and resource efficiency. AI algorithms are utilized to forecast material requirements more accurately, thus reducing waste and overproduction. Concurrently, IoT sensors provide real-time inventory management, ensuring efficient resource use and lessening the environmental impact. Regarding supply chain management, Industry 4.0 introduces enhanced transparency and efficiency through digital connectivity. S-ERP systems facilitate a seamless integration with suppliers, improving communication and collaboration. This digital integration supports more sustainable material sourcing practices, including recycled materials, a fundamental aspect of the circular economy.

IoT-enabled S-ERP systems also contribute significantly to waste reduction through predictive maintenance. By forecasting when equipment maintenance is required, these systems prevent breakdowns and reduce waste, ensuring efficient operation of machinery and supporting sustainable manufacturing practices. Another aspect where S-ERP systems are pivotal is facilitating Product-as-a-Service (PaaS) models. In the Industry 4.0 framework, these systems enable the transition to servitization, where products are offered as services. This encourages the production of durable, repairable, and recyclable products. Real-time tracking of product performance and maintenance needs by S-ERP systems enhances the feasibility and attractiveness of these PaaS models. The real-time data capabilities inherent in Industry 4.0 and the analytical prowess of S-ERP systems lead to improved decision-making. Companies are empowered to swiftly adapt to market changes, regulatory demands, and environmental considerations, aligning their decisions with the principles of the circular economy. Moreover, S-ERP systems in Industry 4.0 settings efficiently manage the processes of returns, disassembly, and recycling. Integration with reverse logistics enables companies to recycle and repurpose products and materials more effectively, thus reducing waste and promoting the circular economy's sustainability ethos.

In the sixth chapter of the present book, we have explored the integration of circular economy with various business functions using S-ERP. This chapter examines how S-ERP systems within Industry 4.0's emerging technologies, like IoT and AI, facilitate the shift from traditional linear business models to a circular economy focusing on sustainability through reuse, repair, and recycling. It discusses integrating these sustainable practices across essential business functions such as procurement and supply chain management, addressing the accompanying challenges, required cultural and process shifts, and the potential benefits for business efficiency and sustainability.

Having delved into integrating circular economy principles with S-ERP systems and their role in transforming traditional business models, the next critical step is understanding the management and utilization of the vast data generated in this process. The subsequent subsection, "Analytics for Harnessing IoT and Industry 4.0 Data Flows," shifts our focus to the pivotal role of analytics. It explores how analyzing the data from IoT and other Industry 4.0 technologies is essential for making informed decisions, optimizing processes, and fully leveraging the capabilities of S-ERP systems in a circular economy. This analysis is about collecting and transforming data into actionable insights that drive efficiency and sustainability in business practices.

1.4.6 Analytics for Harnessing IoT and Industry 4.0 Data Flows in S-ERP

This section delves into how data analytics and cloud technologies are pivotal in this integration. Industry 4.0 revolutionizes manufacturing via connectivity, AI, flexible automation, and cloud technologies. These advancements are about automating processes and enabling machines to provide actionable insights, driving business agility and outcomes. The benefits of adopting Industry 4.0 technologies, as supported by data analytics, are multifaceted. They include data-driven decision-making, improved productivity, faster time to market, enhanced agility and flexibility, better customer and employee experiences, reduced costs, and more significant revenues and profitability. These benefits underscore the transformative impact of Industry 4.0 on business operations. This transformative impact can only be harnessed if proper data analytics techniques are in place in business organizations in S-ERP.

Data analytics significantly improve the entire business cycle, from resource gathering to product delivery, particularly in dynamically adjusting plans in response to unexpected events like supply chain disruptions. This capability is essential for S-ERP systems, which must integrate various data analytic components to support these improvements. An example of this integration is a novel cross-layer data collection approach for active monitoring and control of manufacturing processes in Industry 4.0, which leverages sensor networks' multi-channel and multi-radio architecture. This approach can effectively handle the challenges of industrial environments, ensuring high throughput, low packet loss, and low latency, thus merging the capabilities of Cyber-Physical Systems (CPS) in Industry 4.0 with the system infrastructure in data analytics. In the book's seventh chapter, we explain data analytics to harness the advantages of Industry 4.0 with S-ERP. In the seventh chapter, we discuss the following data analytics technique that can be implemented in S-ERP designed for Industry 4.0:

1. **Real-Time Data Monitoring and Analysis:** Implementing a system for continuous monitoring and real-time data analysis from IoT devices and sensors embedded in manufacturing equipment. This would enable the early detection of potential failures or maintenance needs, optimizing operational uptime and reducing the risk of unplanned downtime.
2. **Predictive Analytics for Maintenance and Operations:** Utilizing predictive analytics algorithms to forecast future trends and potential issues based on historical and real-time data. This approach can predict machine failures, optimize maintenance schedules, and improve overall efficiency in manufacturing processes.
3. **Supply Chain Optimization:** Integrating real-time data analytics into S-ERP's supply chain management component. This would involve using data from intelligent sensors in packaging and shipping to gain accurate, real-time inventory views, thus enabling more agile responses to supply chain disruptions or changes in demand.

4. **Quality Control and Compliance Monitoring:** Deploy AI-driven analytics to provide quality control checkpoints across production lines. This method helps in the early identification of specification non-compliance, reducing waste and ensuring adherence to regulatory standards.
5. **Business Cycle Enhancement:** Extending data analytics beyond manufacturing processes to encompass the entire business cycle, including resource acquisition, component production, product assembly, delivery, and customer relationship management. This holistic approach allows for a more comprehensive optimization of business operations.
6. **Cloud-Based Data Management and Analytics:** Leveraging cloud technology for scalable data storage and advanced analytics. Cloud platforms can handle large volumes of data generated by Industry 4.0 technologies, providing the computational power needed for complex analytics and business intelligence applications.
7. **Cross-Layer Data Collection for CPS:** Implementing a cross-layer data collection approach in the CPS of Industry 4.0, which dynamically switches among different frequency bands and can handle industrial environmental challenges such as noise, signal fading, and electromagnetic interference. This ensures quality of service in terms of high throughput, low packet loss, and low latency.
8. **Customized Analytics for Specific Industry Needs:** Tailoring data analytic methods to meet different industry sectors' specific needs and challenges. This customization ensures the analytics approach aligns with each industry's unique requirements and objectives.

The book's seventh chapter will explore how S-ERP systems, enriched with IoT and Industry 4.0 data analytics, enhance operational efficiency and decision-making. We now turn to the broader context in which these systems operate. The next critical aspect is understanding the integral role of S-ERP in driving digital transformation across organizations. S-ERP, equipped with cutting-edge data analytics, streamlines specific processes and catalyzes comprehensive digital change. This change encompasses all aspects of an organization's operations, from internal processes to customer interactions, and aligns with the evolving digital business landscape.

In discussing the "Role of S-ERP in Digital Transformation" in the following subsection, we will explore how these systems support organizations in adapting to the rapidly changing digital environment. We will examine how S-ERP facilitates the integration of digital technologies into various business areas, transforming traditional business models into more agile, data-driven, and customer-focused entities. Additionally, the following subsection will explore the strategic impact of S-ERP on organizational structure, culture, and the overall value chain, highlighting its significance as a cornerstone in the digital transformation journey.

1.4.7 Role of S-ERP in Digital Transformation

S-ERP plays a pivotal role in digital transformation, particularly in harnessing the benefits of Industry 4.0 technologies. As we transition from discussing the integration of IoT and data analytics in S-ERP systems, it is essential to recognize how S-ERP serves as a foundational element in an organization's digital transformation journey. S-ERP systems play a crucial role in integrating technological advancements of Industry 4.0, acting as a vital link that connects new devices and tools, thus ensuring that organizations remain competitive. These systems are essential for those looking to fully harness the opportunities presented by Industry 4.0, with the risk of being outpaced by competitors who adopt these advancements more effectively. S-ERP systems also significantly enhance operational efficiency; they realize the full potential of Industry 4.0 technologies by providing a platform for tracking and analyzing the extensive data these technologies generate. This capability is critical not only for data protection and network security but also for offering greater process visibility and removing data silos, thereby enabling improved decision-making and efficiency.

Furthermore, modern S-ERP systems are adept at handling large volumes of data, which is crucial for process improvement and cost reduction. They transform this data into accessible dashboards and reports that lay the foundation for enhancing business processes. Integrating AI within these systems further augments their capability by identifying trends and problem areas, streamlining operations, and focusing on critical business aspects. Introducing new technologies also leads to a more decentralized management structure within organizations. S-ERP systems empower employees across all levels, providing them with the tools to gain a comprehensive understanding and control over their departments. This decentralization, balanced with the need for integration, is effectively managed by agile ERP systems, ensuring necessary visibility and connectivity. Additionally, in globalization, cloud computing supported by S-ERP systems enables organizations to monitor and analyze their supply chain comprehensively, optimizing the journey of assets from raw materials to finished products.

Digital transformation, at its core, is about leveraging technology to create more efficient, agile, and customer-centric business models. In this era, sustainability is not just an ethical imperative but also a business necessity. S-ERP systems are at the forefront of this transformation, providing the tools and capabilities necessary for businesses to evolve in an environmentally conscious and digitally advanced landscape. Moreover, the digital transformation facilitated by S-ERP systems extends to developing new, sustainable business models. For example, they can support circular economy models where the lifecycle of products is extended through recycling and repurposing, reducing the need for new resources, and minimizing waste. S-ERP systems help track the journey of products throughout their lifecycle, from production to end-of-life, ensuring that sustainability is embedded in every stage. The digital transformation using S-ERP is explored in the eighth chapter of the present book.

As we have seen, digital transformation facilitated by S-ERP systems is pivotal in creating sustainable, efficient, and customer-focused business models. These systems support the theoretical framework of a circular economy and bring about tangible changes in how businesses operate. However, S-ERP systems' real power and potential in digital transformation are best understood when we examine their integration with Industry 4.0 technologies in practical settings. In the upcoming subsection, "System Integration with Industry 4.0 Technologies: Use Cases," we will explore scenarios where S-ERP systems merge with advanced Industry 4.0 technologies. This exploration will demonstrate how S-ERP, combined with innovations like IoT, AI, and cloud computing, revolutionizes business processes and operations. We will delve into specific cases that illustrate the transformative impact of this integration on different industry sectors, showcasing how these technologies converge to optimize production, enhance supply chain management, and improve overall operational efficiency. This subsection aims to provide a concrete understanding of how S-ERP systems, empowered by Industry 4.0 technologies, are conceptual ideals and active drivers of modernization and efficiency in the business world. By examining these use cases, we will gain insights into the practical applications and benefits of system integration in real-world business environments, thus offering a clearer view of the future possibilities for organizations embracing this digital shift.

1.4.8 System Integration with Industry 4.0 Technologies in S-ERP: Use Cases

In Industry 4.0, particularly regarding S-ERP systems, the system integration process is intricate and multi-dimensional, involving various technologies, methodologies, and human dynamics. Incorporating Industry 4.0 technologies into S-ERP systems aims to improve S-ERP's effectiveness, adaptability, and intelligence capabilities by harnessing the potential of diverse technological fields. In this context, S-ERP under Industry 4.0 is characterized by the amalgamation of recent developments in sectors such as Cyber-Physical Systems, the Internet of Things and Everything (IoE), Cloud Computing, the Internet of Services, Big Data Analysis, Smart Factories, and Augmented Reality, among others. Industry 4.0 marks a novel phase in industrial evolution, posing significant integration challenges that involve various elements of the IoE, including people, data, services, and objects.

Chapter Nine of this book delves into the intricacies of integrating Industry 4.0 technologies within S-ERP systems. The chapter outlines a methodical approach to examining the integration challenges posed by Industry 4.0, utilizing a framework comprising five levels of integration: Connection, Communication, Coordination, Cooperation, and Collaboration, collectively referred to as the "5 C". This book chapter also reviews the latest research and developments in Industry 4.0, structured around these 5 C levels and their interaction with the four primary components of IoE. It focuses explicitly on various research studies, practical applications, use cases, and case studies that address the complex issues of autonomous integration at the more advanced levels of the 5 C model. Additionally, the chapter includes a case study that examines integration from the standpoint of autonomy and self-organization, among other factors, intending to transform conventional industrial operations into intelligent, Industry 4.0-compliant factories.

Having outlined the "5 C" model of integration and its critical role in advancing Industry 4.0 within S-ERP systems, it becomes evident that adapting to these technological advancements requires more than technical integration. The next vital step in this evolutionary process involves fortifying the organizational structure to withstand and thrive amidst these rapid changes. This leads us to an essential aspect of modern S-ERP systems in Industry 4.0: enhancing organizational resilience and agility.

1.4.9 Enhancing Organizational Resilience and Agility in S-ERP for Industry 4.0

We have discussed the concept of organizational resilience and agility in the tenth chapter of the proposed work. This chapter is not just another segment of academic discourse but a critical exploration of how modern organizations can thrive in an era of unprecedented change. At the heart of this chapter lies the concept of organizational resilience—the ability of an organization to anticipate, prepare for, respond to, and adapt to incremental change and sudden disruptions. In the dynamic environment of Industry 4.0, resilience becomes more than just a buzzword; it is a strategic imperative. The chapter opens by defining this concept within the context of S-ERP, shedding light on its pivotal role in maintaining operational continuity and achieving long-term sustainability. It addresses organizations' multifaceted challenges—from market volatility to disruptive technological innovations and pressing environmental concerns.

Alongside resilience, the chapter emphasizes agility—the capacity of an organization to adapt to market and environmental changes productively and cost-effectively rapidly. Agility in the era of Industry 4.0 is about responsiveness and flexibility, which are increasingly becoming decisive factors in an organization's success. The chapter explores how agility can be seamlessly integrated into S-ERP systems, enabling organizations to respond swiftly and effectively to the fast-paced changes of Industry 4.0. The narrative then transitions into a comprehensive analysis of how S-ERP systems can be architecturally designed and functionally implemented to bolster these two critical qualities. It discusses innovative strategies and practical methodologies for building resilience through robust S-ERP frameworks. This involves delving into risk management techniques, sustainable resource allocation, and contingency planning to create a resilient organizational backbone. Simultaneously, the chapter illuminates the pathways to embed agility into S-ERP systems. This includes exploring modular designs, real-time data integration capabilities, and flexible process modeling, all essential for fostering a nimble and adaptable organizational environment.

Finally, the chapter culminates with a forward-looking perspective, proposing a strategic roadmap for organizations aiming to infuse resilience and agility into their S-ERP systems. This roadmap is not just a theoretical construct but a practical guide designed to help organizations navigate the dynamic and challenging terrain of Industry 4.0. It is an invitation to embark on a transformative journey toward creating intelligent, adaptive, and future-ready enterprises. The real-world case studies and practical examples from various industries truly bring this chapter to life. These narratives illustrate the successful integration of resilience and agility into S-ERP systems, showcasing how organizations have not only navigated the complexities of this integration but have also reaped significant benefits regarding sustainability and competitive advantage. These examples inspire and provide practical guidance for organizations aspiring to achieve similar outcomes.

This chapter is a crucial component of the book, providing deep insights and actionable strategies for organizations looking to thrive in the age of Industry 4.0. Focusing on enhancing organizational resilience and agility within S-ERP systems offers a lens through which businesses can view and navigate the complexities of today's industrial landscape, turning challenges into opportunities for sustainable growth and innovation.

The eleventh chapter concludes the discussion on S-ERP. The final eleventh chapter will also discuss using distributed ledger technologies to improve transparency and sustainability and explore new business models and strategies that support sustainability, such as renewable energy. Organizations progressively realize the importance of information systems in supporting sustainability initiatives. Information systems are essential to the success of sustainability initiatives. While many organizations have made significant progress on their sustainability goals, they still need to improve their ability to measure and manage their impact. They also need better tools for sharing information about best practices, monitoring performance over time, and setting new targets. Organizations recognize the importance of business sustainability and integrate sustainability principles into their operations. Enterprises face multiple challenges in implementing effective and sustainable business practices, including managing information resources in a rapidly changing business environment and continuously adapting to environmental and economic changes. One of the critical issues enterprises face today is the need to manage their information and technology resources effectively to ensure the availability of current and reliable data to maximize the effectiveness of business processes and decision-making. The present book aims to explore S-ERP, its key concepts, and challenges, especially for Industry 4.0.

References

1. Akyurt İZ, Kuvvetli Y, Deveci M (2020) Enterprise resource planning in the age of industry 4.0: a general overview. In: Logistics 4.0. CRC Press, p 8. ISBN 978-0-429-32763-6
2. AlMuhayfith S, Shaiti H (2020) The impact of enterprise resource planning on business performance: with the discussion on its relationship with open innovation. J Open Innovat: Technol, Market, Complexity 6(3):87. ISSN 2199-8531. https://doi.org/10.3390/joitmc6030087. https://www.sciencedirect.com/science/article/pii/S2199853122005807
3. Amini M, Abukari AM (2020) ERP systems architecture for the modern age: a review of the state of the art technologies. J Appl Intell Syst Inf Sci 1(2):70–90. ISSN 2821-1987. https://doi.org/10.22034/jaisis.2020.232506.1009. Publisher: FANAP Research Center
4. Anjaria K (2024) Enhancing sustainability integration in sustainable enterprise resource planning (S-ERP) system: application of transaction cost theory and case study analysis. Int J Inf Manag Data Insights 4(2):100243. ISSN 2667-0968. https://doi.org/10.1016/j.jjimei.2024.100243. https://www.sciencedirect.com/science/article/pii/S2667096824000326
5. Anjaria K, Mishra A (2018) Information processing and security analysis of shared system resource based architectures. Int J Cooperat Inf Syst 27(04):1850009. ISSN 0218-8430. https://doi.org/10.1142/S0218843018500090. https://www.worldscientific.com/doi/abs/10.1142/S0218843018500090. Publisher: World Scientific Publishing Co
6. Antoni D, Jie F, Abareshi A. Critical factors in information technology capability for enhancing firm's environmental performance: case of Indonesian ICT sector
7. Ashima R, Haleem A, Bahl S, Nandan D, Javaid M (2023) Automation of AM Via IoT towards implementation of e-logistics in supply chain for industry 4.0. In Pradhan P, Pattanayak B, Das HC, Mahanta P (ed) Recent advances in mechanical engineering. Springer Nature, Singapore, pp 181–189. ISBN 9789811690570
8. Berrah L, Cliville V, Trentesaux D, Chapel C (2021) Industrial performance: an evolution incorporating ethics in the context of industry 4.0. Sustainability 13(16):9209. ISSN 2071-1050. https://doi.org/10.3390/su13169209. https://www.mdpi.com/2071-1050/13/16/9209. Number: 16 Publisher: Multidisciplinary Digital Publishing Institute
9. Chofreh AG, Goni FA, Klemeš JJ (2017) Development of a roadmap for sustainable enterprise resource planning systems implementation (part II). J Cleaner Product 166:425–437. ISSN 0959-6526. https://doi.org/10.1016/j.jclepro.2017.08.037. https://www.sciencedirect.com/science/article/pii/S0959652617317535
10. Chofreh AG, Goni FA, Klemeš JJ (2018) A roadmap for sustainable enterprise resource planning systems implementation (part III). J Cleaner Product 174:1325–1337. ISSN 0959-6526. https://doi.org/10.1016/j.jclepro.2017.10.285. https://www.sciencedirect.com/science/article/pii/S095965261732591X
11. Chofreh AG, Goni FA, Klemeš JJ, Malik MN, Khan HH (2020) Development of guidelines for the implementation of sustainable enterprise resource planning systems. J Cleaner Product 244:118655. ISSN 0959-6526. https://doi.org/10.1016/j.jclepro.2019.118655. https://www.sciencedirect.com/science/article/pii/S0959652619335255
12. Dhirani LL, Armstrong E, Newe T (2021) Industrial IoT, cyber threats, and standards landscape: evaluation and roadmap. Sensors, 21(11):3901. ISSN 1424-8220. https://doi.org/10.3390/s21113901. https://www.mdpi.com/1424-8220/21/11/3901. Number: 11 Publisher: Multidisciplinary Digital Publishing Institute
13. Elbahri FM, Al-Sanjary OI, Ali MAM, Naif ZA, Ibrahim OA, Mohammed MN (2019) Difference comparison of SAP, oracle, and microsoft solutions based on cloud ERP systems: a review. In: 2019 IEEE 15th international colloquium on signal processing & its applications (CSPA), pp 65–70. https://doi.org/10.1109/CSPA.2019.8695976. https://ieeexplore.ieee.org/document/8695976
14. Jimo A, Balaganesh C, Jayasekara DC (2023) Cross-country comparative analysis of digital manufacturing systems. In: Amit RK, Pawar SK, Sundarraj RP, Ratchev S (eds) Advances in digital manufacturing systems: technologies, business models, and adoption. Springer Nature, Singapore, pp 165–196. ISBN 978-981-19707-1-9

15. Kitsantas T (2022) Exploring blockchain technology and enterprise resource planning system: business and technical aspects, current problems, and future perspectives. Sustainability 14(13):7633. ISSN 2071-1050. https://doi.org/10.3390/su14137633. https://www.mdpi.com/2071-1050/14/13/7633. Number: 13 Publisher: Multidisciplinary Digital Publishing Institute
16. Li W, Huynh BH, Akhtar H, Myo KS (2021) Discrete event simulation as a robust supporting tool for smart manufacturing. In: Toro C, Wang W, Akhtar H (eds) Implementing industry 4.0: the model factory as the key enabler for the future of manufacturing. Springer International Publishing, Cham, pp 287–312. ISBN 978-3-030-67270-6
17. Mahmud R, Ramamohanarao K, Buyya R (2020) Application management in fog computing environments: a taxonomy, review and future directions. ACM Comput Surv 53(4):88:1–88:43. ISSN 0360-0300. https://doi.org/10.1145/3403955
18. Martin K, Shilton K, Smith J (2022) Business and the ethical implications of technology: introduction to the symposium. In: Martin K, Shilton K, Smith J (eds) Business and the ethical implications of technology. Springer Nature Switzerland, Cham, pp 1–11. ISBN 978-3-031-18794-0
19. Morawiec P, Soltysik-Piorunkiewicz A (2022) Cloud computing, big data, and blockchain technology adoption in ERP implementation methodology. Sustainability 14(7):3714. ISSN 2071-1050. https://doi.org/10.3390/su14073714. https://www.mdpi.com/2071-1050/14/7/3714. Number: 7 Publisher: Multidisciplinary Digital Publishing Institute
20. Nayernia H, Bahemia H, Papagiannidis S (2022) A systematic review of the implementation of industry 4.0 from the organisational perspective. Int J Product Res 60(14):4365–4396. ISSN 0020-7543. https://doi.org/10.1080/00207543.2021.2002964. Publisher: Taylor & Francis _eprint: https://doi.org/10.1080/00207543.2021.2002964
21. Profanter S, Tekat A, Dorofeev K, Rickert M, Knoll A (2019) OPC UA versus ROS, DDS, and MQTT: performance evaluation of industry 4.0 protocols. In: 2019 IEEE international conference on industrial technology (ICIT), pp 955–962. https://doi.org/10.1109/ICIT.2019.8755050. https://ieeexplore.ieee.org/document/8755050. ISSN: 2643-2978
22. Sazvar Z, Zokaee M, Tavakkoli-Moghaddam R, Al-sadat Salari S, Nayeri S (2022) Designing a sustainable closed-loop pharmaceutical supply chain in a competitive market considering demand uncertainty, manufacturer's brand and waste management. Ann Operat Res 315(2):2057–2088. ISSN 1572-9338. https://doi.org/10.1007/s10479-021-03961-0
23. Tavana M, Hajipour V, Oveisi S (2020) IoT-based enterprise resource planning: challenges, open issues, applications, architecture, and future research directions. Internet Things 11:100262. ISSN 2542-6605. https://doi.org/10.1016/j.iot.2020.100262. https://www.sciencedirect.com/science/article/pii/S2542660520300962
24. Tsalis TA, Malamateniou KE, Koulouriotis D, Nikolaou IE (202) New challenges for corporate sustainability reporting: United Nations' 2030 Agenda for sustainable development and the sustainable development goals. Corporate Soc Responsib Environ Manage 27(4):1617–1629. ISSN 1535-3966. https://doi.org/10.1002/csr.1910. https://onlinelibrary.wiley.com/doi/abs/10.1002/csr.1910. eprint: https://onlinelibrary.wiley.com/doi/pdf/10.1002/csr.1910
25. Tsamados A, Aggarwal N, Cowls J, Morley J, Roberts H, Taddeo M, Floridi L (2022) The ethics of algorithms: key problems and solutions. AI & Society 37(1):215–230. ISSN 1435-5655. https://doi.org/10.1007/s00146-021-01154-8
26. Wu H, Hao Y, Ren S, Yang X, Xie G (2021) Does internet development improve green total factor energy efficiency? Evidence from China. Energy Policy 153:112247. ISSN 0301-4215. https://doi.org/10.1016/j.enpol.2021.112247. https://www.sciencedirect.com/science/article/pii/S0301421521001166

Chapter 2
Foundation of S-ERP in Industry 4.0

2.1 Introduction

Industry 4.0, the disruptive change driven by the proliferation of the Internet of Things (IoT) and Cyber-Physical Systems (CPS) in production and logistics, is a transformation that encompasses innovations such as intelligent products and processes, deeper supply chain integration, and digital technology implementation in production. The primary goal is to enable faster responses to increasingly variable customer demands, enhancing competitiveness in dynamic, volatile markets. Industry 4.0 represents the convergence of advanced technologies to create intelligent, interconnected, and efficient manufacturing and production systems. In this context, S-ERP plays a crucial role in extending the concept of Industry 4.0 by integrating these technologies to achieve operational excellence and environmental and social sustainability. The technologies integrated into S-ERP in Industry 4.0 are shown in Fig. 2.1:

As shown in Fig. 2.1, S-ERP should integrate six technologies.

$$S - ERP = S{-}ERP_{CPS} + S - ERP_{IoT} + S - ERP_{\text{Intelligence}} + S - ERP_{\text{SupplyChain}} + S - ERP_{\text{Responsiveness}} + S - ERP_{\text{volatile markets}}$$

Each component of S-ERP plays a vital role in achieving a holistic and sustainable approach to enterprise resource planning:

- $S - ERP_{CPS}$: Ensures efficient and sustainable operation of physical systems.
- $S - ERP_{IoT}$: Facilitates comprehensive connectivity and data exchange.
- $S - ERP_{\text{Intelligence}}$: Optimizes processes through advanced analytics and automation.
- $S - ERP_{\text{SupplyChain}}$: Enhances supply chain integration for sustainable logistics.

K. Anjaria, *Sustainable Enterprise Resource Planning (S-ERP) for Industry 4.0*, Contributions to Management Science, https://doi.org/10.1007/978-981-96-7734-4_2

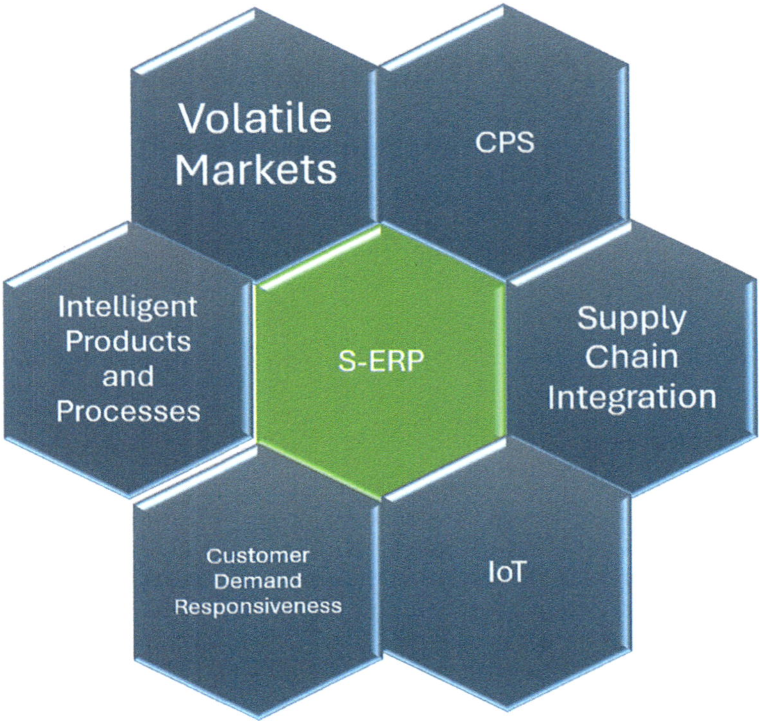

Fig. 2.1 Technologies integrated in S-ERP for Industry 4.0 setup

- $S - ERP_{\text{Responsiveness}}$: Aligns production with real-time customer demands, reducing waste.
- $S - ERP_{\text{volatile markets}}$: Maintains enterprise agility and innovation for competitive advantage.

By formally integrating these elements, S-ERP provides a comprehensive framework that not only enhances operational efficiency but also aligns with broader sustainability goals, ensuring that enterprises can thrive in dynamic and volatile markets while minimizing their negative environmental and social impact. Next, we will understand each integrated component in S-ERP for Industry 4.0.

1. Cyber-Physical Systems (CPS) Integration in S-ERP:

$$\text{S} - \text{ERP}_{CPS} = \sum_{i=1}^{N} \text{CPS}_i$$

where $S - ERP_{CPS}$ represents the integration of multiple CPS within the S-ERP framework. CPS is crucial for real-time monitoring and control of physical processes, enhancing efficiency and reducing waste.

2. Internet of Things (IoT) Connectivity in S-ERP:

$$S - ERP_{IoT} = \sum_{k=1}^{M} IoT_k$$

$S - ERP_{IoT}$ connectivity ensures comprehensive data collection and communication, facilitating sustainable resource management.

3. Intelligent Products and Processes in S-ERP:

$$S - ERP_{\text{Intelligence}} = f\left(\sum_{d=1}^{D} \text{Data}_d, \sum_{a=1}^{A} \text{Analytics}_a, \sum_{o=1}^{o} \text{Automation}_o\right)$$

where $S - ERP_{\text{Intelligence}}$ is the application of data, analytics, and automation within S ERP to create intelligent and adaptive processes that optimize resource use and minimize waste.

4. Supply Chain Integration in S-ERP:

$$\text{S-ERP}_{\text{SupplyChain}} = \sum_{q=1}^{R} (\text{Information Flow and not low}_q + \text{Material Flow}_q + \text{Financial Flow}_q)$$

where $S - ERP_{\text{SupplyChain}}$ ensures that all aspects of the supply chain are integrated, leading to enhanced coordination and sustainability across the entire supply network.

5. Customer Demand Responsiveness in S-ERP:

$$\text{S-ERP Responsiveness} = \frac{1}{\sum_{u=1}^{V} (\text{Lead Time } u + \text{Customization } u)}$$

S-ERP Responsiveness is designed to adapt to customer demands swiftly, ensuring that production aligns with actual market needs and avoiding overproduction and waste.

6. Competitiveness in Volatile Markets via S-ERP:

$$S - ERP_{\text{Competitiveness}} = \sum_{w=1}^{X} (\text{Agility}_w + \text{Flexibility}_w + \text{Innovation}_w)$$

innovation, ensuring that enterprises remain competitive while adhering to sustainable practices.

All these S-ERP components can be integrated using digital twin, machine learning and AI [14]. A digital twin is a virtual representation of a physical object, process, or system that mirrors its real-time status and behaviour. In S-ERP integration, digital twins allow businesses to simulate different scenarios and optimize processes before implementing them in the real world. This ensures efficient resource use and minimal environmental impact. Digital twins help monitor various components' performance by providing real-time data, enabling timely interventions and adjustments. Additionally, digital twins can predict potential failures and maintenance needs, reducing downtime and extending the lifespan of equipment.

Machine learning (ML), a subset of artificial intelligence, involves training algorithms to learn from data and make predictions or decisions without being explicitly programmed. In S-ERP integration, ML algorithms analyze vast amounts of data generated by CPS and IoT devices, identifying patterns and insights that can drive process improvements. ML also automates routine tasks and processes, enhancing efficiency and reducing human error. Furthermore, ML models can predict customer demand more accurately, allowing for better alignment of production schedules and inventory management with market needs.

Artificial intelligence (AI) encompasses a range of technologies that enable machines to perform tasks that typically require human intelligence, such as learning, reasoning, and problem-solving. In the realm of S-ERP integration, AI provides advanced decision support by analyzing complex data sets and offering actionable insights, helping managers make informed decisions. AI-driven automation improves the responsiveness and flexibility of manufacturing and logistics processes, adapting to changes in real-time. Additionally, AI enhances collaboration across the supply chain by facilitating communication and coordination among different stakeholders.

Businesses can seamlessly integrate all S-ERP components by leveraging digital twins, machine learning, and AI, achieving a cohesive and efficient system. These technologies enhance operational efficiency and promote sustainability by optimizing resource use, reducing waste, and enabling proactive decision-making. **Thus, this introduction section lays the foundation and addresses a critical research question that much literature has asked** [1]**: What will be the change in S-ERP if it is to be used in the Industry 4.0 setup?** The clear answer to this question is that the focus of S-ERP in Industry 4.0 will shift from central data collection to the support of mobile, role-based user interactions. Thanks to real-time data access and intelligent data analyses, the sustainability, efficiency, and transparency of process execution can be increased significantly.

The primary objective of S-ERP systems is to function as the central repository for master and transactional data across all business-related processes within organizations, with a particular emphasis on sustainability. S-ERP systems facilitate connectivity to input and output devices such as scanners, illustration, and communication devices, as well as other hardware-software systems within and across organizational boundaries through interfaces. Consequently, S-ERP systems are integral components of the Cyber-Physical Systems (CPS) network and, thus, must be considered in integrated Industry 4.0 frameworks. Nevertheless, the advent of new, flexible processes and moving away from the traditional paradigm that mandates centralized data storage present significant challenges to S-ERP systems. Furthermore, S-ERP underscores sustainable practices by optimizing resource utilization and minimizing environmental impact.

In this chapter, we will explore the integration of S-ERP within the context of Industry 4.0. The chapter is organized into several sections and subsections, each addressing different aspects of this integration. First, we will discuss the goals of Industry 4.0 with S-ERP (Sect. 2.2), outlining the primary objectives and how S-ERP aligns with these goals. Next, we will delve into the technical implementation of S-ERP in an Industry 4.0 environment, covering two critical areas:

- Data Architecture (Sect. 2.3): This section will examine how data is managed, integrated, and utilized within the S-ERP framework.
- Process Architecture (Sect. 2.4): Here, we will focus on the processes involved and how they are optimized using S-ERP.

Following these sections, we will move on to Correlating data and process architectures for S-ERP and Industry 4.0 amalgamation (Sect. 2.5). This section will identify the architectural modifications required to transition from Industry 3.0 to Industry 4.0 (Sect. 2.5). Finally, we will present a Case Study (Sect. 2.6) on implementing S-ERP at GreenTech Industries, providing practical insights and examples of the concepts discussed throughout the chapter. This structure ensures a comprehensive understanding of S-ERP's role and implementation in Industry 4.0, from theoretical foundations to practical applications. The following section discusses the goals of Industry 4.0 that can be achieved using S-ERP. Focusing on these objectives will further illustrate how S-ERP can drive innovation, enhance efficiency, and promote sustainability in a modern industrial context.

2.2 The Goals of Industry 4.0 with S-ERP

Before utilising S-ERP in Industry 4.0, deciding the organisation's goals is crucial. Cote [5] suggested that organizations should set strategic goals. Strategic goals are an organization's measurable objectives that indicate its long-term vision. The strategic goals should have four characteristics. (1) Purpose-driven, (2) Long-term and forward focus, (3) Actionable and (4) Measurable.

The first characteristic involves identifying the company's purpose and values. This guides the development of objectives that are aligned with what you are striving for and why. According to HBS Professor Rebecca Henderson [7], incorporating purpose authentically can enhance business performance by aligning the organization and potentially outperforming competitors. The second characteristic emphasizes long-term and forward focus, suggesting that strategic goals are long-term objectives aligned with the company's values and vision. In contrast, operational goals are daily milestones to achieve them. Ensure that the two are not confused; for example, creating a new marketing strategy is an operational goal that supports the strategic

Table 2.1 The goals the organization should aim to achieve in Industry 4.0 setup with S-ERP'

Characteristics of goals	Goals
Purpose-driven	1. New sustainable business models and services through digitalized products
	2. User-centric work environment through innovative human-machine interaction that promotes sustainability
	3. Incorporating environmentally friendly practices in daily operations
Long-term and forward focus	1. Higher robustness to volatile customer demand
	2. Higher flexibility through dynamic planning, control, and execution
	3. Higher productivity and resource efficiency for individualized products
Actionable	1. Shorter time-to-market for a fast reaction to new customer demands
	2. Increased quality of work through an improved work-life balance
	3. Diversified tasks through the automation of routine jobs
	4. Saving time through more efficient and more transparent production processes
	5. Shorter lead times through the application of intelligent analyses
	6. Different work organizations are organized through mobile process control and execution
Measurable	1. Increased quality through predictive error avoidance

goal of entering a new market segment. Maintain a forward-focused vision to set impactful and challenging objectives.

The third characteristic elucidates that practical strategic goals are actionable and supported by operational goals. The organization should ensure that strategic goals have a clear action plan with achievable operational steps. If not, reconsider their viability to allocate time and resources better. The final characteristic is about crafting strategic goals. Defining how progress and success will be measured to achieve the final characteristic is essential. Based on these characteristics, the goals the organization wants to achieve in Industry 4.0 setup with S-ERP are shown in Table 2.1.

Table 2.1 shows that the rationale behind these goals is to align strategic, organizational, and process improvements with sustainability and environmental responsibility. By incorporating eco-friendly practices, enhancing resource efficiency, and promoting sustainable business models, the goals aim to achieve long-term growth, operational excellence, and reduced environmental impact, ensuring a balance between profitability and ecological stewardship. With these goals in mind, the following section discusses S-ERP's technical implementation details in an Industry 4.0 setup.

2.3 S-ERP's Technical Implementation in an Industry 4.0: Data Architecture

S-ERP's technical implementation consists of two architectural components: (1) Data and (2) process details. This section discusses data architecture in an Industry 4.0 setup. We begin by examining the data architecture-based details. S-ERP facilitates horizontal and vertical data integration within the Industry 4.0 framework, resulting in a tree-like data architecture. Visualize the company as a tree, with branches extending horizontally and roots growing vertically. In this context, horizontal and vertical integration pertains to acquisition strategies. Horizontal integration involves acquiring companies with similar products and customer bases, whereas vertical integration entails incorporating new capabilities and internalizing previously outsourced services. Both integration strategies focus on technologies, processes, and systems that enable effective data collection, collation, communication, and utilization.

Industry 4.0 vertical integration entails connecting all business units and processes within an organization, effectively merging operational technology (OT) at the production level with information technology (IT) at the enterprise level. This integration enables seamless data flow and availability across all business units, including the factory floor, marketing, sales, customer service, purchasing, accounting, human resources, quality control, research and development, and more. Vertical integration extends beyond merely linking enterprise-level systems with the factory floor; it also encompasses opportunities for integration across the entire organizational field layer. For instance, in a batch-size-of-one business model, where a customer requests product customization, the necessary information is readily available in the ERP system, eliminating the need for the sales department to retrieve data from production. Additionally, aligning raw material procurement with market demand optimizes batch production efficiency.

S-ERP supports vertical integration in an Industry 4.0 setup by seamlessly connecting various business units and processes. For example, in a manufacturing company, S-ERP integrates data from production, quality control, supply chain, and customer service. Consider a scenario where the company aims to reduce its carbon footprint. S-ERP collects data on energy consumption, waste generation, and resource utilization from the factory floor. This data is then analyzed and shared with the enterprise management systems. With this integration, the company can identify inefficiencies, such as high energy consumption during certain production phases. Using this information, the company can implement sustainable practices, such as optimizing production schedules to off-peak hours, reducing energy use, and minimizing waste through better resource planning. Additionally, the S-ERP system can align raw material procurement with sustainable suppliers and adjust production plans based on real-time market demand, ensuring efficient batch production and reducing excess inventory. This example illustrates how S-ERP facilitates vertical integration, enabling data-driven decision-making that supports sustainable operations and enhances overall efficiency.

Horizontal data integration in Industry 4.0 pertains to the organization's connectivity with the external environment. It facilitates the seamless data flow across the entire supply chain, encompassing third parties, suppliers, distributors, and customers. This integration ensures consistent connectivity across multi-site operations, thereby maximizing visibility, enhancing production adaptability, and fostering collaboration. S-ERP can achieve horizontal integration by connecting with suppliers to monitor and reduce carbon footprints. For example, an S-ERP system can track the sustainability metrics of raw materials from suppliers, ensuring that only eco-friendly materials are used. This data is shared across the supply chain, enabling real-time collaboration and adjustments to meet sustainability goals, thus enhancing overall supply chain efficiency and environmental responsibility.

S-ERP can adopt the centralization and decentralization of data to achieve horizontal and vertical integration. The S-ERP implements a central data repository within the database system to store and manage essential data from various business units and external partners. It utilises an integrated platform to consolidate data from production, supply chain, sales, marketing, and customer service, ensuring a unified source of truth. Further, the organization ensure real-time data synchronization across the S-ERP system to provide immediate access and updates for all stakeholders, enhancing decision-making and coordination.

Decentralized data in S-ERP can be achieved using localized data storage solutions, such as edge computing and localized servers, to store data closer to its source, like production data at manufacturing sites. Implement distributed S-ERP modules to enable individual business units to manage their data locally while remaining interconnected with the central S-ERP system. Additionally, it leverages cloud-based S-ERP solutions to provide flexibility in storing and accessing data from various geographical locations, facilitating decentralized data management.

Combining centralized and decentralized approaches in S-ERP involves adopting a hybrid data architecture that integrates both centralized and decentralized data management within the system. This ensures critical data is centralized for comprehensive analysis, while operational data is managed locally for efficiency. Utilize APIs and middleware to facilitate seamless data exchange between centralized and decentralized S-ERP modules, ensuring data consistency and interoperability. Establish a robust data governance framework to manage data quality, security, and compliance across both centralized and decentralized environments within the S-ERP system.

Suppose centralized data in the S-ERP system in a manufacturing company could include overall production metrics, financial data, and supply chain information. Decentralized data could involve real-time production data from factory floor sensors and local inventory levels at different warehouses. By combining these approaches, the company can achieve vertical integration (connecting production data with enterprise-level management) and horizontal integration (linking data across the entire supply chain, including suppliers and distributors). By implementing these strategies, organizations can leverage S-ERP's full potential to achieve horizontal and vertical integration, leading to enhanced operational efficiency, better decision-making, and sustainable business practices.

Even though the theory of horizontal and vertical integration with centralized and decentralized data architecture in Industry 4.0 is extensively discussed in the literature [12], there are challenges related to data in implementing S-ERP in the Industry 4.0 setup. In traditional ERP, central data storage is used to plan, control, and execute business processes. However, for S-ERP, it is not feasible to store all data centrally due to the increasing importance of connections to other IT systems, such as sensors, and the need for real-time data usage. As discussed in the previous chapter, S-ERP tracks and processes real-time data to understand sustainable business processes. The challenges lie in capturing and converting all data into information or knowledge in real-time. Researchers have suggested highly responsive user interfaces and solutions based on machine learning and natural language processing (NLP) to address these challenges. Additionally, researchers have recommended using soft computing and fuzzy data instead of imposing hard constraints on the data [2]. After the data architecture, the following section discusses the process architecture.

2.4 S-ERP's Technical Implementation in an Industry 4.0: Process Architecture

The process architecture of S-ERP in Industry 4.0 creates a paradoxical situation as the architecture must support lean management principles and still be flexible. Model-based requirement engineering [10] is used in Industry 4.0 to address this challenge. Model-based requirement engineering in Industry 4.0 is an approach that uses formal models to capture, analyze, and manage requirements for complex systems. It integrates various tools and methodologies to create a comprehensive visual representation of the requirements, facilitating better understanding and stakeholder communication. It reduces inventory-related issues. The inventory, a form of waste in lean management, is also aimed to be reduced in Industry 4.0. It should facilitate the steadily growing range of variants at low cost. A lot size of one is also requested in Industry 4.0.

In traditional manufacturing, the static sequencing and release of production orders, along with the worker and machine being central to production execution (e.g., for quality assurance), support lean management by reducing waste and ensuring efficient workflows [13]. This approach relies on stable, predictable processes and minimal variability. However, Industry 4.0 introduces a need for flexibility. Production orders may be adjusted frequently and released before executing the changed task. This just-in-time release allows for dynamic adjustments based on real-time data and changing customer demands. Consequently, the importance of rough-cut and detailed production planning diminishes as customer orders and requirements can be directly committed to the production line.

The above example explains a paradox: the architecture must support the lean principle of minimizing waste and optimizing efficiency while also being flexible enough to adapt to real-time changes and demands. S-ERP must balance these competing demands by integrating technologies that enable real-time data processing and decision-making while maintaining efficient, streamlined operations.

Moreover, the approach to error avoidance in S-ERP within Industry 4.0 differs significantly from that in traditional Lean Management. In Lean Management, quality is ensured throughout the process, with workers responsible for identifying and cumulatively correcting errors, leading to continuous improvement within the company. However, this process is inherently reactive. Conversely, S-ERP in Industry 4.0 employs technologies proactively to identify errors. Through continuous data collection, data consolidation, and advanced analytics, errors are detected before they disrupt production, enabling predictive maintenance and ensuring more efficient and reliable operations.

To tackle the aforementioned challenges, S-ERP employs decomposed business processes within the context of Industry 4.0. This decomposition framework draws upon insights from Deloitte's white paper [8] and is visualized through Decomposed Business Process Model Diagrams (DBPMDs). These diagrams delineate the segmentation of the business process into six distinct components:

- **Dynamic planning, control, and execution**
- **Integrated end-to-end process**
- **Human-machine interaction**
- **Location-independent control and execution**
- **Efficiency of process steps**
- **Process improvement through intelligent data analyses**

We compare the general/traditional business process model and the decomposed business process model diagram based on the above components. The traditional business process model diagram is shown in Fig. 2.2.

Figure 2.2 is a template of a traditional business process model. It illustrates the activities flow between different users, roles, or departments within an organization. The diagram is divided into horizontal swim lanes, each representing a different user, role, or department (e.g., user/role/department-1 and user/role/department-2). These swim lanes help to visually separate and organize the responsibilities and activities of different participants in the process. Process-1 is the initial process that gets triggered by a start trigger. Process-b is the subsequent process that follows Process-1. Process-n is a later-stage process that is executed based on the outcome of a decision node. Start Trigger initiates Process-1. Arrows between processes indicate activities flow from one process to the next. The end trigger marks the completion of the final process (Process-n). The diamond-shaped node represents a decision point in the process flow. Based on the outcome of the decision, the flow may proceed to different subsequent processes. Database-1, associated with Process-1, indicates that this

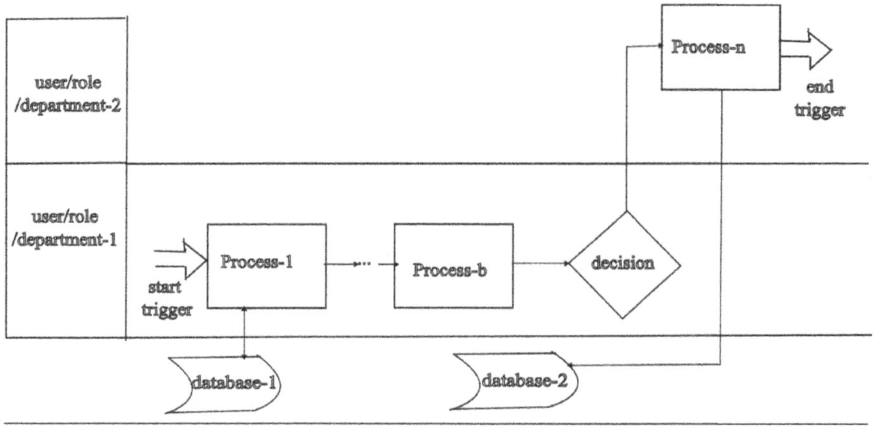

Fig. 2.2 Traditional business process model diagram

process retrieves or stores data in Database-1–database-2, associated with Process-n, indicating data interactions with Database-2. The simple diagram shown in Fig. 2.2 will be decomposed into six components. Readers may note that the decomposed diagrams shown below have not been approved by the Object Management Group (OMG). The decomposed diagram is the modification that we are proposing from our end. OMG provides OCEB 2, which stands for OMG-Certified Expert in BPM 2.0. One track is business-oriented, and the other is technical. OMG intends for BPMN 2.0 to standardize business process modelling like Unified Modelling Language (UML) standardized software modelling.

First, we show and explain **dynamic planning, control, and execution**. It requires seven components in the decomposed business process model diagram shown in Fig. 2.2. The components are: (1) Start, (2) Data Input (Real-time data), (3) Planning (Dynamic adjustments based on data), (4) Execution (Immediate release of production orders), (5) Monitoring (Continuous real-time adjustments) (6) Quality Control (Integrated within the process) and (7) End. These seven components are shown in Fig. 2.3. In Fig. 2.3, we have decomposed process 1.

The interpretation of the BPM diagram shown in Fig. 2.3 is as follows: The triggers have the same meaning as shown in Fig. 2.2. From Database 1, there are two input arrows. These arrows show real-time data input. Since the data used in the process is real-time, the organization has to recheck the data before permanently writing it in the database, as discussed in the previous section of this chapter. From process 1, the arrow goes to the filled rectangle. The filled rectangle depicts dynamic planning done in real time. The concept of a filled rectangle was inspired by Petri Net. The details about Petri Net are available in the literature [4]. Once the arrow leaves the filled rectangle, it depicts the execution of the dynamic planning. Process 1 after the filled rectangle is monitored and quality tested process. It is the combination of the following two parts shown in Table 2.2.

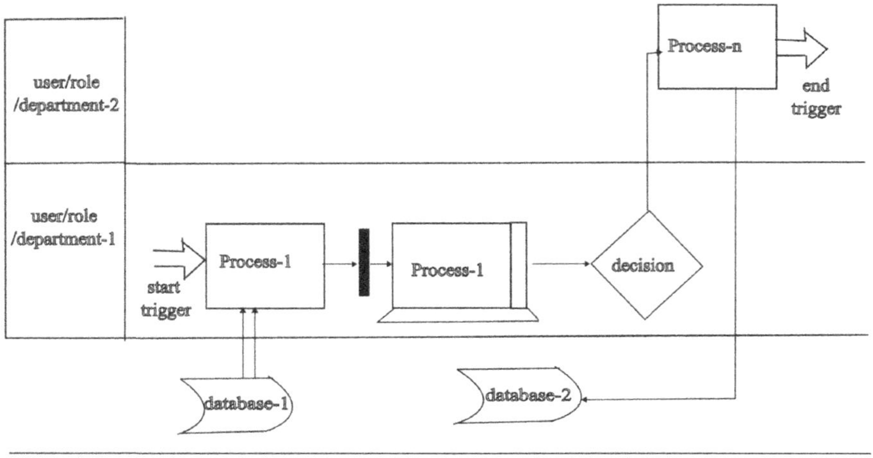

Fig. 2.3 BPM decomposed for dynamic planning, control, and execution

Table 2.2 Meaning of the symbol of Process-1 after the filled rectangle shown in Fig. 2.3

Symbol	Meaning
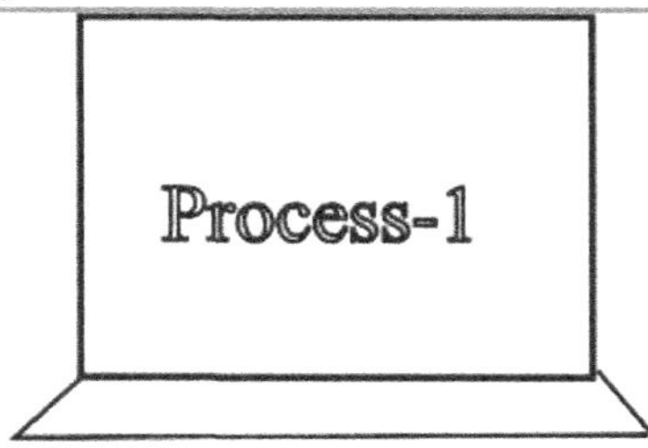	Monitored Process that performs continuous real-time adjustments
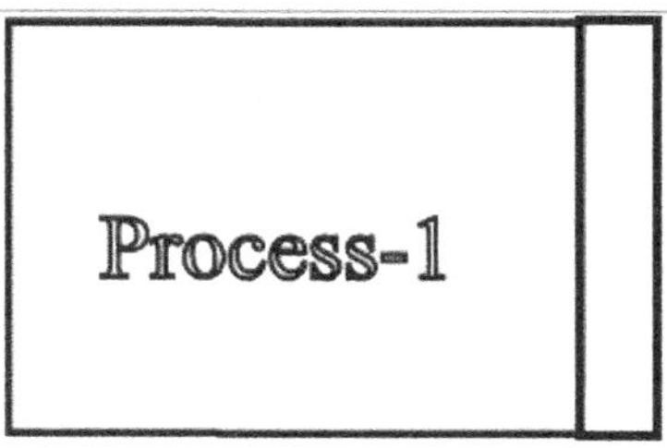	The process has quality control integrated with it.

Example-1: Example of dynamic planning, control, and execution: In an Industry 4.0 environment, a manufacturing company utilizes an S-ERP system to enhance its dynamic planning capabilities. The process begins with a Start trigger initiated by an incoming customer order. Data Input is immediately gathered in real-time from Manufacturing Execution Systems (MES) and Programmable Logic Controllers (PLC), providing up-to-date information on production status, machine performance, and inventory levels. Using this data, the Planning phase involves dynamic adjustments, where the S-ERP system optimizes production schedules and resource allocation to meet the specific requirements of the new order.

The Execution phase follows, where production orders are released immediately based on the optimized plan. This just-in-time release ensures that production can commence without delays. Throughout the process, Monitoring is conducted in real-time, with continuous adjustments as new data is received. For example, if a machine reports a potential issue, the system can reroute tasks to other available machines to maintain efficiency. Quality Control is integrated within the process, with sensors and automated checks ensuring the product meets the required standards at each production stage. This integration allows for immediate detection and correction of defects, further enhancing product quality. Finally, the process reaches its End, with the completed order ready for delivery and all relevant data recorded for future analysis and continuous improvement. This dynamic planning approach ensures that the company can respond swiftly to customer demands, optimize resource use, wastage, and negative impact on the environment and maintain high standards of quality and efficiency.

Next, we demonstrate an **integrated end-to-end process**. It requires eight components in the decomposed business process model diagram. The components are: (1) Start, (2) Supplier Data Integration (real-time), (3) Customer Order Integration (real-time), (4) System processing, (5) Data flow, (6) Execution, (7) Feedback Loop (Continuous improvement), (8) End. These eight components are shown in Fig. 2.4. In Fig. 2.4, we have decomposed process 1.

The interpretation of the diagram shown in Fig. 2.4 is as follows: The start and end trigger concept is similar to Figs. 2.2 and 2.3. The end-to-end integrated process will hold real-time supplier and customer data. Since the database is real-time updated, the double arrows depict that the real-time updates are sent to processes. The data flow and process execution are similar to the traditional process. Process 2 in Fig. 2.4 has implemented the process loop. It sends the feedback to itself. The feedback loop of the process is a concept derived from cybernetics [3].

Example-2: Example of the integrated end-to-end process: In an Industry 4.0 setup, a manufacturing company utilizes S-ERP to create an integrated end-to-end process that promotes sustainability and benefits the environment. The process starts with the reception of real-time data from various sources. Supplier Data Integration is achieved by connecting the S-ERP system with suppliers' databases, providing up-to-date information on raw material availability, delivery schedules, and inventory levels. The S-ERP system prioritises suppliers with eco-friendly practices, ensuring

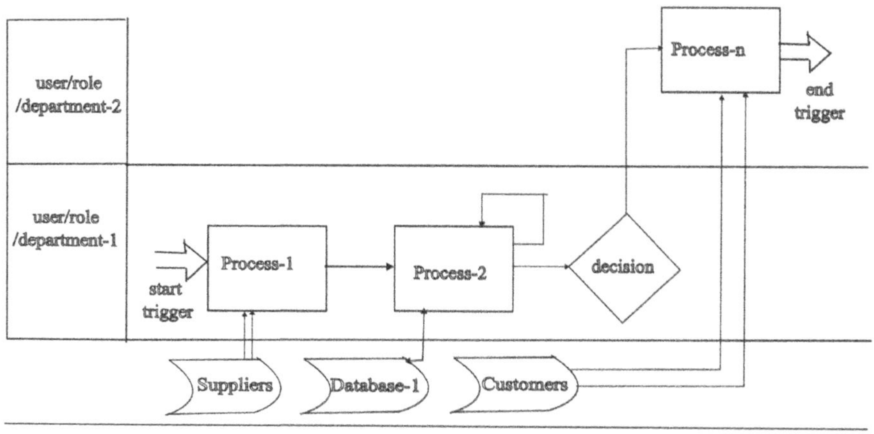

Fig. 2.4 BPM decomposed for an integrated end-to-end process

that raw materials are sourced sustainably. Simultaneously, Customer Order Integration captures real-time customer order data, detailing specifications, quantities, and delivery deadlines. The system encourages customers to opt for environmentally friendly options, such as products with lower carbon footprints or recycled materials.

The ERP System Processing phase uses this comprehensive data to create an optimized production plan, balancing supply and demand while minimizing environmental impact. This involves optimizing transportation routes to reduce fuel consumption, scheduling production runs to minimize energy use during peak hours and reducing waste by accurately forecasting demand. The plan is then communicated seamlessly through Data Flow to Production, where the necessary instructions and schedules are dispatched to the production floor. During the Execution phase, production activities are carried out according to the optimized plan. Real-time monitoring and data collection ensures that deviations or issues are promptly addressed, minimizing waste. Sustainable practices, such as using energy-efficient machinery and recycling production waste, are integrated into the process.

Once production is underway, a Feedback Loop is established, capturing data from every process stage to inform continuous improvement initiatives. This feedback loop includes insights from production metrics, quality control results, and customer feedback, all fed back into the S-ERP system for ongoing optimization. Additionally, the system tracks the environmental impact of each production cycle, allowing the company to refine its sustainability efforts continuously. The process ends with the completed products being ready for delivery and produced efficiently and sustainably. Delivery logistics are optimized to reduce carbon emissions, and customers are informed of the environmental benefits of their purchases. This integrated end-to-end process ensures seamless coordination across the entire supply chain, improving responsiveness, reducing waste, and enhancing overall operational efficiency while promoting sustainability and benefiting the environment.

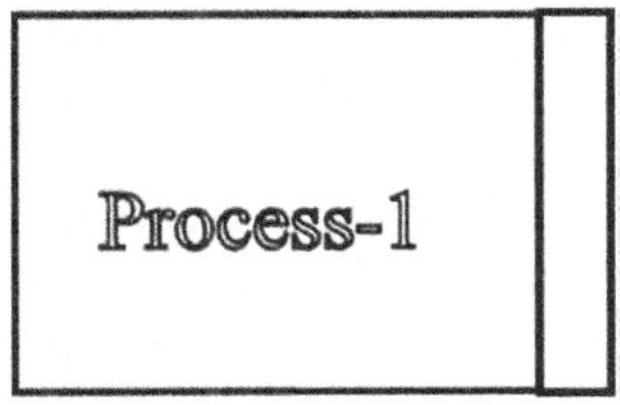

Fig. 2.5 The updated process-1 after error reporting and adjustment

---

Next, we demonstrate **human-machine interaction**. It requires seven components in the decomposed business process model diagram. The components are: (1) Start, (2) user login, (3) interface for input-output, (4) Execution, (5) Error reporting, (6) Adjustment, (7) End. The decomposed BPM diagram for human-machine interaction is the same as shown in Fig. 2.2. Triggers, process-driven input-output, execution, and role-based systems with login mechanisms are the same as the ones shown in Fig. 2.2. The only inclusion in the diagram would be the symbol shown in Fig. 2.5.

Figure 2.5 shows the user can communicate the error and report related issues using processes. The example of human-machine interaction-based decomposition is as follows:

Example-3: An example of human-machine interaction-based decomposition: In an Industry 4.0 environment, a manufacturing company employs an S-ERP system to enhance human-machine interaction, promoting sustainability and benefiting the environment. The process starts with the User Login, where employees access the system through role-based authentication, ensuring that each user has access to relevant data and tools based on their responsibilities. Once logged in, users engage with Data Input/Output through user-friendly interfaces (for example, SAP Fiori UX strategy) designed to simplify complex processes. For example, production managers can easily input data regarding resource usage, while machine operators receive clear instructions for efficient machine operation. These interfaces minimize errors and improve data accuracy, leading to better decision-making.

During the Execution phase, the S-ERP system facilitates seamless coordination between human workers and automated machinery. Real-time data from the shop floor is continuously fed into the system, allowing for precise control and optimization of production processes. This ensures that resources are used efficiently, reducing waste and energy consumption. The Error Reporting feature allows users to provide feedback quickly if there are any discrepancies or issues. For instance, if a machine is not operating within the expected parameters, operators can report the issue immediately through the system. This feedback is crucial for maintaining high standards of quality and efficiency.

The Adjustment phase involves the S-ERP system using the reported errors to make necessary corrections. For example, if a machine error leads to excessive energy use, the system can adjust the settings or schedule maintenance to resolve the issue.

These adjustments are made in real-time, preventing further waste and reducing the environmental impact of production. The process Ends with a continuous loop of improvement, as the feedback and adjustments are recorded and analyzed to refine future operations. This ongoing process enhances productivity and ensures that sustainable practices are consistently implemented.

The S-ERP system significantly contributes to sustainability and environmental benefits by facilitating efficient human-machine interaction. Efficient resource usage reduces waste and energy consumption, while real-time error reporting and adjustments prevent prolonged inefficiencies. Additionally, the system's ability to integrate user feedback ensures that sustainable practices are continually improved, leading to a more environmentally friendly production process.

Next, we demonstrate the decomposition of business processes using **location-independent control and execution**. It requires six components in the decomposed business process model diagram. The components are: (1) Start, (2) Mobile data access, (3) Remote control, (4) Data synchronisation, (5) Monitoring, (6) End. This decomposition doesn't require significant physical changes in the business process model diagram. Most of the changes are logical. The BPM diagram will have start and end triggers. The organization must rethink the data storage mechanism so employees can access mobile data. S-ERP supports various types of storage mechanisms.

To provide data access with mobility and remote control to employees in implementing Industry 4.0, S-ERP allows organizations to leverage cloud-based solutions, enabling real-time data access from any location [11]. S-ERP integrates IoT devices with mobile-friendly applications [15], allowing seamless data collection and monitoring, ensuring employees can make informed decisions on the go. S-ERP utilizes advanced cybersecurity measures like encryption and multi-factor authentication to ensure data integrity and security. As discussed in human-machine interaction-based decomposition, S-ERP implements user-friendly interfaces and training programs to ensure employees can effectively utilize these technologies, enhancing overall productivity and operational efficiency.

To perform data synchronization when implementing Industry 4.0, S-ERP utilizes advanced data integration platforms that support real-time syncing across multiple devices and systems. Leveraging cloud-based storage ensures centralized data access and consistency. Employing middleware solutions and APIs facilitates seamless communication between disparate systems, while edge computing can help manage data locally before syncing with the central repository. Implementing robust data validation and error-checking mechanisms ensures data accuracy during synchronization. Regularly updating synchronization protocols and conducting periodic audits further maintain data integrity and operational efficiency. Finally, the monitoring of the process based on data can be shown with the symbol described in Table 2.2.

Example-4: decomposition of business processes using location-independent control: In an Industry 4.0 setup, a manufacturing company uses an S-ERP system to enable location-independent control, enhancing sustainability and benefiting the environment. The process starts with Mobile Data Access, where managers and operators use a mobile application to access the S-ERP system from anywhere. This feature allows employees to stay connected and informed, regardless of their physical presence on the shop floor. Remote Control is facilitated using the mobile application, enabling users to execute commands and adjust production parameters from their mobile devices. For example, a remote production manager can start or stop machinery, adjust production schedules, or allocate resources based on real-time needs. This flexibility reduces the need for physical travel to the production site, thereby cutting down on transportation-related carbon emissions.

Data Synchronization ensures that all updates and changes made through the mobile application are reflected in real time across the entire S-ERP system. This real-time synchronization allows for immediate adjustments and coordination among different departments. For instance, if a supply chain disruption occurs, the remote manager can quickly reallocate resources or adjust production plans to mitigate the impact, ensuring that production remains efficient and sustainable. Throughout the process, Monitoring is conducted continuously via the mobile application. Real-time data on energy usage, machine performance, and resource consumption is available at the fingertips of remote users. This constant monitoring helps identify inefficiencies and make proactive adjustments to optimize resource usage and reduce waste.

The process Ends with the completion of the production cycle, with all data and adjustments recorded in the S-ERP system for future analysis and continuous improvement. By enabling location-independent control, the S-ERP system supports sustainability and environmental benefits in several ways. Remote access reduces the need for travel, lowering the carbon footprint of the workforce. Real-time monitoring and adjustments ensure optimal resource usage, minimizing waste and energy consumption. Additionally, the flexibility to manage operations from any location enhances overall efficiency, contributing to a more sustainable production process.

Next, we discuss decomposition based on the efficiency of the **process steps**. It requires six components in the decomposed business process model diagram. The components are: (1) Start, (2) Data Processing (Real-time speed monitoring), (3) Execution, (4) Ad-hoc Adjustment (Quick changes based on new data), (5) Monitoring and (6) End. This decomposition doesn't require significant physical changes in the business process model diagram. Most of the changes are logical. The BPM diagram will have start and end triggers. To achieve real-time speed monitoring in data processing when implementing Industry 4.0, S-ERP may employ edge computing to process data close to the source, reducing latency. Integrating IoT sensors with real-time analytics platforms enables continuous data collection and instant processing [9]. Utilizing high-speed data transmission protocols and robust network infrastructure by S-ERP ensures rapid data flow. Implementing machine learning algorithms and game theory can enhance predictive analytics and decision-making [6]. Additionally, leveraging scalable cloud resources to handle large data volumes and employ efficient data compression techniques can further optimise processing speeds and ensure timely insights. The monitoring and adjustment parts in the process would be similar to Fig. 2.3 and Table 2.2.

Example-5: Example of decomposition based on the efficiency of the process: In an Industry 4.0 setup, a manufacturing company utilizes an S-ERP system to enhance the efficiency of process steps. The process starts with Data Processing, where real-time speed monitoring is conducted to gather immediate insights into production performance. The system collects data from various sensors and machines, providing a comprehensive view of operational efficiency. During Execution, the S-ERP system uses this data to optimize production activities. For example, if the system detects a slowdown in a specific machine, it can redistribute tasks to other machines to maintain overall efficiency. This ensures that production continues smoothly without unnecessary delays.

When new data indicates a potential issue or improvement opportunity, ad hoc adjustments are made quickly. For instance, if a sudden increase in energy consumption is detected, the system can adjust machine settings or schedules to reduce energy usage. These quick changes help maintain optimal performance and minimize waste. Throughout the process, Monitoring is ongoing, with the S-ERP system continuously tracking and analyzing performance metrics. This real-time monitoring allows for immediate detection and correction of inefficiencies, ensuring that resources are used effectively. The process Ends with the completion of the production cycle, with all data recorded for future analysis and continuous improvement. By enhancing the efficiency of process steps, the S-ERP system contributes to sustainability and environmental benefits by reducing energy consumption, minimizing waste, and optimizing resource usage. This leads to a more efficient and eco-friendly production process, supporting the company's sustainability goals.

Next, we discuss decomposition based on **the improvement of the process**. It requires seven components in the decomposed business process model diagram. The components are: (1) Start, (2) Data Collection (From various sources), (3) Data Analysis (Using intelligent tools), (4) Predictive Alerts (Maintenance and optimization), (5) Execution, (6) Automated Adjustments (Based on analysis), (7) End. This decomposition also doesn't require significant physical changes in the business process model diagram. Most of the changes are logical. The BPM diagram will have start and end triggers. To achieve data collection from various sources, data analysis using intelligent tools, and predictive alerts for maintenance and optimization when implementing Industry 4.0, organizations plan to integrate IoT devices and sensors in S-ERP to gather data from multiple touchpoints (Example: Embedded analytics functionalities in the core of SAP S/4HANA Enterprise Management). Advanced data analytics platforms powered by AI and machine learning enable comprehensive data analysis and actionable insights. Implementing predictive maintenance systems leverages these insights to generate real-time alerts, facilitating proactive decision-making and optimizing operational efficiency. Combining these elements ensures a seamless flow from data collection to predictive analytics, enhancing overall system performance and reliability. The execution of the business process would be similar to Fig. 2.2.

After studying the data and process architectures required to amalgamate S-ERP and Industry 4.0, we correlate these two architectures in the following section.

2.5 Correlating Data and Process Architectures for S-ERP and Industry 4.0 Amalgamation

This section amalgamates the previous two sections and presents the holistic architecture of S-ERP in Industry 4.0. Here, we point out that process and data architectures should support each other to implement Industry 4.0 and S-ERP successfully.

The first component in the process architecture was Dynamic Planning, Control, and Execution. Dynamic Planning, Control, and Execution involve a comprehensive approach that includes vertical data integration, real-time data exchange, and a focus on cost efficiency. Vertical integration entails connecting ERP systems, MES, and PLCs through steps that ensure seamless communication and coordination among these systems. Real-time data exchange is illustrated by the bi-directional flow of information between production planning systems and on-floor machinery, ensuring that data is continuously updated and shared. To achieve cost efficiency, checkpoints or decision nodes are incorporated throughout the process, allowing for ongoing assessment and adjustment to maintain economic production practices. This holistic approach ensures that all aspects of production are interconnected, data-driven, and optimised for cost savings.

The second component, the Integrated End-to-End Process, addresses delays in responding to customer orders caused by disconnected supply chain data. The lack of integration between supplier data and customer orders creates significant gaps, leading to inefficiencies and slow response times. Real-time data integration highlights these gaps, revealing the need for improved connectivity. By enhancing the real-time integration of supply chain data, businesses can achieve seamless order processing, ensuring timely and efficient responses to customer orders.

The third component, Human-Machine Interaction, currently faces challenges due to high error rates in data input, which are attributed to complex system interfaces. Feedback from workers indicates that they find the current interface difficult to navigate and use effectively. More user-friendly interfaces and robust error-reporting mechanisms are needed to address this issue. These improvements will help reduce errors and enhance the overall efficiency and usability of the system.

The fourth component of the process architecture, Location-Independent Control faces challenges due to inefficient remote monitoring and control. The core issue lies in the lack of real-time updates and synchronization during remote access. To address this, there is a need to enhance mobile application usage, enabling real-time data synchronization and more effective remote control. This improvement will ensure seamless and efficient management of systems from any location.

The fifth component, the efficiency of process steps, is currently hindered by slow responses to changes. Real-time monitoring reveals delays in processing times, highlighting the need for improvement. To address this issue, it is essential to implement ad-hoc adjustments based on real-time data, enabling quicker responses and enhancing overall process efficiency.

The final component, process improvement, needs to shift from reactive to proactive maintenance. Current analysis shows frequent machine breakdowns without prior alerts, leading to inefficiencies and downtime. To address this issue, implementing predictive alerts and automated adjustments will optimise maintenance and enhance overall process execution, ensuring a more reliable and efficient operation. Next, we discuss the algorithm, which helps organizations move from industry 3.0 to 4.0 and implement data architecture and process architectural requirements.

2.5.1 Identify the Requirements for Architectural Modification and Move from Industry 3.0 to 4.0

Organizations should follow a structured approach that involves analyzing current processes, identifying inefficiencies, and leveraging data insights to determine where process improvements are needed. The structured approach is discussed in the form of algorithm 1.

Algorithm 1 Structure Approach to Determine Process Improvement for S-ERP in Industry 4.0

1: Conduct a Process Audit
2: Leverage Real-Time Data and Analytics
3: Engage with Stakeholders
4: Benchmark Against Industry Standards
5: Implement Continuous Improvement Tools
6: Pilot and Validate Improvements

Step 1 of algorithm 1 suggests conducting a process audit. This step starts by documenting and analysing the current business process model. Look for bottlenecks, delays, and areas where resources are not utilized efficiently. For example, in the original diagram, if the data collection phase is taking too long or resulting in errors, this indicates a need for improvement in data input and integration.

Step 2 of algorithm 1 is about leveraging real-time data and analytics. The organization should implement real-time data monitoring tools to gather insights into each process step. Analyze this data to identify patterns and inefficiencies. For instance, during the execution phase, if real-time data from MES and PLC indicates frequent machine downtimes, it suggests a need for predictive maintenance and dynamic planning adjustments.

The third step of the algorithm recommends that organizations engage with stakeholders. Organizations should collect feedback from employees, suppliers, and customers to identify areas where they experience difficulties or delays. For example, if workers report frequent errors during data input in Human-Machine Interaction, it indicates a need for more user-friendly interfaces and better error-reporting mechanisms.

The fourth step of the algorithm is the benchmark against industry standards. In this step, organizations should compare their processes against industry standards and best practices to identify gaps and areas for improvement. For instance, if supplier Data Integration is slower than industry benchmarks, it highlights a need to enhance real-time supply chain data integration.

The fifth step is to implement a continuous improvement tool. For this step, organizations should use Six Sigma and Lean methodologies to identify and eliminate waste systematically. For example, using Six Sigma to analyse data processing times in the Efficiency of Process Steps can help pinpoint stages where delays occur and suggest ad-hoc adjustments.

In **the final step**, organizations look for pilot studies and improvement validation. Before fully implementing changes, organizations should pilot them in a controlled environment to validate their effectiveness. For example, a pilot project for Location-Independent Control can be implemented by enabling a subset of employees to use mobile data access and remote control. Furthermore, organizations should monitor the results to ensure real-time synchronization and improved monitoring.

Consequently, this section has elucidated the technical implementation of S-ERP through the lens of process architecture. The subsequent section will explore the potential outcomes achievable upon fulfilling the technical implementation outlined in the preceding sections. A success story from an organization will be presented, albeit with anonymity maintained and minor adjustments to parametric values, to illustrate the practical application of the discussed concepts.

2.6 Case Study: S-ERP Implementation at GreenTech Industries

GreenTech Industries, a global leader in renewable energy solutions, undertook a significant initiative to enhance operational efficiency and sustainability by implementing a comprehensive S-ERP system. This system integrated various Industry 4.0 technical, driving value across multiple business dimensions, transitioning from Industry 3.0 to Industry 4.0. GreenTech leveraged IoT sensors across its machinery and equipment to collect real-time data on performance and potential failures, enabling predictive maintenance. This approach resulted in a 22% reduction in downtime and a 15% extension in machinery lifespan, translating into substantial cost savings. The S-ERP system also facilitated remote diagnostics and maintenance through augmented reality tools, minimizing the need for on-site visits and reducing maintenance costs by 25%, while enhancing worker safety. The S-ERP system's rapid experimentation and simulation tools allowed GreenTech to test production processes and new product designs virtually, reducing the product development cycle time by 30% and accelerating the launch of innovative renewable energy solutions. Concurrent engineering practices, enabled by the ERP, integrated design, engineering, and manufacturing teams for collaborative project management. This integration improved design quality and reduced errors, resulting in a 15% increase in customer satisfaction.

Customer co-creation and open innovation were fostered through an ERP platform that allowed customers to provide feedback and collaborate on new product ideas, enhancing customer engagement and loyalty and leading to a 10% increase in repeat business. Data-driven design to value, another key feature of the ERP, leveraged big data analytics to optimize product designs based on customer preferences and market trends, improving product-market fit and reducing excess inventory by 20%. Smart energy consumption modules within the S-ERP enabled GreenTech to monitor and optimize energy use, achieving an 18% reduction in energy consumption and contributing to the company's sustainability goals. The deployment of advanced analytics optimized production yields and managed production lots intelligently, enhancing production efficiency and reducing waste and inventory holding costs by 15%.

Through these implementations, GreenTech Industries realized the concept of the "Factory of the Future", characterized by adaptability, resource efficiency, and the integration of customers and business partners in value processes. By leveraging these Industry 4.0 levers through the sustainable S-ERP system, GreenTech Industries achieved significant operational improvements and enhanced its competitive edge in the renewable energy market. The integration of advanced technologies and data-driven decision-making fostered a more resilient, efficient, and customer-centric organization, aligning with GreenTech's sustainability and growth objectives. This comprehensive approach has positioned GreenTech at the forefront of Industry 4.0 adoption, setting a benchmark for the renewable energy sector.

Moreover, implementing Industry 4.0 technologies facilitated the reduction of harmful emissions. Using IoT and AI in production processes increased efficiency, reduced waste, optimized resource use, and lowered carbon emissions per product unit. This shift from mass production to mass customization also contributed to a more sustainable production model by aligning manufacturing outputs more closely with market demands, thus reducing overproduction and waste. Regarding social sustainability, GreenTech's ERP system improved worker safety and satisfaction through remote maintenance capabilities and automation of hazardous tasks. The company also engaged in open innovation and customer co-creation, fostering a collaborative environment that enhanced social responsibility and stakeholder engagement. In summary, the chapter answers the following research questions.

2.7 Research Questions Addressed

Here are the key research questions answered in the chapter:

1. **What will be the change in S-ERP if it is to be used in the Industry 4.0 setup?**

- The chapter discusses how the focus of S-ERP in Industry 4.0 shifts from central data collection to supporting mobile role-based user interactions, leading to increased sustainability, efficiency, and transparency in process execution.

2. **What are the goals of Industry 4.0 that can be achieved using S-ERP?**

- The chapter outlines specific goals such as developing new sustainable business models, achieving higher robustness and flexibility, improving work quality, and ensuring shorter lead times, all of which S-ERP aims to support in an Industry 4.0 context.

3. **How can S-ERP be technically implemented within an Industry 4.0 environment?**

- The chapter answers this question by detailing the data and process architectures required for implementing S-ERP, including how to handle data integration, process optimization, and the challenges posed by decentralized data storage.

4. **How can organizations transition from Industry 3.0 to Industry 4.0 while implementing S-ERP?**

- The chapter discusses the structured approach for transitioning to Industry 4.0, including conducting process audits, leveraging real-time data, engaging stakeholders, benchmarking against industry standards, and implementing continuous improvement tools.

5. **What are the practical outcomes of implementing S-ERP in Industry 4.0?**

- The chapter provides a case study on GreenTech Industries, illustrating the practical benefits of S-ERP in enhancing operational efficiency, sustainability, and customer satisfaction in a real-world scenario.

References

1. Al-Amin Md, Tanjim Hossain Md, Jahidul Islam Md, Biwas SK (2023) History, features, challenges, and critical success factors of enterprise resource planning (ERP) in the era of industry 4.0. Europ Sci J, ESJ 19(6):31–31. ISSN 1857-7431. https://doi.org/10.19044/esj.2023.v19n6p31. https://eujournal.org/index.php/esj/article/view/16475. Number: 6
2. Anjaria K (2022) Dempster-shafer theory and linguistic intuitionistic fuzzy number-based framework for blending knowledge from knowledge repositories: an approach for knowledge management. Expert Syst Appl 199:117142. ISSN 0957-4174. https://doi.org/10.1016/j.eswa.2022.117142. https://www.sciencedirect.com/science/article/pii/S0957417422005383
3. Anjaria K, Mishra A (2017) Relation between cybernetics and information security: from Norbert Wiener's perspectives. Kybernetes 46(10):1654–1673. ISSN 0368-492X. https://doi.org/10.1108/K-04-2017-0129. https://doi.org/10.1108/K-04-2017-0129. Publisher: Emerald Publishing Limited
4. Anjaria KA (2020) Computational implementation and formalism of FAIR data stewardship principles. Data Technol Appl 54(2):193–214. ISSN 2514-9288. https://doi.org/10.1108/DTA-09-2019-0164. https://doi.org/10.1108/DTA-09-2019-0164. Publisher: Emerald Publishing Limited
5. Cote C (2020) How to set strategic planning goals | HBS Online, October 2020. https://online.hbs.edu/blog/post/strategic-planning-goals
6. Hazra T, Anjaria K, Bajpai A, Kumari A (2024) Applications of game theory in deep neural networks. In: Hazra T, Anjaria K, Bajpai A, Kumari A (eds) Applications of game theory in deep learning. Springer Nature Switzerland, Cham, pp 45–67. 978-3-031-54653-2
7. Henderson R, Mitchell W (1997) The interactions of organizational and competitive influences on strategy and performance. Strat Manage J 18(S1):5–14. ISSN 1097-0266. 10.1002/(SICI)1097-0266(199707)18:1+<5::AID-SMJ930>3.0.CO;2-I. eprint: https://onlinelibrary.wiley.com/doi/pdf/10.1002/%28SICI%291097-0266%28199707%2918%3A1%2B%3C5%3A%3AAID-SMJ930%3E3.0.CO%3B2-I
8. Hochmuth C, Bartodziej C, Schwägler C (2017) Industry 4.0: is your erp system ready for the digital era? July 2017. https://doi.org/10.13140/RG.2.2.17725.08169
9. Jayender P, Kundu GK (2022) IoT analytics and ERP interoperability in automotive SCM: ANN-fuzzy logic technique for designing decision support systems. Int J Fuzzy System Appl (IJFSA) 11(3):1–19. ISSN 2156-177X. https://doi.org/10.4018/IJFSA.306282. https://www.igi-global.com/article/iot-analytics-and-erp-interoperability-in-automotive-scm/www.igi-global.com/article/iot-analytics-and-erp-interoperability-in-automotive-scm/306282. Publisher: IGI Global
10. Manoj Kannan S, Suri K, Cadavid J, Barosan I, Van Den Brand M, Alferez M, Gerard S (2017) Towards industry 4.0: gap analysis between current automotive MES and industry standards using model-based requirement engineering. In: 2017 IEEE international conference on software architecture workshops (ICSAW), pp 29–35, April 2017. https://doi.org/10.1109/ICSAW.2017.53. https://ieeexplore.ieee.org/abstract/document/7958432
11. Majstorovic V, Simeunovic V, Mitrovic R, Stosic D, Dimitrijevic S, Miskovic Z (2023) Development of cloud ERP model and its application in smart mining. In: Burduk A, Batako A, Machado J, Kowski RW, Antosz K, Gola A (eds) Advances in production. Springer Nature Switzerland, Cham, pp 28–42. 978-3-031-45021-1
12. Pérez-Lara M, Saucedo-Martínez JA, Marmolejo-Saucedo JA, Salais-Fierro TE, Vasant P (2020) Vertical and horizontal integration systems in Industry 4.0. Wirel Netw 26(7):4767–4775. ISSN 1572-8196. https://doi.org/10.1007/s11276-018-1873-2. https://doi.org/10.1007/s11276-018-1873-2
13. Schulze F, Dallasega P (2023) Barriers to lean implementation in engineer-to-order manufacturing with subsequent assembly on-site: state of the art and future directions. Product Plan & Control 34(1):91–115. ISSN 0953-7287. https://doi.org/10.1080/09537287.2021.1888159. https://www.tandfonline.com/doi/abs/10.1080/09537287.2021.1888159. Publisher: Taylor & Francis

14. Siatras V, Bakopoulos E, Mavrothalassitis P, Nikolakis N, Alexopoulos K (2024) Production scheduling based on a multi-agent system and digital twin: a bicycle industry case. Information 15(6):337. ISSN 2078-2489. https://doi.org/10.3390/info15060337. https://www.mdpi.com/2078-2489/15/6/337. Number: 6 Publisher: Multidisciplinary Digital Publishing Institute
15. Touqeer J, Keyvanzad R (2020) Mobile enterprise resource planning adoption and implementation in norwegian organizations a case study of rambase. Master's thesis, University of Stavanger, Norway, August 2020. https://uis.brage.unit.no/uis-xmlui/handle/11250/3038284. Accepted: 2022-12-16T12:21:36Z

Chapter 3
Change Management and Open Innovation: S-ERP and Industry 4.0 Perspectives

3.1 Introduction

Numerous researchers have observed that businesses faced significant challenges in fostering innovation before the advent of Enterprise Resource Planning (ERP) systems due to deficiencies in expertise, skills, technology, and human resources [27]. Additionally, the demands of dynamic marketplaces necessitate the continuous development of innovative solutions [18]. To address these challenges, there was a critical need for integrated and global data storage solutions, which ERP systems successfully provided. However, in the current context, organizations are increasingly focused on developing innovative products with minimal adverse environmental impacts. Achieving such sustainability objectives requires more than just centralized and integrated production data storage; it necessitates the incorporation of sustainability metrics into this centralized storage framework [3]. This is where S-ERP systems come into play, as they not only integrate sustainability matrices with production data but also play a pivotal role in fostering open innovation, a paradigm that encourages organizations to utilize external and internal ideas and paths to market as they seek to advance their technology and products.

Open innovation is a paradigm that encourages organizations to utilize external and internal ideas and paths to market as they seek to advance their technology and products. Unlike traditional models that rely solely on internal research and development, open innovation leverages the collective intelligence and diverse perspectives of a broader network of stakeholders, including customers, suppliers, academic institutions, and even competitors. This approach accelerates the innovation process by tapping into a wider array of resources and insights, leading to more rapid and cost-effective development cycles. In the context of Industry 4.0, open innovation becomes even more critical as technological advancements and market demands evolve at an unprecedented pace. The integration of S-ERP systems within this framework is particularly beneficial, as these systems are specifically designed to manage and

K. Anjaria, *Sustainable Enterprise Resource Planning (S-ERP) for Industry 4.0*, Contributions to Management Science, https://doi.org/10.1007/978-981-96-7734-4_3

analyze vast amounts of data, a crucial function for identifying innovative opportunities and the collaboration required to exploit them. By aligning sustainability goals with open innovation strategies, businesses can not only drive economic growth but also contribute to environmental and social well-being, ensuring a holistic approach to innovation in the modern industrial landscape.

The present chapter discusses how organizations can leverage S-ERP to achieve great success with open innovation. We argue in the chapter that S-ERP has the architectural advantage that can help businesses achieve open innovation. With sustainability in mind, S-ERP can manage both the channels of idea to achieve open innovation. Effective change management becomes paramount as organizations embark on this journey toward open innovation. Implementing S-ERP systems necessitates significant organizational processes, culture, and mindset changes. The integration of sustainability matrices and open innovation frameworks requires not only technological adaptation but also a comprehensive approach to change management. This ensures the transition is seamless and the organization's workforce is aligned with the new strategic direction. Therefore, understanding the principles and practices of change management in the context of S-ERP implementation is critical for maximizing its potential benefits and achieving long-term success.

The change management activity has two facets: change management due to S-ERP and its associated innovations and change management using S-ERP. The present chapter discusses both approaches. The first facet, where information technology triggers organizational changes that need to be managed, is called "technochange" [21]. Technochange is highly risky yet rewarding. A balance of both facets is required in S-ERP; otherwise, the history of ERP will be repeated, and managers will become involved in technochange inadvertently.

For instance, ERP systems promised substantial strategic advantages and improvements through cross-functional integration and process streamlining. However, they became notorious for their implementation challenges and problematic organizational consequences. Many organizations have implemented ERP systems not to enhance their potential performance but to address concerns related to the year 2000 (Y2K) or to replace outdated legacy systems that were reaching capacity limits. In one such organization, managers could not identify the business benefits of their ERP implementation because they had approached the effort as an IT project rather than a business initiative, delegating full control to IT specialists. In numerous other organizations that had implemented ERP systems, managers were surprised or disappointed when the implementations proved difficult, failed to achieve the desired outcomes, or even failed altogether. By treating technochange situations as IT projects, these organizations had risked encountering unforeseen implementation problems and unintended consequences, thereby missing significant opportunities to benefit from IT-enabled organizational performance improvements.

In this chapter, it is also crucial to explain why open innovation and change management are discussed together. What is common between them? The answer to this question is system usage. Delone and McLean argued that system usage is the main factor in the information system's success [1]. Innovation and change are interrelated with the system's usage [14].

In the context of Industry 4.0, where digital transformation is at the forefront, integrating advanced technologies such as IoT, AI, and big data analytics into S-ERP systems plays a pivotal role. These technologies enhance system usage by providing real-time data, predictive insights, and automated processes, which are essential for fostering open innovation. As organizations adopt these advanced S-ERP systems, they must also navigate the complexities of change management to ensure that employees are adequately trained and that the organizational culture evolves to support new working methods. Moreover, the dynamic nature of Industry 4.0 necessitates continuous innovation and adaptation. Open innovation allows organizations to leverage external ideas and technologies, accelerating the development of new products and services. However, this process requires a robust change management strategy to integrate external innovations seamlessly into existing systems and workflows. Thus, the interplay between system usage, open innovation, and change management becomes critical for achieving sustainable competitive advantage in Industry 4.0. Through effective change management, organizations can maximize the benefits of S-ERP systems, driving innovation while maintaining operational stability and efficiency.

Following the introduction, Sect. 3.2 on Open Innovation with S-ERP in Industry 4.0 is divided into two subsections: the first explores the integration of Web 3.0 technologies with S-ERP, and the second discusses future concepts of Web 4.0 enhancing S-ERP. The next major section, Sect. 3.3, Change Management: S-ERP and Industry 4.0 Perspectives, is divided into three subsections. The first subsection covers strategies for implementing S-ERP, the second focuses on managing change in Industry 4.0, and the third addresses integrating sustainability into change management. This structure outlines the journey from understanding S-ERP's background and technological integration to the practical aspects of managing change within Industry 4.0. The following section discusses open innovation with S-ERP.

3.2 Open Innovation with S-ERP in Industry 4.0

The open innovation process and its contrast with classical closed innovation is shown in Fig. 3.1.

On the left side of Fig. 3.1, the closed innovation model is depicted, where ideas are primarily generated within the confines of the organization, represented by internal ideas (orange circles) moving through a funnel with strict corporate limits, ultimately reaching the market. This model emphasizes internal control and a linear progression of ideas from inception to market deployment. The open innovation model is shown on the right side, characterized by a broader and more flexible funnel that includes both internal (orange circles) and external ideas (blue circles). The funnel extends beyond traditional corporate limits, incorporating inputs from external collaborators and alternative markets. This model facilitates a dynamic exchange of ideas, fostering collaboration and leveraging external resources to enhance innovation, ultimately leading to a diverse and enriched market output. This comparison highlights the shift

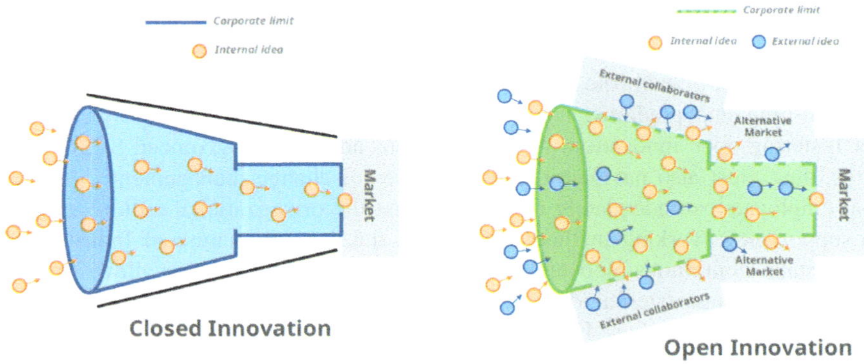

Fig. 3.1 The contrast between closed innovation and open innovation

from a restrictive, internally focused approach to a more inclusive and collaborative paradigm in innovation management.

Ideas are iteratively introduced and filtered through the innovation funnel, with only a select few progressing to the commercialization stage. Technologies play a crucial role in facilitating the expansion of the innovation funnel by enabling interactions that foster the generation of ideas or patents and by supporting the movement of these ideas both within and outside the funnel. Gloor [11] defines "Collaborative Innovation Networks" as "a cyber team of self-motivated individuals with a collective vision, enabled by the Web to collaborate in achieving a common goal through sharing ideas, information, and work." This idea also applies to open innovation through S-ERP due to its architecture.

S-ERP architecture is based on client-server architecture, also known as service-oriented architecture [4] or cloud-based architecture [5]. These architectures allow users to access digital functions and data through web technologies. However, web technologies should be used strategically for open innovation. Suppose there is a blog or a web page where users post ideas. It is not an open innovation. True open innovation involves a collaborative process where diverse internal and external stakeholders actively participate in the idea generation and development process. This includes structured feedback mechanisms, integration of crowd-sourced ideas into the innovation pipeline, and fostering a culture of continuous improvement and co-creation. Effective open innovation leverages web technologies not just for idea submission but for dynamic interaction, collective problem-solving, and iterative refinement of concepts, thereby maximizing the potential for breakthrough innovations.

It is interesting that enterprise and industry versions also progress with web versions. For example, Web 2.0 was used with Enterprise 2.0, Web 3.0 has been used with Enterprise 3.0, and Web 4.0 and Enterprise 4.0, known as Industry 4.0, are also related [24]. We utilize this relation to argue that S-ERP can support open innovation. So, the framework we can use for open innovation is shown in Fig. 3.2.

Fig. 3.2 Framework to study open innovation through S-ERP

Figure 3.2 presents a framework for understanding the application of Web 3.0 and Web 4.0 technologies in the context of open innovation. It is divided into three main stages. The first stage identifies and outlines the essential characteristics of Web 3.0 and Web 4.0 technologies. For Web 3.0, key features include semantic understanding, data interoperability, personalization, decentralization, and artificial intelligence. Web 4.0, while still theoretical, emphasizes ubiquitous connectivity, enhanced AI, human-machine symbiosis, and autonomous agents. The second stage explores how the identified features of Web 3.0 and 4.0 can be leveraged to facilitate and enhance open innovation processes. It involves linking each technological feature to specific aspects of open innovation, such as idea generation, collaboration, evaluation, and implementation. For instance, AI and semantic understanding can be used to analyze and categorize ideas, while decentralization can ensure secure and transparent innovation processes. The final stage describes the various phases of the open innovation process, typically including idea generation, idea evaluation, development, and implementation. It illustrates how Web 3.0 and 4.0 features can support and optimize each of these stages, ensuring a more effective and efficient innovation process. We understand the framework shown in Fig. 3.2 concerning Web 3.0 and 4.0 in the following subsections.

3.2.1 Web 3.0 and Open Innovation in S-ERP

Web 3.0, also known as the Semantic Web, is the third generation of internet services, which focuses on understanding and interpreting the data on the web to provide a more intelligent and connected experience. As discussed above, it has five key features: (1) Semantic understanding, (2) Data interoperability, (3) Personalization, (4) Decentralization, and (5) Utilization of Artificial Intelligence.

The first feature, semantic understanding, can be achieved with Natural Language Processing (NLP). Utilizing natural language processing and machine learning to analyze user-submitted ideas, comments, and feedback enables effective categorization, summarization, and identification of the most promising concepts from large volumes of unstructured data. This approach streamlines the extraction of valuable insights, facilitating a more efficient evaluation of user input. Comprehending the context and semantics behind contributors' inputs enhances the relevance of information shared among collaborators. This ensures that each participant receives pertinent and timely updates, improving the overall effectiveness and coherence of collaborative efforts and ultimately leading to open innovation.

The second feature, data interoperability, is crucial to achieve seamless collaboration. It facilitates data sharing and integration across different platforms and services used by collaborators. Ensure that ideas and feedback from various sources (e.g., social media, internal databases, and third-party innovation platforms) are easily accessible and interoperable. It helps create centralized innovation hubs that integrate data from disparate systems, providing a holistic view of ongoing projects, available resources, and current challenges. It also helps achieve data stewardship requirements for open innovation. Anjaria [6] has discussed stewardship and open innovation in detail.

The third feature is about the personalization. It helps develop personalized innovation portals for users, offering content, challenges, and collaboration opportunities based on their interests, expertise, and past contributions. It requires the use of advanced algorithms to deliver personalized feedback and support to contributors, enhancing their engagement and encouraging continuous participation in the innovation process.

The fourth feature, decentralization, implements blockchain technology to create transparent and secure contributor incentive mechanisms. Smart contracts can automatically reward users for their ideas and contributions, fostering a more collaborative and trust-based innovation environment. It uses decentralized technologies to validate and verify ideas through a distributed network of experts, ensuring unbiased and democratic evaluation processes.

The final feature is artificial intelligence. It incorporates AI-powered virtual assistants to help users navigate the innovation platform, provide insights, and suggest relevant information or collaborators based on their inputs. Further, it uses AI to analyze trends and predict the potential success of new ideas, helping organizations prioritize the most promising innovations and allocate resources effectively. Web 3.0's features align closely with the stages of innovation, enhancing each step from creation to exploitation. The six stages of innovation are creation, filtering, decision-making, acknowledgment, implementation, and exploitation [10].

In the creation stage, semantic understanding and artificial intelligence play pivotal roles. Utilizing NLP and machine learning, organizations can efficiently analyze and categorize user-submitted ideas, comments, and feedback. This enables the identification of the most promising concepts from large volumes of unstructured data. AI-powered tools and virtual assistants further support the creation process by providing insights and suggesting relevant information, fostering a rich environment for generating innovative ideas.

During the filtering stage, semantic understanding and data interoperability are crucial. NLP helps in summarizing and prioritizing ideas based on their potential impact and relevance. Data interoperability ensures that ideas and feedback from various sources, such as social media, internal databases, and third-party platforms, are easily accessible and can be compared seamlessly. This holistic view of data enables better evaluation and selection of ideas, ensuring that only the most viable concepts move forward.

In the decision-making stage, decentralization and artificial intelligence enhance the process. Blockchain technology provides a transparent and secure framework for validating and verifying ideas through a distributed network of experts. This ensures unbiased and democratic evaluation processes. AI helps analyze trends and predict the potential success of new ideas, providing decision-makers with data-driven insights to prioritize innovations effectively.

The acknowledgment stage benefits significantly from personalization and data interoperability. Personalized innovation portals can deliver targeted feedback and recognition to contributors based on their inputs and past contributions. This not only enhances engagement but also ensures that contributors feel valued and motivated. Data interoperability allows for efficient sharing of acknowledgment across platforms, ensuring that recognition is visible and impactful.

During the implementation stage, decentralization and data interoperability facilitate smooth execution. Decentralized technologies, such as blockchain, ensure transparent and secure tracking of the implementation process. Smart contracts can automate the execution of tasks and rewards, fostering a collaborative environment. Interoperability ensures that all relevant data and resources are integrated and accessible, supporting effective project management and implementation.

Finally, in the exploitation stage, artificial intelligence and personalization drive successful commercialization and adoption of innovations. AI can analyze market trends and customer feedback to refine and optimize products or services. Personalized approaches ensure that innovations are tailored to meet the specific needs and preferences of different user segments, enhancing their market acceptance and success.

Thus, the features of Web 3.0—semantic understanding, data interoperability, personalization, decentralization, and artificial intelligence—seamlessly integrate with the stages of innovation, from creation to exploitation. By leveraging these advanced technologies, organizations can enhance their open innovation processes, driving more effective and impactful outcomes. S-ERP can be executed with Web 3.0. So, they support open innovation and stages of open innovation.

Sustainability-related open innovation in business [8] can be significantly enhanced using S-ERP systems. S-ERP integrates sustainability into traditional ERP functions, facilitating the incorporation of environmental, social, and governance (ESG) factors into business processes. By leveraging S-ERP, organizations can drive sustainable innovation through more efficient resource management, compliance tracking, and stakeholder engagement. One key aspect of sustainability-related open innovation using S-ERP is the integration of comprehensive data analytics. S-ERP systems can collect and analyze data from various sources, including energy consumption, waste management, and supply chain operations. S-ERP enables businesses to identify opportunities for reducing environmental impact and enhancing resource efficiency by providing real-time insights into these areas. For instance, data analytics can highlight patterns of excessive energy use or waste generation, prompting innovative solutions such as energy-efficient processes or circular economy practices.

Furthermore, S-ERP facilitates seamless collaboration and data sharing among stakeholders, which is crucial for open innovation. Sustainable business practices often require input and cooperation from a wide range of stakeholders [2], including suppliers, customers, regulatory bodies, and nongovernmental organizations (NGOs). S-ERP systems enable secure and transparent data sharing, ensuring that all stakeholders have access to the information needed to contribute to sustainable innovation. This collaborative approach fosters the development of innovative solutions that are not only environmentally sustainable but also economically viable and socially responsible.

S-ERP also supports compliance with sustainability standards and regulations, which is essential for fostering open innovation in this area. By integrating regulatory requirements into business processes, S-ERP ensures that sustainability initiatives are aligned with legal and industry standards. This alignment not only mitigates the risk of non-compliance but also encourages the adoption of best practices and innovative approaches to sustainability. For example, S-ERP can automate the tracking and reporting of carbon emissions, enabling businesses to meet regulatory requirements and identify opportunities for carbon reduction initiatives. Additionally, S-ERP systems enhance the ability to track and measure the impact of sustainability initiatives, providing a clear framework for continuous improvement and innovation. By monitoring key performance indicators (KPIs) related to sustainability, businesses can assess the effectiveness of their initiatives and make data-driven decisions to enhance their sustainability strategies. This ongoing evaluation process encourages a culture of continuous improvement and innovation, as businesses can identify areas for further innovation and refine their approaches to achieve better sustainability outcomes.

Incorporating stakeholder feedback is another crucial element of sustainability-related open innovation using S-ERP. These systems can facilitate the collection and analysis of feedback from various stakeholders, including employees, customers, and community members. By understanding stakeholder perspectives and concerns, businesses can develop more inclusive and impactful sustainability initiatives. S-ERP systems can also support the creation of forums or platforms for stakeholders to share ideas and collaborate on sustainability projects, further driving innovation through diverse perspectives and collective problem-solving.

Thus, the final argument would be that sustainability-related open innovation in business can be significantly enhanced using S-ERP systems. By integrating comprehensive data analytics, facilitating seamless collaboration, ensuring compliance with sustainability standards, tracking and measuring impact, and incorporating stakeholder feedback, S-ERP enables businesses to drive sustainable innovation effectively. This approach enhances environmental performance and supports economic and social sustainability, ultimately leading to more resilient and forward-thinking businesses. Through the strategic use of S-ERP, organizations can transform sustainability challenges into opportunities for innovation and growth, fostering a more sustainable future for all. Thus, this subsection has followed the steps described in Fig. 3.2. The following subsection discusses Web 4.0 and open innovation in Industry 4.0.

3.2.2 *Web 4.0 and Open Innovation in S-ERP*

Web 4.0 is a theoretical concept representing the future evolution of the web, often referred to as the Symbiotic Web. While it is still largely speculative, it envisions a highly intelligent and interconnected web where machines and humans interact seamlessly. Web 4.0 has four features: (1) Ubiquitous connectivity, (2) Enhanced AI, (3) Human-Machine Symbiosis, (4) Autonomous Agents.

The first feature, ubiquitous connectivity, is the pervasive internet access provided by advanced networks and IoT devices, ensuring that innovators and stakeholders can collaborate anytime and anywhere. This constant connectivity allows for real-time communication and collaboration, breaking geographical barriers. It facilitates the continuous flow of ideas and feedback, ensuring all participants are engaged and can contribute to the innovation process. Furthermore, IoT devices can collect and share vast amounts of data from different environments, providing valuable insights to drive innovation.

The second feature, enhanced AI, is a more sophisticated AI capable of understanding and anticipating human needs and can be used to analyze and synthesize vast amounts of data generated during the innovation process. AI can identify trends, patterns, and emerging opportunities that may not be immediately apparent to humans. It can also provide personalized recommendations to innovators, helping them refine their ideas and solutions. AI-driven tools can automate repetitive tasks, freeing human resources to focus on more creative aspects of innovation. Additionally, AI can facilitate better decision-making by providing data-driven insights and predictive analytics.

The third feature, human-machine symbiosis, is close integration between humans and machines, enabling intuitive and seamless interactions, which is crucial for open innovation. This symbiosis allows for the efficient use of human creativity and machine precision. Innovators can leverage advanced interfaces like augmented reality (AR) and virtual reality (VR) to visualize and prototype their ideas in immersive environments. Machines can assist humans by performing complex calculations, simulations, and analyses, enhancing the accuracy and feasibility of innovative solutions. This collaboration between humans and machines fosters a more dynamic and interactive innovation process.

The fourth feature of Web 4.0 is that autonomous agents are highly autonomous systems that can make decisions and perform tasks on behalf of users, which can streamline and accelerate the innovation process. Autonomous agents can manage routine tasks, coordinate activities among stakeholders, and ensure the innovation process remains on track. These agents can also gather and analyze data, identify potential issues, and propose solutions, reducing the time and effort required for manual oversight. By handling administrative and operational aspects, autonomous agents allow innovators to focus on generating and developing new ideas. Thus, the features of Web 4.0—ubiquitous connectivity, enhanced AI, human-machine symbiosis, and autonomous agents—can significantly enhance open innovation by fostering continuous collaboration, providing valuable insights, enabling seamless human-machine interactions, and automating routine tasks. By leveraging these advanced technologies, organizations can create a more efficient, dynamic, and inclusive innovation ecosystem that drives continuous improvement and breakthrough innovations. Web 4.0's features align closely with the stages of innovation, enhancing each step from creation to exploitation. The six stages of innovation are creation, filtering, decision-making, acknowledgment, implementation, and exploitation.

In the creation stage, ubiquitous connectivity and enhanced AI are crucial. Ubiquitous connectivity, provided by advanced networks and IoT devices, ensures that innovators and stakeholders can collaborate anytime and anywhere. This constant connectivity allows for real-time communication and collaboration, breaking geographical barriers and facilitating the continuous flow of ideas and feedback. Enhanced AI, capable of understanding and anticipating human needs, can analyze vast amounts of data, identify trends, and generate insights that help in the creation of innovative ideas.

During the filtering stage, enhanced AI and data interoperability are vital. AI helps in summarizing and prioritizing ideas based on their potential impact and relevance. It can also provide personalized recommendations to innovators, helping them refine their ideas and solutions. Ubiquitous connectivity ensures that data and feedback from various sources are easily accessible and interoperable, allowing for a seamless comparison and evaluation of ideas. This holistic view of data enables better evaluation and selection of ideas, ensuring that only the most viable concepts move forward.

In the decision-making stage, human-machine symbiosis and enhanced AI enhance the process. Human-machine symbiosis allows for intuitive and seamless interactions between humans and machines, leveraging human creativity and machine precision. This collaboration facilitates more accurate and informed decision-making. Enhanced AI provides data-driven insights and predictive analytics, helping decision-makers prioritize the most promising innovations and allocate resources effectively.

The acknowledgment stage benefits significantly from personalization and ubiquitous connectivity. Personalized innovation portals can deliver targeted feedback and recognition to contributors based on their inputs and past contributions. This not only enhances engagement but also ensures that contributors feel valued and motivated. Ubiquitous connectivity allows for efficient sharing of acknowledgment across platforms, ensuring that recognition is visible and impactful.

During the implementation stage, decentralization and human-machine symbiosis facilitate smooth execution. Decentralized technologies, such as blockchain, ensure transparent and secure tracking of the implementation process. Smart contracts can automate the execution of tasks and rewards, fostering a collaborative environment. Human-machine symbiosis ensures that all relevant data and resources are integrated and accessible, supporting effective project management and implementation.

Finally, in the exploitation stage, enhanced AI and personalization drive successful commercialization and adoption of innovations. AI can analyze market trends and customer feedback to refine and optimize products or services. Personalized approaches ensure that innovations are tailored to meet the specific needs and preferences of different user segments, enhancing their market acceptance and success. Thus, the features of Web 4.0—ubiquitous connectivity, enhanced AI, human-machine symbiosis, and autonomous agents—seamlessly integrate with the stages of innovation, from creation to exploitation. By leveraging these advanced technologies, organizations can enhance their open innovation processes, driving more effective and impactful outcomes.

Sustainability-related open innovation in business can be significantly enhanced using S-ERP systems combined with the advanced features of Web 4.0. S-ERP integrates sustainability into traditional ERP functions, facilitating the incorporation of environmental, social, and governance (ESG) factors into business processes. Leveraging the features of Web 4.0—ubiquitous connectivity, enhanced AI, human-machine symbiosis, and autonomous agents –alongside S-ERP can drive sustainable innovation through more efficient resource management, compliance tracking, and stakeholder engagement. One key aspect of sustainability-related open innovation using S-ERP and Web 4.0 is the integration of comprehensive data analytics. S-ERP systems, enhanced by Web 4.0's ubiquitous connectivity and IoT devices, can collect and analyze data from various sources, including energy consumption, waste management, and supply chain operations. By providing real-time insights into these areas, S-ERP enables businesses to identify opportunities for reducing environmental impact and enhancing resource efficiency. For instance, enhanced AI can highlight patterns of excessive energy use or waste generation, prompting innovative solutions such as energy-efficient processes or circular economy practices.

Furthermore, S-ERP, supported by Web 4.0's seamless collaboration capabilities, facilitates data sharing among stakeholders, which is crucial for open innovation. Sustainable business practices often require input and cooperation from a wide range of stakeholders, including suppliers, customers, regulatory bodies, and non-governmental organizations (NGOs). S-ERP systems, with the help of Web 4.0's ubiquitous connectivity, enable secure and transparent data sharing, ensuring that all stakeholders have access to the information needed to contribute to sustainable innovation. This collaborative approach fosters the development of innovative solutions that are environmentally sustainable, economically viable, and socially responsible.

S-ERP, augmented by Web 4.0's autonomous agents, supports compliance with sustainability standards and regulations, which is essential for fostering open innovation. By integrating regulatory requirements into business processes, S-ERP ensures that sustainability initiatives are aligned with legal and industry standards. This alignment mitigates the risk of non-compliance and encourages the adoption of best practices and innovative approaches to sustainability. For example, autonomous agents can automate the tracking and reporting of carbon emissions, enabling businesses to meet regulatory requirements and identify opportunities for carbon reduction initiatives.

Additionally, S-ERP systems, enhanced by Web 4.0's human-machine symbiosis, enhance the ability to track and measure the impact of sustainability initiatives, providing a clear framework for continuous improvement and innovation. By monitoring key performance indicators (KPIs) related to sustainability, businesses can assess the effectiveness of their initiatives and make data-driven decisions to enhance their sustainability strategies. This ongoing evaluation process encourages a culture of continuous improvement and innovation, as businesses can identify areas for further innovation and refine their approaches to achieve better sustainability outcomes.

Incorporating stakeholder feedback is another crucial element of sustainability-related open innovation using S-ERP and Web 4.0. These systems can facilitate the collection and analysis of feedback from various stakeholders, including employees, customers, and community members. By understanding stakeholder perspectives and concerns, businesses can develop more inclusive and impactful sustainability initiatives. Web 4.0's features, such as enhanced AI and ubiquitous connectivity, can support the creation of forums or platforms for stakeholders to share ideas and collaborate on sustainability projects, further driving innovation through diverse perspectives and collective problem-solving.

In conclusion, sustainability-related open innovation in business can be significantly enhanced using S-ERP systems combined with the advanced features of Web 4.0. By integrating comprehensive data analytics, facilitating seamless collaboration, ensuring compliance with sustainability standards, tracking and measuring impact, and incorporating stakeholder feedback, S-ERP and Web 4.0 enable businesses to drive sustainable innovation effectively. This approach enhances environmental performance and supports economic and social sustainability, leading to more resilient and forward-thinking businesses. Through the strategic use of S-ERP and Web 4.0, organizations can transform sustainability challenges into opportunities for innovation and growth, fostering a more sustainable future for all.

Throughout the section, we have followed Fig. 3.2 by first outlining the key features of Web 3.0 and Web 4.0, including semantic understanding, data interoperability, personalization, decentralization, ubiquitous connectivity, enhanced AI, human-machine symbiosis, and autonomous agents. We then related these features to open innovation by explaining how each can enhance collaboration, data sharing, decision-making, and implementation processes. Finally, we mapped these features to the stages of open innovation—creation, filtering, decision-making, acknowledgment, implementation, and exploitation—demonstrating their impact at each stage. After the discussion on open innovation, we discuss change management with and due to S-ERP and Industry 4.0.

3.3 Change Management: S-ERP and Industry 4.0 Perspectives

Change management is critical to modern organizational strategies, particularly with the advent of Industry 4.0 and the integration of S-ERP systems. Industry 4.0, characterized by the fusion of advanced technologies such as the IoT, AI, and big data analytics, demands a significant shift in traditional business processes and workflows. Implementing S-ERP systems facilitates this transformation by offering a more agile, efficient, and environmentally responsible approach to resource planning and management. Effective change management practices ensure that organizations can seamlessly transition to these advanced systems, minimizing disruption and maximizing the benefits of technological advancements. This involves technological integration and addressing the human and organizational factors essential for a successful transformation. By focusing on S-ERP and Industry 4.0 perspectives, organizations can create a cohesive strategy supporting innovation, enhancing productivity, ensuring sustainability, and maintaining competitive advantage in an increasingly digitalized and eco-conscious world. The present section discusses the four aspects of change management. These aspects are shown in Fig. 3.3.

We discuss each aspect of change management shown in Fig. 3.3 in the following subsections. The literature discusses the generic frameworks of change management like Lewin's change management model [15] and McKinsey's 7S model [30]. In the following subsections, we will discuss specific change management frameworks critical for S-ERP in Industry 4.0 instead of generic ones.

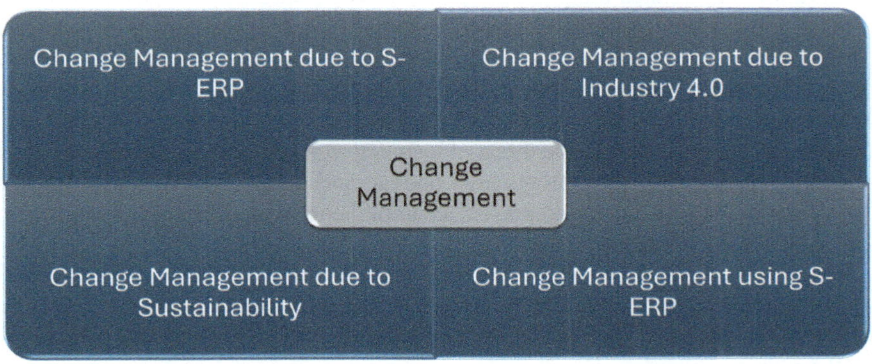

Fig. 3.3 Aspects of change management

3.3.1 Change Management in the Organization Due to the Adoption of S-ERP

Jing and Qiu have clearly stated that implementing S-ERP, or any ERP in general, is complex, expensive, time-consuming, and arduous [16]. Till 2007, companies used to spend at least four times as much on ERP implementations as they had on software licenses. Many researchers have suggested various success factors of ERP implementation. For example, Nah stated 11 success factors: Appropriate Business and IT legacy Systems, Business plan and vision, Business Process Reengineering (BPR), Change Management Culture and Program, Effective communication, ERP teamwork and composition, Monitoring and evaluation of Performance, Project Champion, Project management, Software Development, Testing and Troubleshooting, Top Management Support [22]. Salas provided the following five success factors of ERP: top management support, business case, change management, project management, and training [23].

Thus, in each of these studies, change management is a common factor that is considered an important element. Change management is essential to prepare users for implementing new systems, mitigate resistance to these systems, and positively influence user attitudes toward the change. Many studies have shown that companies that perform proper change management activities feel they ultimately control ERP implementation [17]. The change management exercise combines psychology, human resources, and strategy making [13]. Furthermore, change management is a continuous process that starts at the beginning of the project and is constant throughout the project lifecycle.

In technology-related things like ERP, S-ERP, and Industry 4.0, change management becomes more complex as the organization's employees may or may not be ready to change the technological platform. On top of that, they will resist the change in organizational or business processes [25]. In the case of S-ERP, both would be necessary. Researchers have tried to identify the reason behind the resistance and found many reasons, such as the threat of losing power, jobs, extra training or project schedules to adopt the technology, high budget, and many more. We have studied these reasons, and we attempt to present a comprehensive model based on that study. The model is shown in Fig. 3.4.

Figure 3.4 shows the framework based on the fact that S-ERP systems integrate sustainability into traditional ERP functions, ensuring that business operations align with environmental, social, and economic sustainability goals. Implementing an S-ERP requires comprehensive change management to ensure smooth transitions and lasting benefits.

Conducting a thorough assessment of current ERP systems, technology and sustainability practices is the **first step**. Identifying gaps and areas needing improvement to align with sustainability goals is crucial. Developing a detailed implementation plan, outlining steps, timelines, and resources required, follows this assessment. Securing commitment from top leadership to champion the S-ERP initiative and engaging key stakeholders early, including employees, customers, suppliers, and

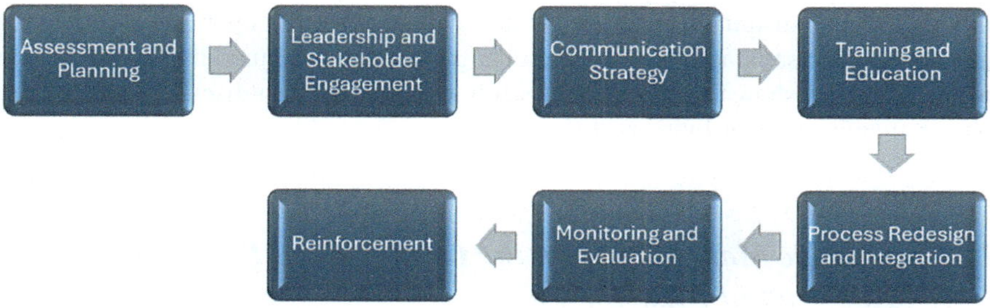

Fig. 3.4 Change management model required due to S-ERP

regulatory bodies, are vital steps. Establishing a steering committee to oversee the project and ensure alignment with sustainability objectives enhances this process. These tasks are achieved in the **second step**, shown in Fig. 3.4.

Developing a comprehensive communication plan to inform all stakeholders about the benefits and impacts of S-ERP is essential, as shown in the **third step** of Fig. 3.4. Using multiple channels such as meetings, emails, and workshops to disseminate information and gather feedback helps ensure transparent communication, which builds trust and mitigates resistance. Providing extensive training programs for employees at all levels to understand and use the new S-ERP system is crucial. Incorporating sustainability principles into the training modules to highlight the system's environmental and social benefits and offering ongoing support and refresher courses ensures proficiency and adaptation. The importance of the training is incorporated in the **fourth step**.

Redesigning business processes to integrate S-ERP functionalities effectively and ensuring that new methods are aligned with sustainability goals and regulatory requirements are necessary. The importance of redesigning is in the **fifth step** of Fig. 3.4. Testing the redesigned processes in pilot programs before full-scale implementation ensures their effectiveness. Establishing metrics to measure the performance of the S-ERP system against predefined sustainability objectives and conducting regular audits and evaluations to ensure compliance and identify improvement areas are also essential. This monitoring is achieved in the **sixth step**. Using feedback mechanisms to improve the system and address any issues continuously and promptly is vital. Recognizing and rewarding individuals and teams that contribute to the successful implementation of S-ERP fosters a culture of continuous improvement and sustainability within the organization. Maintaining momentum by setting new sustainability targets and updating the S-ERP system as needed is crucial. This reinforcement is achieved in the **seventh step** of Fig. 3.4.

The key takeaway is that implementing an S-ERP system requires a well-structured change management model that focuses on thorough planning, strong leadership, stakeholder engagement, effective communication, comprehensive training, process integration, continuous monitoring, and reinforcement of changes. By following these steps, organizations can ensure that their S-ERP implementation not only meets their sustainability goals but also enhances overall operational efficiency and stakeholder satisfaction. Clavert's change management model [9] inspires the proposed model shown in Fig. 3.4. The following subsection discusses the change management model for Industry 4.0.

3.3.2 *Change Management in the Organization Due to the Adoption of Industry 4.0*

Like in the previous subsection, we study the success factors of influential Industry 4.0 implementation from the literature. Zhou et al. suggested five success factors for Industry 4.0 implementation: (1) top management support, (2) technology competence, (3) supplier support, (4) pilot projects, and (5) organizational change management [29]. Kumar et al. suggested six success factors for Industry 4.0 implementation for sustainability: (1) environmental regulations for sustainability, (2) adequate labor laws for the workforce, (3) Effective change management and restructuring of the organization, (4) Support from all the stakeholders, (5) Strategic roadmap, and (6) branding of the green image [19]. Babel indicated five elements for the success of Industry 4.0: (1) collaborative innovation, (2) comprehensive integration, (3) data-driven, (4) organizational change, and (5) internet of everything [7]. In all the literature, change management is the only common thing endorsed by most researchers. This subsection presents the common framework of change management for Industry 4.0. The framework is presented in Fig. 3.5.

Step 1 of Fig. 3.5 involves aligning the technological advancements with the needs and expectations of individuals, the company, and public policies. It ensures that integrating new technologies considers the human element, organizational goals, and regulatory requirements. This alignment is crucial for creating a cohesive strategy that supports innovation while maintaining compliance and addressing stakeholder concerns. **Step 2** focuses on decentralizing functions within the organization. Instead of centralized control, various functions are distributed across different departments or teams. This approach promotes agility, enhances collaboration, and allows for more efficient decision-making processes. Distributed functionalities enable the organization to respond rapidly to changes and leverage diverse expertise.

Fig. 3.5 Change management model required in the organization due to adoption of Industry 4.0

In **Step 3** of Fig. 3.5, the organization restructures its traditional hierarchical teams into a more dynamic network of teams. This transformation emphasizes interconnectivity and collaboration among teams, breaking down silos and fostering a culture of teamwork. A network of teams enhances flexibility, innovation, and the ability to tackle complex challenges by leveraging the collective strengths of diverse team members. **Step 4** involves providing extensive training to employees to equip them with the necessary skills and knowledge to adapt to new technologies and processes. Effective communication strategies keep all stakeholders informed and engaged throughout the change process. Continuous monitoring ensures that the implementation progresses as planned, identifying any issues early and allowing for timely interventions. This step is critical for maintaining momentum and ensuring everyone is on the same page.

The final step, **Step 5** focuses on evaluating the outcomes of the change initiatives against predefined objectives. It involves assessing the performance, identifying areas for improvement, and reinforcing successful practices. Recognizing and rewarding contributions to successfully implementing changes helps sustain the momentum and fosters a culture of continuous improvement. Evaluation and reinforcement ensure the changes are embedded within the organization and drive long-term benefits. By following these steps, organizations can effectively manage the transition to Industry 4.0, ensuring that technological advancements are seamlessly integrated into their operations while enhancing overall efficiency and stakeholder satisfaction. The following subsection discusses the change management required due to sustainability aspects.

3.3.3 Change Management in the Organization Due to the Adoption of Sustainable Practices

For several reasons, change management is essential when an organization is moving toward sustainability. Sustainability initiatives require changes across various elements of the company, including operations, management, strategy, procurement, marketing, and communication. Change management helps ensure that these efforts are integrated holistically rather than compartmentalized [20]. The decision to pursue sustainability often stems from internal motivations and external pressures. Change management helps balance these forces, aligning internal operations with external expectations and standards, such as those from the Global Reporting Initiative (GRI) [26]. Sustainability is not a one-time effort but a continuous process.

Change management facilitates ongoing improvement and adaptation, helping organizations transition from the current state to a more sustainability-oriented state (MSOS) and making these changes part of the new status quo. Organizational change often encounters resistance from employees and other stakeholders. Change management provides strategies to overcome this resistance, ensuring smoother implementation of sustainability initiatives. Moving toward sustainability usually requires a shift in corporate culture, including changes in values, visions, policies, and employee behaviors. Change management helps orchestrate these cultural changes, fostering an environment that supports sustainability goals. Effective sustainability reporting and efforts require active stakeholder engagement, including employees, customers, suppliers, and regulators.

Change management ensures that these stakeholders are involved and that their feedback is incorporated into sustainability strategies. Change management helps set up systems for monitoring sustainability performance and aligning these efforts with the company's strategic goals. This ensures that sustainability initiatives contribute to the overall success of the organization. Properly managed change toward sustainability can enhance an organization's reputation and provide external validation of its efforts. Change management ensures that sustainability reporting accurately reflects the company's efforts and achievements, improving transparency and trust. The document emphasizes that change management and sustainability reporting have reciprocal reinforcing relationships. Sustainability reporting acts as both a catalyst for change and a reflection of the changes made, driving a continuous cycle of improvement and adaptation toward sustainability. The framework for the change management is shown in Fig. 3.6.

Step 1 of the change management model for sustainability, as shown in Fig. 3.6, is holistic integration. Sustainability initiatives require changes across various elements of the company, including operations, management, strategy, procurement, marketing, and communication. Change management ensures these efforts are integrated holistically rather than in a compartmentalized manner [28]. In **Step 2**, internal and external drivers, the decision to pursue sustainability often stems from a combination of internal motivations and external pressures. Change management helps balance these forces, aligning internal operations with external expectations and

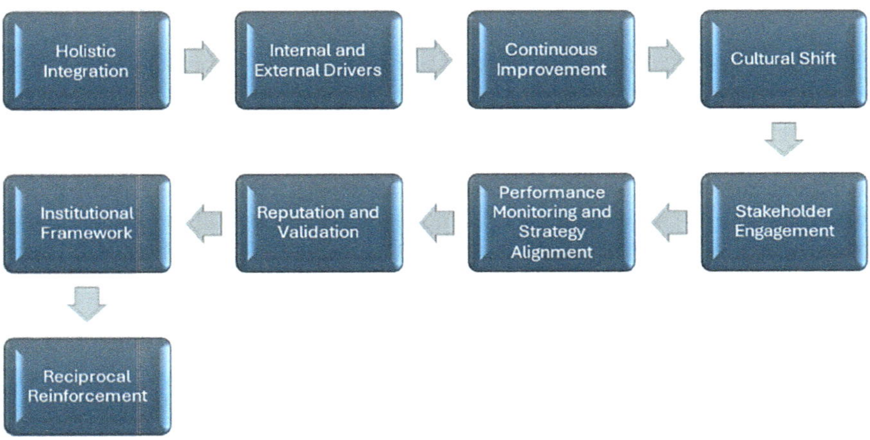

Fig. 3.6 Change management model required in the organization due to adoption of sustainable practices

standards, such as those from the Global Reporting Initiative (GRI). **Step 3** suggests that Sustainability is a continuous process. Change management facilitates ongoing improvement and adaptation, helping organizations transition from the current state to a more sustainability-oriented state (MSOS) and then making these changes part of the new status quo. Organizational change often encounters resistance from employees and other stakeholders. Change management provides strategies to overcome this resistance, ensuring smoother implementation of sustainability initiatives.

Step 4 discusses cultural shift. Moving toward sustainability often requires a shift in corporate culture, including changes in values, visions, policies, and employee behaviors. Change management helps orchestrate these cultural changes, fostering an environment that supports sustainability goals. Effective sustainability reporting and efforts require active engagement with stakeholders, including employees, customers, suppliers, and regulators. Change management ensures that these stakeholders are involved and that their feedback is incorporated into sustainability strategies. **Step 5**, titled "Stakeholder Engagement," elucidates that effective sustainability reporting and efforts require active engagement with stakeholders, including employees, customers, suppliers, and regulators. Change management ensures that these stakeholders are involved and that their feedback is incorporated into sustainability strategies. **Step 6** suggests that change management should incorporate performance monitoring and strategy alignment. Change management helps set up systems for monitoring sustainability performance and aligning these efforts with the company's strategic goals, ensuring that sustainability initiatives contribute to the overall success of the organization.

Properly managed change toward sustainability can enhance an organization's reputation and provide external validation of its efforts, as shown in **Step 7**. Change management ensures that sustainability reporting accurately reflects the company's efforts and achievements, thereby improving transparency and trust. The Institutional Framework of **Step 8** helps maintain stability during changes and facilitates corporate sustainability (CS) institutionalization. Planning organizational changes while engaging with different organizational levels can help overcome resistance to change and integrate sustainability efforts holistically, including technological, managerial, and organizational changes. Sustainability Reporting (SR) and Organizational Change Management for Sustainability (OCMS) have reciprocal reinforcing relationships that **Step 9** suggests. SR acts as both a catalyst for change and a reflection of the changes made, driving a continuous cycle of improvement and adaptation toward sustainability. The next subsection discusses S-ERP as a tool to manage the change.

3.3.4 *Change Management in the Organization Using S-ERP*

Change Management using ERP or S-ERP is not a widely used and explored concept in the literature. Gunjal, in 2019, discussed ERP as a change management tool [12]. After that, no other research on this aspect of ERP. However, we believe that S-ERP can be used to manage the change in the organization.

We believe that S-ERP can be used as a change management tool as it provides a holistic view of the organization's operations, enabling better planning and coordination of change initiatives. It supports change management by providing a centralized, integrated, and flexible platform that enhances communication, standardizes processes, and enables real-time monitoring and decision-making. These capabilities make S-ERP an indispensable tool for managing organizational change. Organizations can follow the steps shown in Fig. 3.7 to use S-ERP as a change management tool.

S-ERP systems are scalable, allowing organizations to expand their sustainability efforts as they grow. This scalability ensures that sustainability remains a core focus even as the organization evolves. Implementing S-ERP for change management in Industry 4.0 involves several steps, as shown in Fig. 3.7. In **Step 1**, assess current sustainability practices by evaluating existing ones using S-ERP's data analytics tools, identifying areas for improvement, and setting sustainability goals. In **Step 2**, a comprehensive change management plan should be developed integrating Industry 4.0 technologies with S-ERP capabilities, defining clear objectives, timelines, and roles for implementing sustainable practices. Communication and education are essential; use S-ERP's communication tools to inform employees about the sustainability goals and changes and provide training on new processes and technologies to ensure smooth adoption as described in **Step 3**. Monitoring and adjusting are crucial elements shown in **Step 4**; utilize real-time monitoring and reporting features of S-ERP to track progress, gather feedback, and make necessary adjustments to the change management plan. Collaboration, as described in **Step 5**, with stakeholders

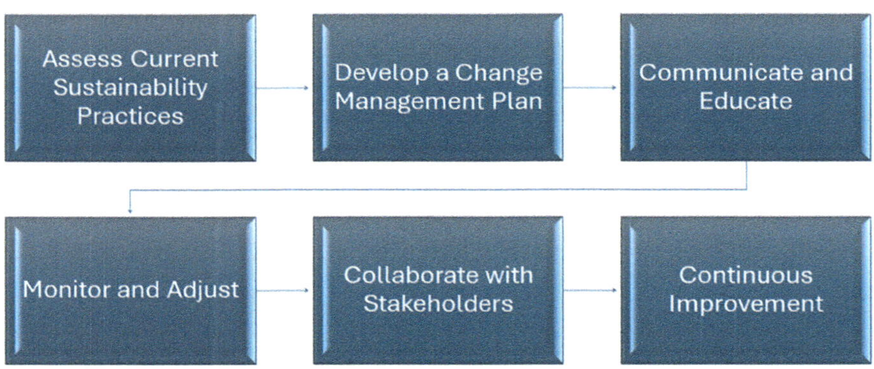

Fig. 3.7 Steps to use S-ERP as a change management tool

is also important; engaging with suppliers, partners, and other stakeholders using S-ERP's collaboration tools to ensure sustainability practices are aligned across the supply chain. Finally, **Step 6** fosters a culture of continuous improvement by regularly reviewing and updating sustainability practices, using insights from S-ERP's data analytics to drive further enhancements. For example, a manufacturing company in Industry 4.0 can use S-ERP to optimize its energy consumption by integrating IoT sensors throughout its production line. The S-ERP system analyzes the real-time data collected by these sensors to identify inefficiencies and suggest adjustments. The company can significantly reduce its energy usage by standardizing these adjustments across all production sites, lowering costs and minimizing its environmental impact. We believe that further research is required to test the above steps. Further, the execution of the steps should be in which sequence also needs to be studied. In summary, the chapter answers the following research questions.

Research Questions Addressed

The chapter titled "Change Management and Open Innovation with S-ERP in Industry 4.0" addresses the following key research questions:

1. **How can S-ERP systems facilitate open innovation in Industry 4.0?**

- The chapter explores the integration of Web 3.0 and Web 4.0 technologies with S-ERP systems, discussing how these integrations can enhance open innovation processes, data interoperability, personalized innovation, and decentralized validation.

2. **What are the critical success factors for implementing change management in S-ERP systems within Industry 4.0?**

- The chapter delves into change management models explicitly tailored for S-ERP and Industry 4.0, highlighting key factors such as top management support, effective communication, continuous training, and stakeholder engagement.

3. **How do S-ERP systems align sustainability goals with open innovation strategies?**

- The chapter examines how S-ERP systems incorporate sustainability metrics into traditional ERP functions, enabling organizations to drive sustainable innovation through better resource management, compliance tracking, and stakeholder engagement.

4. **What are the challenges and strategies for managing change due to implementing S-ERP systems and Industry 4.0 technologies?**

- The chapter discusses the complexities of change management required by the technological advancements of Industry 4.0 and the introduction of S-ERP systems, offering a comprehensive change management model to address these challenges.

5. **How can S-ERP be used as a tool for change management in the context of Industry 4.0?**

- The chapter proposes that S-ERP systems can be employed as a change management tool, providing a centralized, integrated, and flexible platform that enhances communication, standardizes processes, and enables real-time monitoring and decision-making.

These research questions guide the exploration of the role of S-ERP in fostering innovation, managing change, and supporting sustainability within the dynamic environment of Industry 4.0.

References

1. Delone WH, McLean ER (2003) The DeLone and McLean model of information systems success: a ten-year update. J Manage Informat Syst 19(4):9–30. ISSN 0742-1222. https://doi.org/10.1080/07421222.2003.11045748. Publisher: Routledge _eprint: https://doi.org/10.1080/07421222.2003.11045748
2. Andersson S, Svensson G, Otero-Neira C, Laurell H, Lindgren J, Karlsson NPE (2023) Sustainable development considerations in supply chains: Firms' relationships with stakeholders in their business sustainability practices – A triangular comparison. Business Strategy Environ 32(4):1885–1899. ISSN 1099-0836. https://doi.org/10.1002/bse.3225. https://onlinelibrary.wiley.com/doi/abs/10.1002/bse.3225. _eprint: https://onlinelibrary.wiley.com/doi/pdf/10.1002/bse.3225
3. Anjaria K (2024) Enhancing sustainability integration in sustainable enterprise resource planning (S-ERP) system: application of transaction cost theory and case study analysis. Int J Inf Manage Data Insights 4 (2):100243. ISSN 2667-0968. https://doi.org/10.1016/j.jjimei.2024.100243. https://www.sciencedirect.com/science/article/pii/S2667096824000326
4. Anjaria K, Mishra A (2017) Quantitative analysis of information leakage in service-oriented architecture-based Web services. Kybernetes 46(3):479–500. ISSN 0368-492X. https://doi.org/10.1108/K-07-2016-0178. Publisher: Emerald Publishing Limited
5. Anjaria K, Patel N (2021) Attainment of green computing in cloudlet-based mobile cloud computing model using squirrel search algorithm. In: Mahapatra RP, Panigrahi BK, Kaushik BK, Roy S (eds) Proceedings of 6th international conference on recent trends in computing. Springer, Singapore, pp 457–470
6. Anjaria KA (2020) Computational implementation and formalism of FAIR data stewardship principles. Data Technol Applicat 54(2):193–214. ISSN 2514-9288. https://doi.org/10.1108/DTA-09-2019-0164. Publisher: Emerald Publishing Limited
7. Babel W (2022) Industry 4.0, China 2025, IoT (internet of things), AI. In: Babel W (ed) Industry 4.0, China 2025, IoT: the hype around the world of automation. Springer Fachmedien, Wiesbaden, pp 7–46. ISBN 978-3-658-37852-3
8. Bengtsson NJ (2020) Open innovation for enhancing sustainability: a case study on the sustainability-related implications of open innovation projects. https://urn.kb.se/resolve?urn=urn:nbn:se:uu:diva-415470
9. Calvert C (2006) A change-management model for the implementation and upgrade of ERP systems. In: ACIS 2006 proceedings. https://aisel.aisnet.org/acis2006/18
10. Carbone F, Contreras J, Hernández JZ, Gomez-Perez JM (2012) Open innovation in an enterprise 3.0 framework: three case studies. Expert Syst Applicat 39(10):8929–8939. ISSN 0957-4174. https://doi.org/10.1016/j.eswa.2012.02.015. https://www.sciencedirect.com/science/article/pii/S095741741200259X

11. Gloor PA (2006) Swarm creativity: competitive advantage through collaborative innovation networks. Oxford University Press. ISBN 978-0-19-804136-8. Google-Books-ID: 5b3L8jP_IqEC
12. Gunjal S (2019) Enterprise Resource Planning (ERP) as a change management tool. https://papers.ssrn.com/abstract=3525989
13. Hagl C, Kanitz R, Gonzalez K, Hoegl M (2024) Change management interventions: taking stock and moving forward. Human Resourc Manage Rev 34(1):101000. ISSN 1053-4822. https://doi.org/10.1016/j.hrmr.2023.101000. https://www.sciencedirect.com/science/article/pii/S1053482223000530
14. Hekkert MP, Suurs RAAA, Negro SO, Kuhlmann S, Smits REHM (2007) Functions of innovation systems: a new approach for analysing technological change. Technol Forecast Soc Change 74(4):413–432. ISSN 0040-1625. https://doi.org/10.1016/j.techfore.2006.03.002. https://www.sciencedirect.com/science/article/pii/S0040162506000564
15. Hussain ST, Lei S, Akram T, Haider MJ, Hussain SH, Ali M (2018) Kurt Lewin's change model: a critical review of the role of leadership and employee involvement in organizational change. J Innovat Knowl 3(3):123–127. ISSN 2444-569X. https://doi.org/10.1016/j.jik.2016.07.002. https://www.sciencedirect.com/science/article/pii/S2444569X16300087
16. Jing R, Qiu X (2007) A study on critical success factors in ERP systems implementation. In 2007 international conference on service systems and service management, pp 1–6
17. Kemp MJ, Low GC (2008) ERP innovation implementation model incorporating change management. Business Process Manage J 14(2):228–242. ISSN 1463-7154. https://doi.org/10.1108/14637150810864952. Publisher: Emerald Group Publishing Limited
18. Kreiterling C (2023) Digital innovation and entrepreneurship: a review of challenges in competitive markets. J Innovat Entrepreneurship 12(1):49. ISSN 2192-5372. https://doi.org/10.1186/s13731-023-00320-0
19. Kumar V, Vrat P, Shankar R (2022) Factors influencing the implementation of industry 4.0 for sustainability in manufacturing. Glob J Flexible Syst Manage 23(4):453–478. ISSN 0974-0198. https://doi.org/10.1007/s40171-022-00312-1
20. Lozano R, Nummert B, Ceulemans K (2016) Elucidating the relationship between sustainability reporting and organisational change management for sustainability. J Cleaner Product 125:168–188. ISSN 0959-6526. https://doi.org/10.1016/j.jclepro.2016.03.021. https://www.sciencedirect.com/science/article/pii/S0959652616300816
21. Markus ML (2004) Technochange management: using IT to drive organizational change. J Inf Technol 19(1):4–20. ISSN 0268-3962. https://doi.org/10.1057/palgrave.jit.2000002. Publisher: SAGE Publications Ltd
22. Nah FF-H, Zuckweiler KM, Lau JL-S (2003) ERP implementation: chief information officers' perceptions of critical success factors. Int J Human – Comput Interact 16(1). ISSN 1044-7318
23. Salas WH (2023) Model to improve an ERP implementation based on agile best practice: a Delphi study. Proc Comput Sci 219:1785–1792. ISSN 1877-0509. https://doi.org/10.1016/j.procs.2023.01.474. https://www.sciencedirect.com/science/article/pii/S1877050923004866
24. Sheth A, Thirunarayan K (2013) Semantics empowered web 3.0: managing enterprise, social, sensor, and cloud-based data and services for advanced applications. Synthesis lectures on data management. Springer International Publishing, Cham. ISBN 978-3-031-00766-8 978-3-031-01894-7. https://doi.org/10.1007/978-3-031-01894-7
25. Singh S, Singh S, Misra SC (2022) Post-implementation challenges of ERP system in pharmaceutical companies. Int J Quality Reliab Manage 40(4):889–921. ISSN 0265-671X. https://doi.org/10.1108/IJQRM-10-2020-0333. Publisher: Emerald Publishing Limited
26. Sroufe R (2017) Integration and organizational change towards sustainability. J Cleaner Product 162: 315–329. ISSN 0959-6526. https://doi.org/10.1016/j.jclepro.2017.05.180. https://www.sciencedirect.com/science/article/pii/S0959652617311320
27. Liem NT, Khuong NV, Khanh THT (2019) Firm constraints on the link between proactive innovation, open innovation and firm performance. J Open Innovat: Technol, Market, Complexity 5(4):88. ISSN 2199-8531. https://doi.org/10.3390/joitmc5040088. https://www.mdpi.com/2199-8531/5/4/88. Number: 4 Publisher: Multidisciplinary Digital Publishing Institute

28. Warner B (2024) Organization change, July 2024. https://uk.sagepub.com/en-gb/eur/organization-change/book259161
29. Zhou H, Zhou B, Nie Z, Zheng L (2024) Identifying key success factors for industry 4.0 implementation: an empirical analysis using SEM and fsQCA. Appl Sci 14(12):5244. ISSN 2076-3417. https://doi.org/10.3390/app14125244. https://www.mdpi.com/2076-3417/14/12/5244. Number: 12 Publisher: Multidisciplinary Digital Publishing Institute
30. Zincir O, Tunç AÖ (2020) An imagination of organizations in the future: rethinking McKinsey's 7S model. In: Natural language processing: concepts, methodologies, tools, and applications. IGI Global, pp 1667–1685. ISBN 978-1-79980-951-7. https://doi.org/10.4018/978-1-7998-0951-7.ch077. https://www.igi-global.com/chapter/an-imagination-of-organizations-in-the-future/www.igi-global.com/chapter/an-imagination-of-organizations-in-the-future/240008

Chapter 4
Sustainable Enterprise Resource Planning (S-ERP) Roadmapping in Industry 4.0

4.1 Introduction

Roadmapping is a strategic planning technique to outline the future path and evolution of a project, technology, product, or business. It visually represents the steps, milestones, goals, and deliverables over time. Roadmaps help organizations present the strategic direction and priorities to all stakeholders, ensuring that everyone involved is working toward the same goals and objectives. They also highlight potential challenges and dependencies that could impact progress and provide a framework for making informed decisions about resource allocation, scheduling, and priority setting.

S-ERP in an Industry 4.0 setup involves integrating advanced technologies such as IoT, AI, machine learning, and big data analytics into ERP systems to enhance efficiency, sustainability, and adaptability. Roadmapping is essential in this context for several reasons. It helps in planning and coordinating the integration of new technologies into the existing ERP system, aligning the ERP implementation with sustainability goals to ensure that the system supports environmentally friendly and sustainable practices. Roadmapping also assists in managing the organizational change required to adopt new technologies and processes, ensures that the ERP system is scalable and adaptable to future technological advancements and business needs, and helps effectively allocate financial and human resources to different phases of the ERP implementation.

Roadmapping and change management are both essential components of strategic planning but serve different purposes. Roadmapping concentrates on the strategic planning and visualization of future projects, technologies, or business developments. It is broad and long-term, covering the overall direction, milestones, and goals over an extended period. The purpose of roadmapping is to provide a high-level view of what needs to be done, when, and why, aligning all stakeholders towards a common vision. On the other hand, change management deals with the human side of change, ensuring that individuals and teams successfully adopt and adapt to new changes. Its

K. Anjaria, *Sustainable Enterprise Resource Planning (S-ERP) for Industry 4.0*,
Contributions to Management Science, https://doi.org/10.1007/978-981-96-7734-4_4

scope is narrower and more focused on the short to medium term, addressing specific changes and their impacts on the organization. The purpose of change management is to ensure a smooth transition and adoption of new processes, systems, or behaviors, minimizing resistance and maximizing engagement.

Effective roadmapping for S-ERP in Industry 4.0 must integrate both strategic planning and change management to ensure seamless transitions and successful implementations [21]. The strategic roadmap provides the blueprint for technological advancements and sustainability initiatives, detailing the timeline, key milestones, and expected outcomes. Concurrently, change management ensures that the human element is addressed by preparing, supporting, and guiding individuals through the transition. This dual approach helps mitigate resistance, fosters a culture of adaptability, and enhances overall engagement. By aligning the technical and human aspects, organizations can achieve a more cohesive and resilient integration of S-ERP systems [1]. Continuous feedback loops and iterative processes are critical in this context, as they allow for real-time adjustments and refinements to the roadmap, ensuring it remains responsive to both internal dynamics and external market shifts. Additionally, leveraging data-driven insights and predictive analytics can further optimize resource allocation, risk management, and decision-making processes, ultimately driving the organization toward its long-term strategic goals. This holistic approach ensures that S-ERP implementations not only meet immediate operational needs but also align with the broader vision of sustainability and technological innovation in Industry 4.0. The chapter explains the Roadmapping of S-ERP in Industry 4.0 in four segments. The segments are shown in Fig. 4.1

The four segments illustrated in Fig. 4.1 are examined as distinct individual sections in this chapter. Sections 4.2 and 4.3 address the strategic alignment and operationalization of roadmapping, respectively. Section 4.4 explores practical tools for

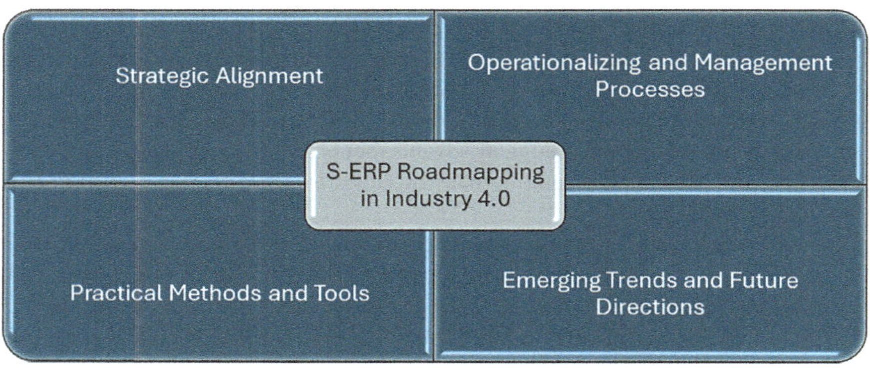

Fig. 4.1 Segments of S-ERP roadmapping in industry 4.0

roadmapping, including knowledge graphs, interaction scenarios, agile roadmapping, design structure matrices, and a multi-layer roadmapping framework. The final section, Sect. 4.5, delves into emerging trends and future research directions related to roadmapping within the context of S-ERP in Industry 4.0.

4.2 Strategic Alignment

S-ERP roadmapping and strategic alignment are crucial components for organizations navigating the complexities of Industry 4.0. This integration ensures that business processes, technology, and innovation are synchronized to achieve sustainable growth and competitive advantage.

Strategic alignment refers to the process of ensuring that all organizational activities, from marketing and technology functions to product development or service delivery, are coordinated and aligned with the overall business strategy. This alignment is essential in Industry 4.0, where rapid technological advancements and market dynamics require a cohesive approach to strategy execution. There are three types of strategic alignment in any business: (1) Horizontal alignment (also known as functional alignment, (2) Vertical alignment (also known as hierarchical alignment), and (3) Temporal alignment (also known as synchronization alignment).

The **horizontal alignment** involves aligning different functions within the organization, such as marketing, technology, and product development, to ensure they work together seamlessly. Miscommunication and conflict between these functions can hinder innovation and strategic goals. Roadmaps serve as a visual tool to bridge these gaps, providing a clear picture of how various functions interrelate and contribute to the overall strategy. The **vertical alignment** ensures that strategies and activities at lower organizational levels are consistent with higher-level strategic goals. It involves integrating strategic plans across different hierarchical levels within the organization, from top management to operational teams. Roadmapping supports this by providing a structured framework that aligns lower-level activities with high-level strategic objectives. The **temporal alignment** focuses on aligning activities and developments over time. It ensures that technological advancements, product development, and market introductions occur in a synchronized manner, meeting commercial expectations and strategic goals. Temporal synchronization in roadmapping helps organizations plan and time their initiatives effectively, ensuring readiness and market alignment.

S-ERP roadmapping involves creating a structured plan that aligns enterprise resource planning systems with sustainability goals and Industry 4.0 requirements. The roadmapping process helps organizations visualize their path toward integrating sustainable practices into their core business processes using advanced technologies.

Various strategic fit frameworks, such as the ESCO, and SAM models, provide foundational principles for roadmapping. These frameworks emphasize aligning organizational structure, strategy, systems, style, skills, staff, and superordinate goals with the external environment, strategy, core competencies, and organizational configuration. The visual nature of roadmaps supports effective communication within and between organizations. Roadmaps provide a layered representation of key stakeholder views, reducing information asymmetries and facilitating better decision-making. This structured approach supports hierarchical integration, functional alignment, and temporal synchronization, which are critical for strategic fit. Phaal and Muller discuss the visual strategy for roadmapping in detail [14]. Roadmapping integrates with various strategic management processes, including knowledge management, change management, innovation management, design thinking, and systems engineering. By linking these areas, roadmapping helps bridge the gap between theory and practice, enhancing the effectiveness of strategic planning and execution in Industry 4.0 environments. Various strategic fit frameworks, such as the ESCO, and SAM models, provide foundational principles for roadmapping. These frameworks emphasize aligning organizational structure, strategy, systems, style, skills, staff, and superordinate goals with the external environment, strategy, core competencies, and organizational configuration. The visual nature of roadmaps supports effective communication within and between organizations. Roadmaps provide a layered representation of key stakeholder views, reducing information asymmetries and facilitating better decision-making. This structured approach supports hierarchical integration, functional alignment, and temporal synchronization, which are critical for strategic fit. Phaal and Muller discuss the visual strategy for roadmapping in detail [14]. Roadmapping integrates with various strategic management processes, including knowledge management, change management, innovation management, design thinking, and systems engineering. By linking these areas, roadmapping helps bridge the gap between theory and practice, enhancing the effectiveness of strategic planning and execution in Industry 4.0 environments.

S-ERP roadmapping and strategic alignment are vital for organizations seeking to thrive in the dynamic landscape of Industry 4.0. By leveraging strategic fit frameworks and a structured, visual approach, organizations can ensure that their business processes, technological advancements, and sustainability initiatives are aligned and synchronized. This alignment not only supports efficient resource management but also drives innovation and competitive advantage in a rapidly evolving market. The importance and implementation of alignment and roadmapping are discussed in the following case.

4.2.1 Case Study: Implementing S-ERP Roadmapping and Strategic Alignment in an Industry 4.0 Manufacturing Firm

TechManufacture Inc. is a mid-sized manufacturing company specializing in producing smart home devices. With the advent of Industry 4.0, TechManufacture Inc. aims to integrate advanced technologies such as IoT, AI, and robotics into their manufacturing processes. Additionally, they are committed to sustainability and seek to implement S-ERP systems to align their operations with environmental goals.

TechManufacture Inc. faces the challenge of synchronizing its technological advancements with its strategic business objectives. The company needs a cohesive plan that integrates various functions (marketing, technology, product development) and ensures that all activities are aligned with their sustainability goals and Industry 4.0 requirements. To address this challenge, TechManufacture Inc. decides to implement an S-ERP roadmapping strategy to ensure strategic alignment across all organizational levels.

1. Horizontal alignment TechManufacture Inc. creates a cross-functional team involving members from marketing, technology, and product development. This team collaborates to develop a roadmap that aligns their functions. For instance, the marketing team provides insights into market demands, the technology team identifies the required technological advancements, and the product development team integrates these insights into their product designs. The roadmap visually depicts how these functions interrelate, ensuring clear communication and reducing conflicts.
2. Vertical alignment The roadmap is structured to align activities at all hierarchical levels within TechManufacture Inc. Top management defines high-level strategic goals, such as increasing market share in smart home devices and reducing carbon footprint by 30% within five years. These goals are broken down into actionable plans at the operational level. For example, the production team focuses on implementing energy-efficient manufacturing processes, while the R&D team works on developing eco-friendly product materials.
3. Temporal alignment The roadmap includes a timeline that synchronizes the development and introduction of new products with market demands and technological advancements. For instance, the roadmap schedules the launch of a new smart thermostat with integrated AI features to coincide with the peak home improvement season. This synchronization ensures the product is ready for market introduction when demand is highest, maximizing commercial success.

TechManufacture Inc. adopts the McKinsey 7S framework to ensure that all organizational elements are aligned. They assess their current structure, strategy, systems, style, skills, staff, and superordinate goals to identify areas that need adjustment. For example, they implement a new training program to enhance the skills of their staff in using advanced manufacturing technologies. Using the roadmap, TechManufacture Inc. creates a visual representation of its strategic plan. The roadmap includes layers representing market trends, product development timelines, and technological advancements. This visual tool is used in meetings to communicate the strategic plan to all stakeholders, ensuring everyone is on the same page. The roadmapping process is integrated with TechManufacture Inc.'s strategic management activities. They use the roadmap to guide their innovation management efforts, ensuring that new product developments are aligned with market needs and sustainability goals. The roadmap also helps in managing changes within the organization, such as the adoption of new technologies and processes.

Outcome: By implementing S-ERP roadmapping and ensuring strategic alignment, TechManufacture Inc. achieves several key outcomes:

- **Enhanced Collaboration:** Improved communication and collaboration between marketing, technology, and product development teams.
- **Strategic Coherence:** Clear alignment of all activities with high-level strategic goals, leading to more efficient operations.
- **Market Readiness:** Products are developed and launched in sync with market demands, enhancing commercial success.
- **Sustainability:** Implementing sustainable manufacturing practices reduces the company's carbon footprint.

TechManufacture Inc.'s case demonstrates how S-ERP roadmapping and strategic alignment can help organizations navigate the complexities of Industry 4.0. By using structured visual tools and aligning all organizational activities with strategic goals, companies can achieve sustainable growth and maintain a competitive edge in the market. The following section discusses operationalizing and management processes.

4.3 Operationalizing and Managing Processes

This section focuses on operationalizing processes using roadmaps for S-ERP in Industry 4.0. In the context of Industry 4.0, the operationalization of S-ERP systems involves the seamless integration of strategic goals with the day-to-day operations of an organization. Technology RoadMaps (TRM) serve as critical instruments in this process, aligning innovation strategies with operational management to ensure the effective implementation of S-ERP systems. This alignment is vital for maintaining competitiveness in a rapidly evolving technological landscape. The process of operationalization involves four phases. These phases are shown in Fig. 4.2.

The first phase discusses one of the primary challenges in operationalizing S-ERP in Industry 4.0, i.e., the transition from broad strategic goals to specific, actionable projects. This involves decomposing complex strategic objectives into a limited number of key priorities that can be addressed through targeted innovation projects. For instance, a major Russian oil and gas company utilized a roadmap for

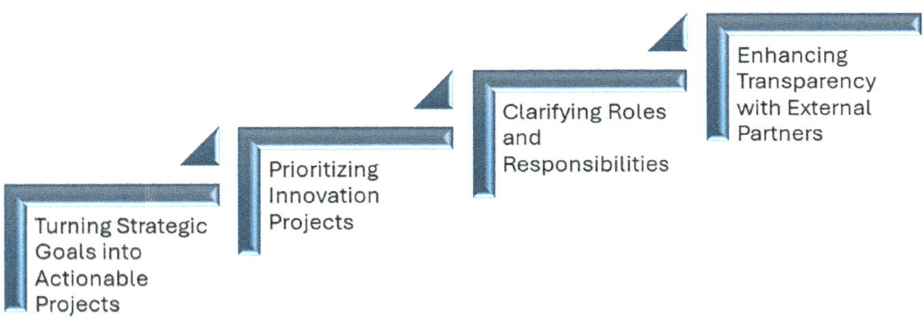

Fig. 4.2 The operationalizing processes of S-ERP in Industry 4.0

developing local technologies for subsea production [5]. This roadmap facilitated aligning strategic goals with operational tasks by breaking down the development process into manageable projects, each with clear milestones and timelines.

The second phase is about the prioritization of innovation projects. It is another critical aspect of operationalizing S-ERP. This process ensures that the most crucial projects, which are aligned with the strategic goals, receive the necessary resources and attention. The prioritization can be based on a combination of financial criteria (e.g., cost-to-income ratio) and non-financial criteria (e.g., market opportunities). For example, a large Russian aircraft manufacturer prioritized technology platforms based on market forecasts and technological readiness levels, ensuring that resources were allocated to projects with the highest strategic impact.

The third phase establishes clear roles and responsibilities within projects, which is essential for operationalizing processes. This clarity helps in avoiding overlaps and gaps in task execution, ensuring that each team member understands their specific contributions to the overall strategic goals. The experience of large Russian companies demonstrated that technology roadmaps could provide a structured representation of project works, establishing clear roles and ensuring transparency in project performance. This structured approach enhances accountability and commitment among project participants.

The fourth phase advocates for the involvement of external partners in the projects and introduces additional complexity in the operationalization of S-ERP. Effective technology roadmaps help improve transparency regarding the contributions of these partners. By providing a clear overview of the roles and performance of external collaborators, TRMs ensure that the integration of external knowledge and resources is managed effectively. This approach mitigates the risks associated with external partnerships, such as misalignment of objectives and lack of control over external activities.

The operationalization of S-ERP in Industry 4.0 relies heavily on the effective use of technology roadmaps. These roadmaps facilitate the transition from strategic goals to specific projects, prioritize resources, clarify roles, and enhance collaboration transparency. By addressing these critical areas, TRMs ensure that S-ERP systems are not only aligned with strategic objectives but also effectively implemented at the operational level, thereby supporting sustainable innovation and competitiveness in the dynamic landscape of Industry 4.0. Next, we discuss establishing management processes for S-ERP in Industry 4.0.

In the context of Industry 4.0, integrating S-ERP through establishing the management process is crucial for maintaining competitive advantage and achieving strategic goals. This integration involves not only the adoption of advanced technologies but also the effective management of Key Performance Indicators (KPIs). KPIs are essential for measuring the performance of business operations and ensuring alignment with strategic goals. Effective KPI management involves defining, monitoring, and refining KPIs to ensure they remain relevant and accurately reflect the performance metrics crucial for business success. The process begins with the identification of stakeholders and understanding their expectations through gap analysis. Gap analysis identifies discrepancies between the current state of KPI management and the expectations of stakeholders. This analysis is vital for understanding where improvements are needed and how they can be achieved through technology roadmapping. Key elements of KPI management include ensuring KPIs are Specific, Measurable, Assignable, Realistic, and Time-related (SMART) [17].

Engaging stakeholders from various business units is critical for successfully implementing KPI management processes. It ensures that KPIs are relevant and agreed upon across the organization, facilitating better decision-making and performance tracking. Stakeholder identification and interviews help recognize gaps and expectations, forming the basis for developing a comprehensive technology roadmap. The operationalizing management process through KPI has several advantages, such as internal quality assurance, central communication channels, KPI documentation, and dashboard enhancements. The following section discusses practical methods and tools of roadmapping for S-ERP in Industry 4.0.

4.4 Practical Methods and Tools of Roadmapping for S-ERP in Industry 4.0

This section discusses tools or methods to implement S-ERP in Industry 4.0. The first tool is the digitalization of the roadmapping process. The digitization of S-ERP roadmapping has four stages, as shown in Fig. 4.3. This framework integrates activity-based and psychosocial perspectives, ensuring comprehensive and effective roadmapping processes.

The first phase, named organize, involves forming dedicated roadmapping teams that include coordinators and facilitators. These teams are tasked with defining time and resource requirements for the process. This involves creating an activity plan, specifying required participants, gathering necessary information, and designing the

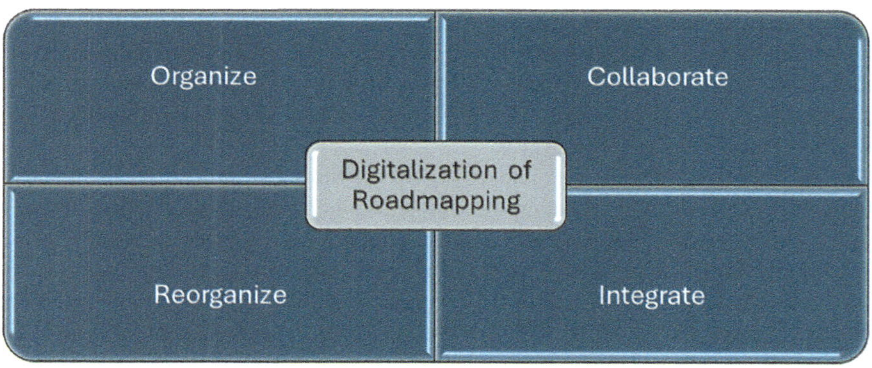

Fig. 4.3 Phases to digitalize the roadmapping

roadmap architecture. **The second phase**, i.e., collaborate, utilizes digital collaborative whiteboards to consolidate topic roadmaps in one place, thereby optimizing sharing and reducing the number of files and templates needed. Digital formats facilitate immediate communication and sharing of results with participants, eliminating the need for time-consuming transcription of physical notes.

The third phase, i.e., the reorganize phase focuses on data technology that enhances data organization and information convergence, leading to fewer gaps and doubts during the roadmapping process. The digital format allows for easy reorganization of ideas and connections within the roadmap, creating more consistent strategic narratives compared to physical methods. **The final phase**, i.e., integrate supports both remote and co-located digital roadmapping, with remote applications often providing empirical insights from industrial projects. By merging activity-based methods with cognitive and social interaction perspectives, the framework ensures robust strategizing processes that engage all participants effectively. The advantages of digitalization of roadmapping are enhanced productivity, improved quality of roadmaps, effective stakeholder engagement and immediate implementation [13]. The following subsection discusses the tools for S-ERP Roadmapping in Industry 4.0, such as knowledge graphs and interaction scenarios.

4.4.1 Knowledge Graph and Interaction Scenario

In the context of Industry 4.0, integrating S-ERP systems with advanced technologies such as IoT, AI, and robotics necessitates a structured approach to strategic planning. A technology roadmapping framework based on a technology knowledge graph and expert interaction is crucial for achieving this integration effectively. This framework facilitates the development, implementation, and management of S-ERP systems, ensuring they are aligned with the technological and strategic goals of an organization.

A technology knowledge graph is a structured representation of technical knowledge that reflects the relationships between various technologies within a specific field [18]. This graph combines data analysis results with expert knowledge to provide a comprehensive view of the technological landscape. For S-ERP in Industry 4.0, the technology knowledge graph helps in several ways:

1. **Technical Field Division:** It provides a clear categorization of technical fields relevant to S-ERP, such as IoT, AI, robotics, and sustainable technologies. This categorization helps in identifying the key areas where S-ERP can be integrated.
2. **Technical Relationships:** The graph outlines the relationships between different technologies, helping experts understand how various components of S-ERP interact with each other and with other technologies in Industry 4.0.
3. **Data and Expert Integration:** By integrating data analysis with expert knowledge, the graph ensures that the roadmap is based on both empirical data and the insights of experienced professionals. This combination enhances the reliability and relevance of the roadmap.

The interaction scenario involves the collaboration between domain experts and support personnel, facilitated by the technology knowledge graph. This collaboration follows a structured process to develop and refine the technology roadmap for S-ERP in Industry 4.0:

1. **Trend Analysis:** Experts use the technology knowledge graph to analyze current trends in relevant fields. This analysis helps in identifying emerging technologies and their potential impact on S-ERP systems.
2. **Technology Foresight List:** Based on trend analysis, experts create a list of critical technologies that need to be integrated into the S-ERP system. This list is refined through multiple rounds of expert feedback and data analysis, ensuring it is comprehensive and aligned with strategic goals.
3. **Expert Questionnaire and Interview:** Experts are consulted through questionnaires and interviews to gather diverse perspectives on the technology foresight list. This process helps in validating the list and identifying any gaps or additional opportunities.
4. **Technology Roadmapping:** The final step involves developing the actual roadmap based on the refined technology foresight list. The roadmap outlines the strategic steps needed to implement and manage S-ERP systems, including timelines, resource allocation, and performance metrics.

The application of a technology knowledge graph and interaction scenario in developing a technology roadmap for S-ERP in Industry 4.0 ensures a systematic and data-driven approach. This framework enhances the ability of organizations to integrate advanced technologies into their S-ERP systems effectively, aligning operational processes with strategic goals and maintaining competitiveness in a rapidly evolving technological landscape. The following subsection discusses agile roadmapping as a tool.

4.4.2 Agile Roadmapping for S-ERP in Industry 4.0

For agile Roadmapping, the first step is to establish a roadmap architecture and set an appropriate time horizon of 6 to 12 months, considered long-term in a dynamic environment. The roadmap layers should include market/external environment, product/service, process, and resources. The decision is required to determine who should be involved in each layer's discussion, ensuring key stakeholders from various functions participate. Engage in small meetings instead of large workshops to effectively improve participation and manage resource constraints.

In the next step, use SCRUM methodology for roadmapping [11]. Utilization of SCRUM methodology to guide short-term actions at tactical-operational levels. This involves planning micro sprints, which are detailed planning sessions that occur periodically throughout the year, usually every three months. Implement ARMA-SCRUM for both technical and business teams. This involves adapting the traditional

SCRUM process to ensure that both types of teams can work effectively with the roadmap. Technical teams should maintain the standard SCRUM structure but adjust inputs and outputs to align with roadmap requirements. For business teams, adapt SCRUM to accommodate the more dynamic nature of business activities, with frequent reflections and adjustments. It is crucial to establish a routine for continuous updating of the roadmap based on feedback from SCRUM activities. This ensures the roadmap remains a living document that evolves with the project's progress and external changes.

In agile methodology, one can also use a mixed approach of hand-made and digital maps. Applying a combination of hand-drawn and digital maps during meetings to capture and translate ideas quickly and efficiently is beneficial. Tools like Trello can digitize the maps immediately, visually representing the roadmap as a Kanban board [8]. This approach helps prioritize tasks and identify gaps effectively. Using digital tools that support agile principles, such as intuitive search, easy updating, and prioritization functionalities, will be a valuable exercise. These tools should facilitate virtual meetings, require minimal training, and be customizable to meet specific needs without incurring high costs. Finally, the business should not forget to address the external pressure for long-term planning from investors and other stakeholders by maintaining a six-month horizon for strategic levels while allowing flexibility for short-term operational adjustments. Leave blank spaces in long-term roadmaps to accommodate future changes and uncertainties.

The agile roadmapping will also change the alignments discussed in Section 2 and subsection 2.1. In agile roadmapping, a cross-functional team develops a roadmap for horizontal alignment that aligns marketing insights, technological advancements, and product development, ensuring clear communication and reduced conflicts. In vertical alignment of agile roadmapping, the team breaks down high-level strategic goals into actionable plans at all hierarchical levels. Finally, the roadmap includes a timeline synchronizing product development with market demands and technological advancements in temporal alignment. Enhanced collaboration, strategic coherence, market readiness, and sustainability practices, demonstrating the effectiveness of agile roadmapping for S-ERP in Industry 4.0. This approach ensures that the roadmapping process for S-ERP is agile, continuously updated, and effectively integrates strategic and operational activities, leading to better alignment with Industry 4.0 requirements and sustainability goals. The following section discusses roadmapping tools for sustainable practices achieved using S-ERP in Industry 4.0.

4.4.3 Roadmapping Tools for Sustainable Practices Achieved Using S-ERP in Industry 4.0

In the context of Industry 4.0, S-ERP systems are pivotal in achieving sustainability goals. Integrating sustainability into business practices requires a strategic approach that aligns with advanced technological frameworks. Roadmapping for sustainability through S-ERP involves leveraging tools and methods that ensure environmental, social, and economic benefits while maintaining competitive advantage. Technology roadmaps are essential for operationalizing sustainability strategies. They provide a structured approach to planning and implementing sustainability initiatives by aligning technological advancements with business objectives. Technology roadmaps help identify gaps, set goals, and define the steps necessary to achieve sustainability targets.

Faludi et al. [10] presented the algorithmic steps as tools for roadmapping for sustainability. In the present work, we have molded the algorithmic steps for S-ERP in Industry 4.0. The algorithm is as follows:

Algorithm 1 Roadmapping S-ERP in Industry 4.0

1: **Step 1:** Establish a clear vision and set ambitious goals.
2: **Step 2:** Assess the current state of sustainability within the organization.
3: **Step 3:** Identify the gaps between the current state and the desired vision.
4: **Step 4:** Develop strategic initiatives that address the identified gaps.
5: **Step 5:** Create a detailed implementation plan that outlines the steps required to execute the strategic initiatives.
6: **Step 6:** Establish mechanisms for monitoring and evaluating the progress of sustainability initiatives.

In the above algorithm, the **first step** in creating a sustainability roadmap is establishing a clear vision and setting ambitious goals. These should align with the Sustainable Development Goals (SDGs), particularly SDG 12 (Responsible Consumption and Production) and SDG 9 (Industry, Innovation, and Infrastructure). The vision should encompass long-term sustainability strategies deeply integrated into product development and business practices. The **second step** helps assess the current state of sustainability within the organization. This involves evaluating existing practices, identifying strengths and weaknesses, and understanding the environmental impact of current operations. The baseline provides a foundation for measuring progress and identifying areas for improvement. The **third step** identifies the gaps between the current state and the desired vision. This step involves determining the barriers to achieving sustainability goals and understanding the needs and expectations of stakeholders. Gap analysis helps prioritize actions and allocate resources effectively.

The **fourth step** of the above algorithm develops strategic initiatives that address the identified gaps. These initiatives should focus on integrating sustainability into core business processes, such as product design, supply chain management, and customer engagement. Initiatives may include adopting eco-design principles, improving resource efficiency, and enhancing product lifecycle management. The **fifth step** creates a detailed implementation plan that outlines the steps required to execute the strategic initiatives. This plan should include timelines, responsibilities, and performance indicators. Effective implementation requires collaboration across different departments and continuous monitoring to ensure alignment with sustainability goals. The **sixth step**, the final one, is to establish mechanisms for monitoring and evaluating the progress of sustainability initiatives. This involves setting up Key Performance Indicators (KPIs) to measure the impact of actions and using feedback loops to make necessary adjustments. Regular reporting and transparent communication are essential to maintain stakeholder engagement and accountability. Roadmapping for sustainability through S-ERP in Industry 4.0 involves a strategic approach that aligns technological advancements with sustainability goals. By setting a clear vision, conducting thorough assessments, identifying gaps, and implementing targeted initiatives, organizations can achieve significant environmental, social, and economic benefits. The following subsection explains the tool called the design structure matrix for roadmapping.

4.4.4 Design Structure Matrix for S-ERP Roadmapping in Industry 4.0

The Design Structure Matrix (DSM) is a powerful tool used to model, analyze, and optimize the flow of information, tasks, and dependencies within complex systems [9]. Unlike traditional project management tools focusing on sequential and parallel processes, DSM emphasizes interdependencies, feedback, and iterations, which are crucial for complex product development and organizational processes. In the context of Industry 4.0, applying DSM to S-ERP can significantly enhance the planning, execution, and management of complex, interdependent tasks and information flows.

Implementation of DSM is a five-step process. The steps are as follows:

Step 1: Visualization of Dependencies

Using DSM, S-ERP systems can map out the intricate web of dependencies between different tasks, teams, and processes. This visualization helps identify key areas where tasks intersect and depend on each other, which is essential for optimizing resource allocation and scheduling. For instance, in a manufacturing setting, the DSM can illustrate how the design phase depends on market analysis and how production planning relies on design and supply chain logistics.

Step 2: Clustering and Chunking

DSM helps streamline communication and reduce unnecessary information exchange by clustering related tasks into chunks. In S-ERP, this means grouping functions such as procurement, inventory management, and production scheduling, which often need to interact closely. This chunking reduces the complexity of communication paths and enhances the efficiency of inter-departmental workflows.

Step 3: Sequencing and Scheduling

DSM enables the sequencing of tasks in a way that minimizes iterations and feedback loops. For S-ERP, this translates to more efficient scheduling of production runs, maintenance activities, and supply chain operations. For example, the DSM can help determine the optimal sequence for machinery maintenance, ensuring that downtime is minimized and production schedules are not disrupted.

Step 4: Enhancing Collaboration and Communication

In an S-ERP system, various departments, such as R&D, production, and quality control, need to collaborate effectively. DSM provides a framework for enhancing this collaboration by making communication paths explicit and optimizing them for efficiency. This can lead to better alignment of strategic goals with operational tasks, ensuring that all departments work toward common objectives in a coordinated manner.

Step 5: Feedback and Iteration Management

One of the strengths of DSM is its ability to manage feedback and iterations effectively. In the context of S-ERP, this means continuously improving processes based on feedback from various stages of production and supply chain activities. For example, feedback from the production floor can be quickly integrated into the planning phase, ensuring that any issues are addressed promptly and do not cause significant delays. The following subsection explains DSM using the above steps.

4.4.4.1 Example of DSM Implementation for S-ERP in Industry 4.0

In the context of S-ERP within Industry 4.0, understanding the interdependencies between various tasks is crucial for efficient planning and execution. As suggested in the **first step**, we have identified some tasks and drawn a dependency graph for DSM implementation. The graph is shown in Fig. 4.4. Figure 4.4 shows the visualization of the dependencies of the functions.

The graph in Fig. 4.4 is a Directed Acyclic Graph (DAG) [12], as the graph does not form any cycle. Ideally, the dependency graph should not have any cycle as it creates a dependency chain, and we end up getting a task dependent on itself [6]. So, we try to derive some of the insights from DAG, as shown in Fig. 4.4.

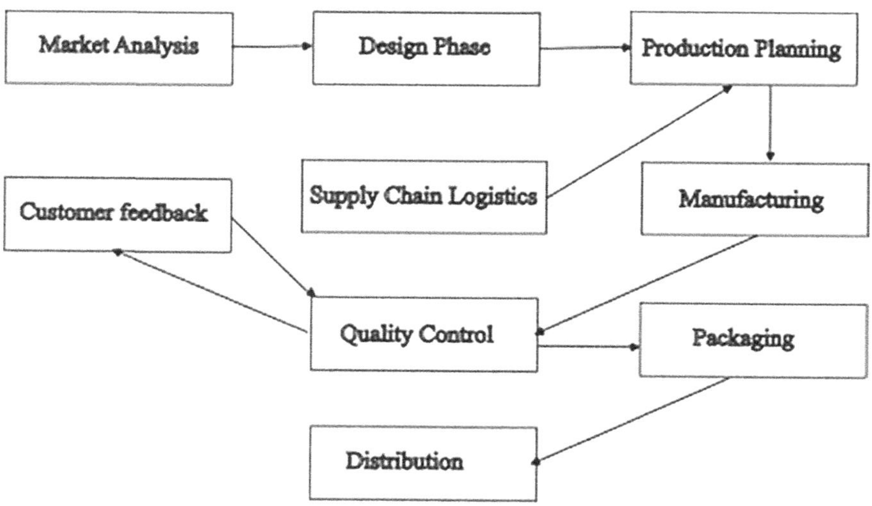

Fig. 4.4 DSM implementation (Step 1 visualization of dependencies)

The longest path in a DAG is the critical path, determining the minimum project completion time. The graph reveals the critical path [3]: Market Analysis → Design Phase → Production Planning → Manufacturing → Quality Control → Packaging → Distribution. This sequence represents the most time-sensitive tasks. Prioritize resource allocation, risk mitigation, and optimization along this path to ensure timely product delivery. Further, in-degree measures incoming edges (tasks dependent on completion), while out-degree measures outgoing edges (tasks impacted). *Quality Control* has a high in-degree, indicating it's a crucial bottleneck as many tasks rely on its completion [7]. Streamlining this phase is vital. *Production Planning* has a high out-degree, impacting multiple downstream tasks. Ensure its quality to prevent cascading issues.

After the first step, shown in Fig. 4.4, **the second step** is clustering [16]. Clustering groups related tasks that frequently interact or depend on each other. In the given S-ERP tasks shown in Fig. 4.4, the clusters can be identified based on their functional areas. The clusters are shown in Fig. 4.5

Figure 4.5 shows four clusters. C1 is the market and design cluster. The second cluster, C2, is the planning and supply chain cluster. The third cluster, C3, is the manufacturing and quality cluster. The final one, C4, is the packaging and distribution cluster. The **third step** is to perform the sequencing of the clusters. The final order of the cluster is C1, C3, C2, C3, C4. By scheduling the clusters in this order, we ensure that all dependencies are respected and tasks are completed efficiently. This structured approach helps optimize resources, minimize delays, and maintain a smooth workflow across the S-ERP system in Industry 4.0.

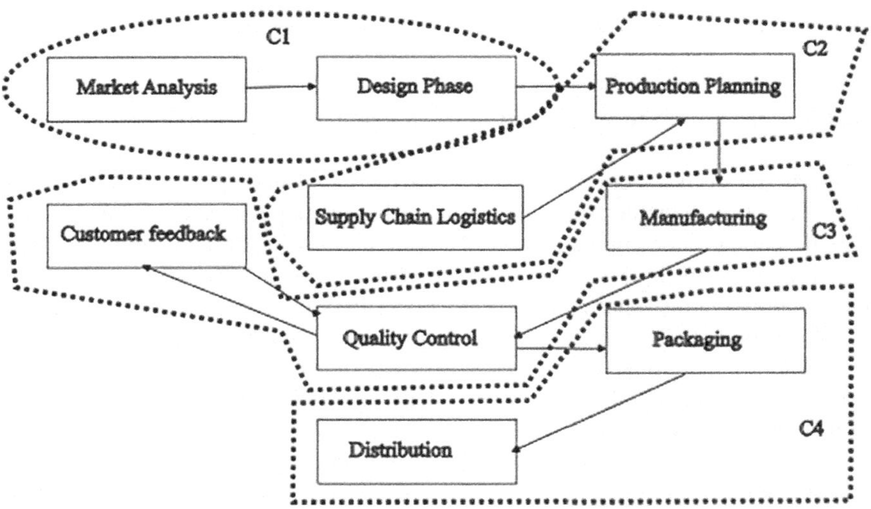

Fig. 4.5 Clustering of the task

Next, to enhance collaboration, organizations can take several steps. It may form cross-functional teams that include members from market analysis, design, production planning, supply chain logistics, manufacturing, quality control, packaging, and distribution. This ensures diverse perspectives and expertise are brought together to solve problems and make decisions. Further, organization can utilize centralized communication platforms to streamline information sharing. Implement project management tools like Trello, Asana, or Jira [4, 19], where team members can post updates, share documents, and track progress. Organizations can implement the following feedback mechanisms for communication and continuous monitoring as discussed in **steps 4 and 5**:

- **Continuous Feedback Loops:** Establish continuous feedback loops where feedback is collected at every process stage. For example, quality control can provide immediate feedback to manufacturing, which can be used to make real-time adjustments.
- **Customer Feedback Integration:** Collect and integrate customer feedback into the product development cycle. Use customer feedback to inform market analysis, design improvements, and quality control measures.
- **Iterative Development Cycles:** Adopt iterative development cycles (e.g., Agile, SCRUM) where products and processes are continuously improved based on feedback. Each iteration should aim to refine and enhance the product, incorporating learnings from the previous cycle.
- **Performance Metrics and KPIs:** Define and monitor Key Performance Indicators (KPIs) that reflect the performance and impact of each iteration. Regularly review these metrics to assess progress and identify areas for improvement.

Next, we compare DSM with a Multi-Layer Roadmap Framework.

4.4.4.2 DSM and Multi-Layer Roadmap Framework: Comparison

As discussed in this section, DSM is a matrix representation used to model, analyze, and optimize the flow of information, tasks, and dependencies within complex systems. It includes components or tasks, with matrix entries indicating dependencies or interactions. It is used for dependency management, scheduling, and process optimization. On the other hand, a Multi-Layer Roadmap Framework is a strategic planning tool that aligns various business functions (market, business, product, service, system, technology, science, resources) over time to support integrated strategic and innovation planning [15]. Figure 4.6, derived from Phaal's work [14], shows the Multi-Layer Roadmap Framework.

Layers representing different functional perspectives, a time axis (short term, medium term, long term), and key questions (Where do we want to go? Where are we now? How can we get there?). The multi-layer roadmap framework guides strategic planning, innovation, and alignment of organizational goals over time.

DSM and the Multi-Layer Roadmap Framework visualize dependencies among tasks or components, facilitating better understanding and management of complex

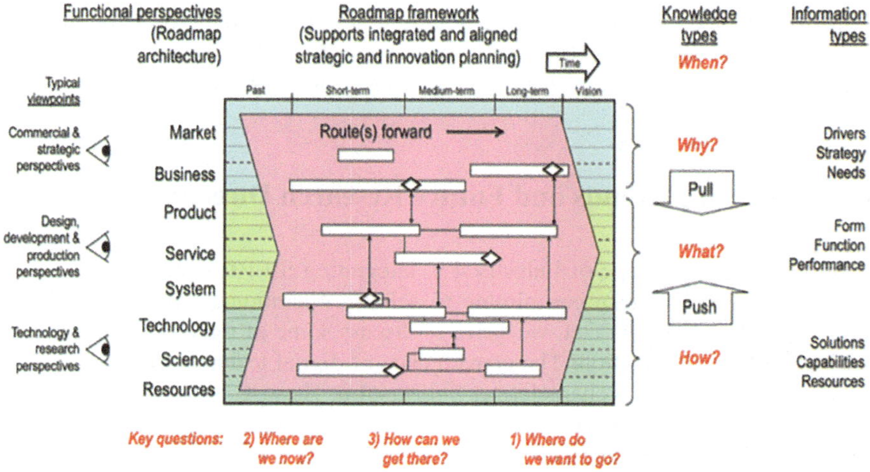

Fig. 4.6 Multi-layer roadmap framework derived from [14]

systems. Both tools help align various functions and activities with strategic goals, ensuring cohesive and efficient operations. Both approaches emphasize continuous improvement and optimization based on feedback and iterative processes.

DSM and the Multi-Layer Roadmap Framework differ in representation, scope, focus, and complexity management. In the case of representation, DSM uses a matrix format to represent dependencies, making it easier to see direct interactions and feedback loops between specific components. On the contrary, the Multi-Layer Roadmap Framework uses a layered format with a time axis, highlighting how different business functions evolve and align over time. In the case of scope and focus, DSM focuses on the dependencies and interactions within a system or project, often used for process optimization and detailed scheduling. The Multi-Layer Roadmap Framework focuses on strategic planning and long-term vision, integrating various business functions and ensuring alignment with overarching goals. Finally, regarding complexity management, DSM deals directly with the complexity of interactions within a system, often used for engineering and product development processes. The Multi-Layer Roadmap Framework manages complexity by aligning different functional perspectives and timeframes, and it is used for high-level strategic and innovation planning.

Both DSM and the Multi-Layer Roadmap Framework are valuable tools for managing complexity and aligning activities with strategic goals. While DSM is more suited for detailed dependency management and process optimization, the Multi-Layer Roadmap Framework excels in strategic planning and aligning various business functions over time. The choice between these tools depends on the specific needs of the organization and the nature of the projects being managed. For comprehensive planning, organizations may benefit from integrating both approaches, using

DSM for detailed process management and the Roadmap Framework for strategic alignment and long-term vision. The following section discusses emerging trends and future research directions for the roadmapping of S-ERP in Industry 4.0.

4.5 Emerging Trends and Future Research Directions

In the evolving landscape of Industry 4.0, integrating S-ERP systems has become a critical factor for businesses aiming to achieve operational efficiency and sustainability. Based on this fact, we strive to discuss some of the emerging trends in roadmapping in this section. The convergence of digital technologies such as IoT, AI, and big data analytics characterizes Industry 4.0. These technologies enhance the capabilities of S-ERP systems by providing real-time data, predictive analytics, and automated decision-making processes. The shift toward cloud-based S-ERP solutions offers scalability, flexibility, and cost-effectiveness. Cloud computing reduces the need for extensive on-premises infrastructure, enabling organizations to allocate resources more efficiently and focus on core business activities. The switching from on-premises to cloud forces organizations to change the roadmaps as the cloud modifies the focus from managing physical infrastructure to leveraging cloud services, optimizing cost, ensuring security within a shared responsibility framework, and enabling faster deployment and scalability.

Incorporating green technology and sustainability will also force organization to modify the roadmap. The new roadmap should prioritize eco-friendly practices, reduce carbon footprints, and comply with environmental regulations. This shift involves adopting sustainable materials, energy-efficient processes, and renewable energy sources, ultimately transforming product development, supply chains, and operational strategies.

Introducing regulatory mechanisms, enhanced governance, and improved situation awareness [2] will necessitate organizations to revise their roadmaps. This includes ensuring compliance with new regulations, strengthening governance frameworks, and adopting advanced real-time monitoring and decision-making tools. These changes enhance accountability, risk management, and adaptability to evolving market conditions and legal requirements.

User-centered design in an organization compels a shift in the roadmap by prioritizing the needs and preferences of end-users throughout the product development process [20]. This approach requires organizations to gather extensive user feedback, conduct usability testing, and iteratively refine their products based on user experiences. Consequently, the focus shifts from merely meeting technical specifications to enhancing user satisfaction and engagement. This change often leads to reallocating resources, adopting new design methodologies, and a more agile development process that can quickly respond to user insights and evolving market demands. Overall, embracing user-centered design ensures that the products are more intuitive, accessible, and aligned with the actual needs of the users, driving a significant transformation in the organization's strategic planning and execution.

The integration of big data, machine learning, deep learning, and AI in an organization necessitates a fundamental shift in the roadmap, as these technologies offer powerful tools for data analysis, predictive modeling, and automation. Leveraging big data allows organizations to process vast amounts of information and extract valuable insights, leading to more informed decision-making. Machine learning and deep learning enable the development of sophisticated algorithms that can identify patterns, make predictions, and improve over time with minimal human intervention. AI enhances automation and operational efficiency, transforming how tasks are performed and services are delivered. As a result, organizations must re-evaluate their strategic goals, invest in new infrastructure and talent, and adopt an agile mindset to rapidly implement and scale these technologies. This technological shift drives innovation, enhances competitive advantage, and necessitates continuous adaptation of the organization's roadmap to stay aligned with the evolving digital landscape.

Managing changes in roadmaps for S-ERP within the context of Industry 4.0 involves a dynamic and iterative approach that accommodates the rapid advancements in technology and evolving business needs. Effective change management in S-ERP roadmaps requires continuous monitoring and adaptation to ensure alignment with strategic goals and market demands. This process begins with establishing a robust governance framework that oversees the implementation and operation of the S-ERP system, ensuring compliance with international standards and regulations. Engaging stakeholders across all levels of the organization is crucial, as it fosters collaboration and facilitates the identification of emerging challenges and opportunities. Incorporating feedback mechanisms allows for the proactive identification of potential issues and the development of timely solutions. Utilizing advanced technologies such as AI, IoT, and big data analytics provides real-time insights and predictive capabilities, enabling organizations to anticipate and respond to changes effectively. Furthermore, maintaining a flexible and modular approach to S-ERP system design allows for seamless integration and upgrading of components, ensuring the system can evolve with changing business requirements. Continuous training and support for users ensure they remain proficient in utilizing the S-ERP system, thereby enhancing overall efficiency and responsiveness. By embracing these strategies, organizations can manage changes in their S-ERP roadmaps effectively, ensuring sustained competitiveness and alignment with Industry 4.0 objectives.

Another crucial aspect of managing changes in S-ERP roadmaps is emphasizing continuous innovation and improvement. Organizations must foster a culture that encourages experimentation and the adoption of new technologies and methodologies. This involves investing in research and development to explore emerging trends and innovations that can further enhance the capabilities of S-ERP systems. By staying abreast of the latest technological advancements, businesses can integrate cutting-edge solutions that drive efficiency and sustainability. Collaboration with external partners, including technology providers, academic institutions, and industry consortia, can provide valuable insights and resources to support ongoing innovation. Regularly revisiting and updating the roadmap ensures that it remains relevant and aligned with the organization's strategic vision and the dynamic landscape of Industry 4.0. This proactive approach addresses current challenges and anticipates future opportunities, positioning the organization as a leader in sustainable and efficient enterprise resource planning. In summary, the chapter answers the following research questions.

Research Questions Addressed The chapter titled "Sustainable Enterprise Resource Planning (S-ERP) Roadmapping in Industry 4.0" addresses the following key research questions:

1. **How does roadmapping serve as a strategic planning tool for integrating advanced technologies like IoT, AI, machine learning, and big data analytics into S-ERP systems?**

- The chapter explores how roadmapping aids in visualizing and coordinating the integration of these advanced technologies into S-ERP systems within Industry 4.0, enhancing efficiency, sustainability, and adaptability.

2. **What are the critical elements of strategic alignment in S-ERP roadmapping within Industry 4.0?**

- The chapter discusses the importance of aligning business processes, technology, and innovation with sustainability goals and how strategic alignment ensures that organizational activities support sustainable growth and competitive advantage.

3. **How can organizations operationalize S-ERP systems in Industry 4.0 through roadmapping?**

- The chapter outlines the phased approach to transitioning from broad strategic goals to specific, actionable projects, focusing on the prioritization of innovation projects, role clarity, and the involvement of external partners.

4. **What practical tools and methods can be used in S-ERP roadmapping for Industry 4.0?**

- The chapter provides a detailed exploration of practical tools like digitalization of roadmapping processes, knowledge graphs, interaction scenarios, agile roadmapping, and the design structure matrix (DSM) for managing complex interdependencies and enhancing collaboration within organizations.

5. **What are the emerging trends and future research directions in roadmapping for S-ERP in Industry 4.0?**

- The chapter discusses emerging trends such as the convergence of digital technologies, the shift toward cloud-based solutions, the integration of green technologies, regulatory mechanisms, user-centered design, and the incorporation of big data, machine learning, and AI into S-ERP systems.

These research questions guide the comprehensive exploration of S-ERP roadmapping as a strategic tool in Industry 4.0, focusing on both the technological and managerial aspects necessary for achieving sustainable and competitive advantages.

References

1. Anjaria K (2024) Enhancing sustainability integration in Sustainable Enterprise Resource Planning (S-ERP) system: application of transaction cost theory and case study analysis. Int J Informat Manage Data Insights 4(2):100243. ISSN 2667-0968. https://doi.org/10.1016/j.jjimei.2024.100243. https://www.sciencedirect.com/science/article/pii/S2667096824000326
2. Anjaria K, Mishra A (2017) Relating Wiener's cybernetics aspects and a situation awareness model implementation for information security risk management. Kybernetes 47(1):58–79. ISSN 0368-492X. https://doi.org/10.1108/K-06-2017-0226. https://doi.org/10.1108/K-06-2017-0226. Publisher: Emerald Publishing Limited
3. Aquilano NJ, Smith DE (1980) A formal set of algorithms for project scheduling with critical path scheduling/material requirements planning. J Operat Manage 1(2):57–67. ISSN 0272-6963. https://doi.org/10.1016/0272-6963(80)90013-3. https://www.sciencedirect.com/science/article/pii/0272696380900133
4. Arya S, Kulkarni MS (2024) A comprehensive study of agile project management tools. In: 2024 4th international conference on innovative practices in technology and management (ICIPTM), pp 1–6. https://doi.org/10.1109/ICIPTM59628.2024.10563875. https://ieeexplore.ieee.org/abstract/document/10563875
5. Bereznoy A, Snegirev A (2023) Technology roadmaps as an instrument for operationalizing innovation strategies of large corporations. In: Daim TU, Phaal R, Meissner D, Kerr C (eds) Next generation roadmapping: establishing technology and innovation pathways towards sustainable value. Springer International Publishing, Cham, pp 25–40. 978-3-031-38575-9
6. Bärmann A, Gemander P, Merkert M (2020) The clique problem with multiple-choice constraints under a cycle-free dependency graph. Discrete Appl Math 283:59–77. ISSN 0166-218X. https://doi.org/10.1016/j.dam.2019.12.015. https://www.sciencedirect.com/science/article/pii/S0166218X1930558X
7. Darvish M, Yasaei M, Saeedi A (2009) Application of the graph theory and matrix methods to contractor ranking. Int J Project Manage 27(6):610–619. ISSN 0263-7863. https://doi.org/10.1016/j.ijproman.2008.10.004. https://www.sciencedirect.com/science/article/pii/S0263786308001488
8. de Souza ML, de Souza WC, Freitas JS, de Melo Filho LD, Bagno RB (2022) Agile roadmapping: a management tool for digital entrepreneurship. IEEE Trans Engin Manage 69(1):94–108. ISSN 1558-0040. https://doi.org/10.1109/TEM.2020.3027918. https://ieeexplore.ieee.org/abstract/document/9247459. Conference Name: IEEE Transactions on Engineering Management
9. Eppinger SD, Browning TR (2012) Design structure matrix methods and applications. MIT Press. 978-0-262-01752-7. Google-Books-ID: MPQoumoGXHYC
10. Faludi J, Hoffenson S, Kwok SY, Saidani M, Hallstedt SI, Telenko C, Martinez V (2020) A research roadmap for sustainable design methods and tools. Sustainability 12(19):8174. ISSN 2071-1050. https://doi.org/10.3390/su12198174. https://www.mdpi.com/2071-1050/12/19/8174. Number: 19 Publisher: Multidisciplinary Digital Publishing Institute
11. Fernández-Valderrama P, Antón D, MaríÃ‚Ân-GarcíÃ‚Âa D, Carretero-Ayuso MJ, (2024) Adapting agile scrum methodology for the subject of building project design in university teaching. In: Bienvenido-Huertas D, de la Hoz-Torres ML, Aguilera AJA (eds) Teaching innovation in architecture and building engineering: challenges of the 21st century. Springer Nature Switzerland, Cham, pp 275–291. 978-3-031-59644-5
12. Park N-J, Kim Y-U, Ahn H-S (2024) Fixed node determination and analysis in directed acyclic graphs of structured networks. J Franklin Inst 361(13):106995. ISSN 0016-0032. https://doi.org/10.1016/j.jfranklin.2024.106995. https://www.sciencedirect.com/science/article/pii/S0016003224004162
13. Paul M, Upadhyay P, Dwivedi YK (2020) Roadmap to digitalisation of an emerging economy: a viewpoint. Transf Government: People, Process Policy 14(3):401–415. ISSN 1750-6166. https://doi.org/10.1108/TG-03-2020-0054. https://doi.org/10.1108/TG-03-2020-0054. Publisher: Emerald Publishing Limited

14. Phaal R, Muller G (2009) An architectural framework for roadmapping: towards visual strategy. Technol Forecasting and Social Change 76(1):39–49. ISSN 0040-1625. https://doi.org/10.1016/j.techfore.2008.03.018. https://www.sciencedirect.com/science/article/pii/S0040162508000632
15. Phaal R, Simonse L, Ouden ED (2008) Next generation roadmapping for innovation planning. Int J Technol Intell Plan 4(2):135–152. ISSN 1740-2832. https://doi.org/10.1504/IJTIP.2008.018313. https://www.inderscienceonline.com/doi/abs/10.1504/IJTIP.2008.018313. Publisher: Inderscience Publishers
16. Ren Y, Pu J, Yang Z, Xu J, Li G, Pu X, Yu PS, He L (2024) Deep clustering: a comprehensive survey. IEEE Trans Neural Netw Learn Syst 1–21. ISSN 2162-2388. https://doi.org/10.1109/TNNLS.2024.3403155. https://ieeexplore.ieee.org/abstract/document/10585323. Conference Name: IEEE Transactions on Neural Networks and Learning Systems
17. Robert A, Daim T (2023) Technology roadmapping: KPI management process. In: Daim TU, Phaal R, Meissner D, Kerr C (eds) Next generation roadmapping: establishing technology and innovation pathways towards sustainable value. Springer International Publishing, Cham, pp 41–70. 978-3-031-38575-9
18. Sarica S, Song B, Luo J, Wood K, Technology knowledge graph for design exploration: application to designing the future of flying cars. https://doi.org/10.1115/DETC2019-97605. https://dx.doi.org/10.1115/DETC2019-97605
19. Sravan NK, Vuyyyru VA, Gottapu P, Theeda A, Peddineni L (2024) Agile management tools: technological evaluations and future archetype. In 2024 4th international conference on pervasive computing and social networking (ICPCSN), pp 914–918. https://doi.org/10.1109/ICPCSN62568.2024.00154. https://ieeexplore.ieee.org/abstract/document/10607751
20. Unger R, Chandler C (2023) A project guide to UX design: for user experience designers in the field or in the making. New Riders. ISBN 978-0-13-818806-1. Google-Books-ID: wsHWEAAAQBAJ
21. Vukman K, Klarić K, Greger K, Perić (2024) Driving efficiency and competitiveness: trends and innovations in ERP systems for the wood industry. Forests 15(2):230. ISSN 1999-4907. https://doi.org/10.3390/f15020230. https://www.mdpi.com/1999-4907/15/2/230. Number: 2 Publisher: Multidisciplinary Digital Publishing Institute

Chapter 5
Ethical and Security Aspects of Sustainable Enterprise Resource Planning (S-ERP) in Industry 4.0

5.1 Introduction

The study of ethical and security aspects of Sustainable Enterprise Resource Planning (S-ERP), a system that integrates sustainability into enterprise resource planning in Industry 4.0, is critical due to the profound impact these systems have on business operations, data management, and societal values. S-ERP ensures that business processes are efficient, resource-conscious, and environmentally friendly. The ethical and security dimensions become even more significant in the context of Industry 4.0, characterized by advanced technologies such as the Internet of Things (IoT), artificial intelligence (AI), and big data analytics. These technologies, for instance, loT, enable the interconnection of devices, AI facilitates predictive analytics, and big data analytics allows for processing large volumes of data. Ethical considerations involve ensuring that deploying these technologies does not lead to unfair labor practices, data privacy violations, or environmental degradation. Security aspects are equally crucial as S-ERP systems handle vast amounts of sensitive data, making them prime targets for cyberattacks. Robust security measures are necessary to protect this data from breaches that could lead to financial loss, operational disruptions, and reputational damage. Moreover, addressing these aspects helps build trust among stakeholders, including customers, employees, and regulators, thereby fostering a sustainable and resilient business environment. Therefore, a thorough understanding and implementation of ethical and security principles in S-ERP systems are fundamental to achieving the goals of sustainability and technological advancement in Industry 4.0.

The study of ethical and security aspects of S-ERP differs significantly from that of traditional classical Enterprise Resource Planning (ERP) systems due to the broader scope and deeper integration of sustainability principles in S-ERP [4]. Traditional ERP systems primarily focus on optimizing business processes for efficiency, cost reduction, and productivity enhancement [13]. Ethical and security considerations in classical ERP systems are often limited to ensuring data integrity, user privacy,

K. Anjaria, *Sustainable Enterprise Resource Planning (S-ERP) for Industry 4.0*, Contributions to Management Science, https://doi.org/10.1007/978-981-96-7734-4_5

and compliance with relevant regulations. In contrast, S-ERP encompasses a holistic approach that integrates environmental, social, and economic sustainability into its core framework. This integration necessitates a more comprehensive examination of ethical issues, including the impact of business operations on the environment, the ethical sourcing of materials, which involves ensuring that materials are obtained in a responsible and sustainable manner, and the promotion of fair labor practices throughout the supply chain. Furthermore, the security landscape in S-ERP is more complex due to the incorporation of advanced technologies characteristic of Industry 4.0, such as IoT and AI, which introduce new vulnerabilities and require sophisticated security protocols. The ethical dimension in S-ERP also extends to the responsible use of AI and data analytics, ensuring that these technologies are used transparently and do not perpetuate biases or inequality. Consequently, the study of ethical and security aspects in S-ERP is more dynamic and multifaceted, requiring an ongoing assessment of how these systems interact with broader societal goals and technological advancements.

Sophisticated cybersecurity is necessary for S-ERP systems due to the intricate and interconnected nature of these platforms within the framework of Industry 4.0. S-ERP systems, by design, integrate various advanced technologies such as the Internet of Things (IoT), artificial intelligence (AI), and big data analytics to enhance sustainability and operational efficiency [35]. While offering substantial benefits, these technologies also expand the attack surface, making S-ERP systems more vulnerable to cyber threats. Cybersecurity in this context must address the increased complexity and scale of potential risks, including data breaches, ransomware attacks, and cyber espionage. Since S-ERP systems manage sensitive data across multiple domains, including financial information, supply chain logistics, and customer data, a robust cybersecurity framework is essential to protect this information from unauthorized access and manipulation. Furthermore, the interconnectedness of S-ERP systems with external partners and suppliers amplifies the risk of cyberattacks, as a breach in one part of the network can have cascading effects throughout the entire supply chain. Effective cybersecurity measures are therefore critical not only to safeguard data integrity and confidentiality but also to maintain operational continuity and trust among stakeholders. Advanced cybersecurity protocols, such as multi-factor authentication, encryption, and real-time threat detection and response, are imperative to defend against sophisticated cyber threats. These measures ensure that S-ERP systems can function securely and reliably, supporting the overarching goals of sustainability and technological innovation in Industry 4.0.

In addition to sophisticated cybersecurity measures, integrating ethical principles into S-ERP systems necessitates reevaluating how these technologies are deployed and managed. Implementing AI and data analytics within S-ERP should be guided by ethical frameworks prioritising fairness, transparency, and accountability. This includes ensuring that AI algorithms are free from biases that could lead to discriminatory outcomes, as well as maintaining transparency in data collection and processing practices to uphold user privacy and trust. Ethical considerations also extend to the environmental impact of S-ERP systems, where sustainable practices must be embedded into the lifecycle of technology deployment, from development and implementation to maintenance and decommissioning. This holistic approach ensures that the environmental footprint of S-ERP systems is minimized, aligning with broader corporate sustainability goals. Furthermore, fostering a culture of ethical awareness within organizations is crucial, involving continuous training and education of employees on ethical and security best practices. By integrating these ethical considerations alongside robust security measures, organizations can ensure that their S-ERP systems not only drive operational efficiency and sustainability but also adhere to high standards of ethical responsibility and resilience in the face of evolving technological challenges in Industry 4.0.

Studying the ethics and security aspects of S-ERP after examining change management and roadmapping within the context of Industry 4.0 is essential for a comprehensive understanding of these systems' holistic impact and implementation. Change management and roadmapping provide the strategic framework and structured approach necessary to successfully adopt and integrate S-ERP systems. These initial steps focus on aligning organizational objectives with technological advancements, identifying potential barriers, and developing a phased plan to achieve desired outcomes. However, S-ERP systems' long-term success and sustainability hinge on addressing the ethical and security implications that arise as these technologies are deployed. Without thoroughly considering ethics, deploying S-ERP could lead to unintended negative consequences such as data privacy violations, unfair labor practices, and environmental harm. Similarly, robust security measures are imperative to protect against cyber threats compromising system integrity, data confidentiality, and operational continuity. By studying ethics and security aspects after the foundational stages of change management and roadmapping, organizations can ensure that S-ERP is strategically sound, ethically responsible, and secure, thereby fostering a sustainable, resilient, and trustworthy business environment in Industry 4.0.

The following section discusses the ethics of S-ERP in Industry 4.0.

5.2 Ethics of S-ERP in Industry 4.0

In the context of S-ERP in Industry 4.0, various ethical frameworks can guide the development and implementation of sustainable practices. Utilitarianism can help evaluate the overall impact of S-ERP systems on society by maximizing benefits such as increased efficiency and reduced environmental footprint while minimizing negative outcomes like resource depletion [18]. Virtue ethics emphasizes fostering an organizational culture that values sustainability, encouraging individuals to develop character traits like responsibility and stewardship toward the environment [28]. Deontological ethics ensures that S-ERP practices adhere to established rules and principles, such as regulatory compliance and ethical labor practices, regardless of the consequences [23]. Relativism allows for flexibility in implementing S-ERP solutions that are culturally sensitive and tailored to specific regional needs, acknowledging that sustainability practices may vary across different contexts [41]. Egoism can be leveraged to highlight the long-term benefits of sustainable practices for the organization's self-interest, such as cost savings and brand reputation. Care ethics underscores the importance of nurturing relationships with all stakeholders, including employees, customers, and the community, by prioritizing their well-being through sustainable initiatives. Contractarianism supports the idea that sustainable practices can be viewed as part of a social contract where businesses commit to ethical standards for the common good [25]. Natural law ethics can guide organizations to align their S-ERP strategies with fundamental human and ecological principles, ensuring harmony with the natural world. Existentialist ethics encourages organizations to take proactive and authentic actions toward sustainability, fostering a sense of individual and collective responsibility [22]. Pragmatism emphasizes the practical implications of S-ERP, focusing on the tangible benefits and effectiveness of sustainable practices [21]. Feminist ethics brings attention to gender equality and inclusivity within sustainable practices, ensuring that the benefits of Industry 4.0 advancements are equitably distributed [33]. Consequentialism, broader than utilitarianism, also considers the outcomes of S-ERP practices but may weigh various types of consequences differently. Principlism, particularly in the bioethical realm, provides a balanced approach by integrating principles like autonomy, beneficence, non-maleficence, and justice into the decision-making process for sustainable initiatives [19]. Lastly, postmodern ethics challenges the notion of universal sustainability solutions, advocating for a more nuanced and context-sensitive approach that recognizes the diversity of perspectives and the complexity of ethical considerations in S-ERP within Industry 4.0 [11].

The aforementioned ethical frameworks have been thoroughly examined and are directly relevant to S-ERP. Consequently, this chapter will focus on ethical considerations specific to algorithms, artificial intelligence (AI), sensors, Industry 4.0, computer vision, sustainability, and environmental impact, which are pivotal for S-ERP within the Industry 4.0 context and less explored in the literature. These principles are shown in Fig. 5.1. Each of the subsequent subsections is dedicated to exploring these critical ethical principles pertaining to S-ERP in Industry 4.0.

First, we discuss the ethics of algorithms used in S-ERP for Industry 4.0 in the following subsection.

5.2.1 Ethics of Algorithms Used in S-ERP for Industry 4.0

The deployment of algorithms within S-ERP in Industry 4.0 raises several critical ethical concerns that are well-documented in the broader discourse on algorithmic ethics. According to Mittelstadt et al. [24], algorithms mediate various operations and decisions, influencing business transactions, social interactions, and how data is interpreted and utilized. This mediation is value-laden, where operational parameters defined by developers and configured by users often privilege specific values and interests over others, leading to ethically problematic outcomes.

Epistemic concerns with algorithms revolve around the quality of evidence they produce. In S-ERP, algorithms convert data into actionable insights that drive

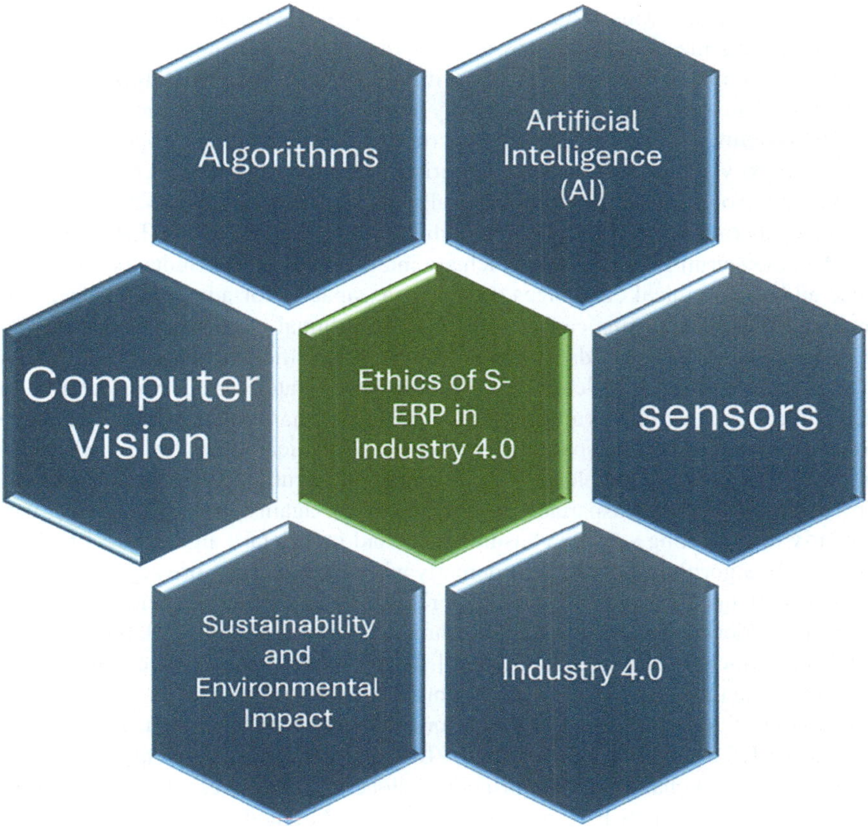

Fig. 5.1 The ethical principles discussed in the subsequent subsections related to S-ERP in Industry 4.0

decision-making processes. The risk of producing inconclusive, inscrutable, or misguided evidence is significant. For instance, inconclusive evidence reflects the inherent uncertainty in algorithmic outputs, which may not be reliable enough to justify specific actions. Inscrutable evidence concerns the opacity of algorithmic decision-making processes, making it difficult for stakeholders to understand how conclusions are derived. Misguided evidence stems from biases in the input data, which propagate through the algorithm and affect the outputs, potentially leading to flawed decision-making in S-ERP systems.

Normative concerns address the ethical acceptability of actions taken by algorithms. In the context of S-ERP, algorithms that result in unfair outcomes can perpetuate discrimination or inequity, impacting marginalized groups disproportionately. Furthermore, the transformative effects of algorithms are profound, as they can reontologize the world by altering how businesses conceptualize and interact with their data and operations. This can lead to ethical dilemmas when algorithmic decisions restructure organizational practices in ways that may not be ethically neutral. An overarching concern in the ethics of algorithms is traceability, which involves determining the cause and responsibility for any harm caused by algorithmic decisions. In S-ERP systems, ensuring traceability is crucial for accountability, especially given the complexity and opacity of many algorithms. Traceability issues complicate the identification of responsible parties when algorithms cause harm, necessitating robust mechanisms for oversight and accountability in the deployment of S-ERP systems.

Machine learning algorithms, which are integral to S-ERP in Industry 4.0, introduce additional ethical challenges due to their capacity for autonomous decision-making. These algorithms can modify their operational parameters and decision-making rules based on new data, making their actions difficult to predict and explain. The autonomy of machine learning algorithms necessitates careful consideration of their ethical implications, as unforeseen and potentially harmful decisions can emerge from their operation. Transparency is essential for ethical algorithmic governance in S-ERP, allowing stakeholders to understand and scrutinize the decision-making processes. However, transparency must be balanced against privacy concerns, particularly in protecting sensitive business data and proprietary algorithms. Ensuring that S-ERP algorithms operate transparently while safeguarding privacy is a delicate ethical challenge that requires thoughtful regulation and implementation.

Some readers may contend that algorithms can serve as a framework for enforcing rules and principles, as discussed in [26]. In the context of S-ERP, certain ethical principles can indeed be operationalized through the use of algorithms. While ethical principles provide a foundational framework, they alone are insufficient to guarantee ethical S-ERP systems in the context of Industry 4.0. This argument is grounded in several critical challenges that principles alone cannot address. Let's understand this fact by comparing S-ERP with the systems used in medicine. Unlike medicine, which has a unified goal of promoting patient health, S-ERP systems are developed in diverse, often conflicting, industrial contexts where the aims of developers, users, and stakeholders do not align. This absence of a shared goal complicates ethical decision-making and transforms it into a competitive rather than cooperative process, making it difficult to balance public and private interests.

Secondly, a relative lack of professional history and established norms in the field of S-ERP—medicine benefits from a long history of ethical standards and professional codes that have evolved over centuries. In contrast, S-ERP algorithm development lacks a cohesive historical foundation and the associated professional culture. This absence makes establishing and enforcing consistent ethical norms and behaviors challenging.

Thirdly, proven methods to translate high-level ethical principles into practical applications are lacking in S-ERP algorithm development. While medicine has developed robust mechanisms, such as ethics review committees and licensing schemes, to operationalize ethical principles, S-ERP algorithm development lacks comparable empirically tested methods. The environmental rules and regulations are also different and non-uniform in different countries. The process of translating abstract principles into actionable guidelines and requirements for S-ERP systems is complex and context-dependent, often leading to inconsistent implementation across different sectors. Therefore, while ethical principles are necessary, they are not sufficient to address the multifaceted ethical challenges posed by S-ERP in Industry 4.0. A comprehensive approach that includes well-defined aims, established professional norms, proven translation methods, and robust accountability mechanisms is essential to ensure the ethical development and deployment of these systems.

So, the question is, how do we make S-ERP algorithms ethical in Industry 4.0 setup? The question of how to make S-ERP algorithms ethical within an Industry 4.0 setup does not have a single, straightforward answer. Nonetheless, we propose that organizations, upon implementing S-ERP, consider the following five steps. These steps are not required to be executed in the exact sequence presented, nor are they exhaustive. Thus, they should be applied with caution. The steps are as follows:

Algorithm 1 Ethical Integration in S-ERP for Industry 4.0

Step 1: Integrating Ethical Principles
Step 2: Human-in-the-Loop Systems
Step 3: Transparency and Explainability
Step 4: Cultural and Social Sensitivity
Step 5: Continuous Ethical Training and Education
Step 6: Value-Sensitive Design
Step 7: Regulatory and Ethical Frameworks

The explanation of the above seven steps are as follows:

Step 1. Integrating Ethical Principles:

S-ERP algorithms should be designed based on core ethical principles such as beneficence, non-maleficence, autonomy, and justice. These principles should guide the decision-making processes embedded within the algorithms, ensuring that they promote well-being, avoid harm, respect user autonomy, and ensure fairness and equity.

Step 2. Human-in-the-Loop Systems:

Incorporating human-in-the-loop systems ensures that human judgment and ethical considerations are part of the algorithmic decision-making process as discussed in [20]. This approach can help address unforeseen ethical dilemmas and maintain a check on automated decisions, providing a balance between efficiency and ethical responsibility.

Step 3. Transparency and Explainability:

We stress the importance of explainability in algorithms. S-ERP systems should be designed to be transparent, providing clear explanations for their decisions and actions. This transparency helps in building trust and accountability, ensuring that users and stakeholders understand how decisions are made and can scrutinize them for ethical compliance.

Step 4. Cultural and Social Sensitivity:

Ethical S-ERP algorithms should consider the cultural, social, and economic contexts in which they operate. This includes understanding and respecting different stakeholders' diverse values and norms and ensuring that the algorithms do not inadvertently reinforce biases or inequalities.

Step 5. Continuous Ethical Training and Education:

Developers and users of S-ERP systems should be continuously educated and trained on ethical issues related to algorithm design and implementation. This education should cover the cultural history of consciousness and its implications for technology, fostering a deep understanding of the ethical impact of their work.

Step 6. Value-Sensitive Design:

Adopting a value-sensitive design approach ensures that the moral values and ethical considerations of all stakeholders are embedded in the S-ERP algorithms from the early stages of design. This approach helps in creating systems that are aligned with the ethical expectations and values of the society they serve.

Step 7. Regulatory and Ethical Frameworks:

Establishing robust regulatory and ethical frameworks can provide guidelines for the ethical development and deployment of S-ERP algorithms. These frameworks should be dynamic, allowing for continuous improvement and adaptation to new ethical challenges as they arise.

By integrating these ethical considerations into S-ERP algorithms, the technology can be aligned with the broader goal of creating a sustainable and socially responsible Industry 4.0, ensuring that technological advancements contribute positively to society and human well-being. The following subsection relates AI ethics and S-ERP from Industry 4.0 perspectives.

5.2.2 AI Ethics and S-ERP in Industry 4.0

Artificial intelligence (AI) ethics plays a pivotal role in S-ERP within the framework of Industry 4.0. Integrating AI into S-ERP systems promises significant advancements in efficiency, data management, and decision-making capabilities. However, this integration also raises several ethical concerns that need to be addressed to ensure the sustainability and social responsibility of these technologies. Researchers have used a variety of approaches for ethical AI.

AI ethics involves the study and application of ethical principles to AI systems, focusing on issues such as privacy, transparency, accountability, and fairness [3]. In the context of S-ERP, these ethical principles are essential for building systems that not only enhance business operations but also uphold the values and rights of all stakeholders involved. Privacy is a critical concern, as S-ERP systems often handle vast amounts of sensitive data, including personal and organizational information. Ensuring that this data is collected, stored, and processed in compliance with privacy regulations is paramount. AI systems must be designed to protect against data breaches and misuse, thereby maintaining the trust of users and stakeholders.

Transparency is another crucial aspect of AI ethics in S-ERP. The decision-making processes of AI algorithms must be understandable and explainable to users, as discussed in Sect. 5.2.1. This transparency helps in building trust and allows stakeholders to understand how decisions are made, ensuring that the AI systems are used in a fair and unbiased manner. It also facilitates accountability, where clear documentation and understanding of AI processes enable organizations to identify and address any issues that may arise. Fairness and non-discrimination are essential to ensure that AI systems in S-ERP do not perpetuate existing biases or create new ones. This involves rigorous testing and validation of AI algorithms to detect and mitigate any biases in the data or decision-making processes. Ensuring fairness helps in promoting equality and preventing any negative societal impacts that could arise from biased AI systems.

Moreover, the sustainability aspect of AI ethics in S-ERP involves considering the environmental impact of AI technologies. AI systems require substantial computational resources, which can lead to high-energy consumption. Developing energy-efficient AI solutions and using renewable energy sources can help mitigate the environmental footprint of these technologies, aligning with the goals of sustainable development. Incorporating ethical considerations into S-ERP systems also means addressing the potential impacts on employment and the workforce. Automation and AI can lead to significant changes in job roles and may result in job displacement. Ethical AI development in S-ERP should include strategies for workforce transition, reskilling, and ensuring that the benefits of AI are equitably distributed across the organization.

Overall, integrating AI ethics into S-ERP within Industry 4.0 is vital for creating systems that are not only technologically advanced but also socially responsible and sustainable. By addressing ethical concerns such as privacy, transparency, fairness, sustainability, and workforce impacts, organizations can develop S-ERP systems that contribute positively to business operations and society as a whole. This ethical approach ensures that the advancements in AI and S-ERP lead to long-term benefits and trust among all stakeholders involved. So, the question is how to embed AI ethics in S-ERP, especially in the context of Industry 4.0.

To achieve ethical and risk-free AI for S-ERP in the context of Industry 4.0, it is essential to incorporate a transdisciplinary approach that merges cognitive sciences, computational ethics, and the humanities. This involves creating frameworks that foster ethical awareness and responsibility among developers and scientists. The first model is about the human-centered design of AI in S-ERP. This model incorporates human values and ethical principles, ensuring that AI systems are designed to benefit individuals and society. This includes respecting privacy, ensuring fairness, and preventing bias. Since S-ERP is a central technological structure with a massive amount of sensitive information, privacy, and fairness through AI and data analytics algorithms are crucial.

To achieve fairness and transparency, transdisciplinary training about the AI and ML algorithms to all the systems' stakeholders (within and outside of the organization) is critical. Businesses should develop educational programs that integrate humanities with technical training, enabling developers to understand the broader societal impacts of their work. This should include courses on the cultural history of consciousness and ethical theories related to technology. In this training, the roles, responsibilities, and accountabilities of executing the AI-based systems should also be conveyed. To do this, organizations should ensure that AI systems are transparent and their decision-making processes can be understood and audited. This helps maintain accountability and trust in AI systems. AI systems in S-ERP should enhance human autonomy and well-being. They should be designed to assist rather than control, empowering users to make informed decisions. Sometimes, users feel that S-ERP controls them in the sensory environment that drives Industry 4.0. So, here, the related ethical aspect of AI system design is vital. Finally, organizations should encourage collaboration between technologists and ethicists to embed ethical considerations into AI development. This collaborative effort ensures that ethical guidelines are practical and relevant.

By embedding these principles into the development and implementation of AI systems in S-ERP, it is possible to create technology that not only drives efficiency and innovation but also upholds ethical standards and mitigates risks associated with AI in Industry 4.0. After discussing ethical AI, the following subsection discusses the ethical aspects of sensors.

5.2.3 The Amalgamation of Sensors and Ethics in S-ERP for Industry 4.0

In the realm of S-ERP for Industry 4.0, ethics in sensors plays a pivotal role, primarily revolving around privacy and user consent. As sensors increasingly collect and analyze emotional data, critical privacy concerns arise. The ethical framework within which these sensors operate is shaped by the designer's moral decisions regarding which emotions are recognized, who has access to this data, and how it is utilized. These decisions must align with ethical theories such as contractualism, which emphasizes mutual agreement and respect for individual rights. Users generally prefer systems that provide transparency and control over their data through explicit ethical contracts. These contracts detail the handling, usage, and accessibility of emotional data, thereby enhancing user trust and comfort. Therefore, ethics in sensors are essential to ensure that emotional data collection does not infringe on privacy and that users feel respected and secure, fostering broader acceptance and responsible use of affective computing technologies within S-ERP systems in Industry 4.0.

At this point, it may still be confusing to the readers. Is it possible that ethical concerns can arise due to sensors? How? In the context of S-ERP within Industry 4.0, sensors can potentially violate ethical behavior in several ways. Sensors embedded in wearable devices or workplace equipment could collect emotional or physiological data without explicit consent from the employees. For instance, if sensors monitor stress levels to optimize productivity without employees' knowledge or approval, it breaches privacy and ethical standards. Even if data is collected with consent, it could be used in ways that were not originally agreed upon. For example, emotional data intended to enhance employee well-being might instead be used to make unfair employment decisions, such as targeting individuals for layoffs based on perceived emotional instability. When the purpose and scope of data collection are not transparently communicated, it leads to ethical violations. If sensors in a smart factory collect detailed movement and biometric data without clear disclosure of how this data will be used, employees might feel their privacy is invaded. Constant monitoring of employees using various sensors can create a surveillance environment that infringes on personal privacy and autonomy. For example, sensors tracking every movement, gesture, or facial expression can lead to a feeling of being constantly watched, which is ethically problematic and can adversely affect employee morale and mental health. Sensors could be used to manipulate behavior in unethical ways. For example, if sensors detect signs of fatigue and companies use this data to push employees to work harder instead of allowing them to rest, it can lead to exploitation and health risks. When the ethical implications of sensor data use are not clearly outlined, it can lead to a lack of accountability. For instance, if a company uses sensor data to implement productivity measures that negatively impact employee well-being but does not take responsibility for these outcomes, it represents an ethical lapse. These examples highlight the importance of establishing clear ethical guidelines and robust privacy policies to ensure that the use of sensors in S-ERP systems aligns with ethical standards and respects the rights and dignity of individuals.

So, the question is how to achieve an ethical sensory system, especially S-ERP, in the context of Industry 4.0. To achieve an ethical sensory system, especially within S-ERP in the context of Industry 4.0, several critical steps and considerations must be addressed based on the principles outlined in the attached document. First, transparency and informed consent are paramount. Users should be clearly informed about what data is being collected by sensors, how it will be used, and who will have access to it. This can be achieved through explicit privacy policies or ethical contracts, which detail the scope of data collection and usage. For example, in an S-ERP system, employees should be made aware if their emotional or physiological data is being monitored to optimize productivity and how this data will impact their work environment.

Second, the principles of contractualism should be applied. This involves forming a mutual agreement between the system designers and the users. Ethical decisions embedded in the technology should reflect a consensus on what is acceptable and respectful of user privacy. For instance, if affective sensors are used to monitor employee stress levels, it should be done with the understanding and agreement of the employees, ensuring that the data is used solely for enhancing their well-being rather than for punitive measures. Third, data protection and security are crucial to maintaining ethical standards. Collected data must be securely stored and processed to prevent unauthorized access and misuse. Implementing robust encryption, regular security audits, and access controls are necessary measures. In the context of Industry 4.0, where vast amounts of data are collected and processed, ensuring the security of this data helps maintain trust and prevents ethical breaches.

Fourth, addressing potential biases in sensor algorithms is essential to avoid discriminatory practices. Algorithms should be carefully designed and tested to ensure they accurately interpret data across diverse user groups. For example, emotional recognition sensors should be trained on diverse datasets to avoid misinterpretation of emotions across different cultures. Fifth, providing users with control and feedback mechanisms enhances ethical acceptability. Users should have the ability to access, review, and delete their data if they choose. Additionally, providing feedback on how their data is used can increase transparency and trust. In S-ERP systems, employees could have dashboards where they can see what data is being collected and adjust their privacy settings accordingly. Finally, continuous evaluation and improvement of ethical practices are necessary. Regularly soliciting user feedback and conducting ethical reviews of the sensory systems can help identify and address emerging concerns. For instance, periodic surveys and focus groups involving employees can provide insights into their comfort and acceptance levels of the sensory systems in place.

Thus, achieving an ethical sensory system in S-ERP for Industry 4.0 requires a multifaceted approach involving transparency, mutual agreement, data protection, bias mitigation, user control, and ongoing evaluation. These measures ensure that the use of sensors respects user privacy, maintains trust, and aligns with ethical standards. The following subsection discusses the ethics of Industry 4.0, which is critical to understanding S-ERP implementation in the Industry 4.0 context.

5.2.4 Ethics in Industry 4.0 from S-ERP Perspectives

Ethics is essential in Industry 4.0, particularly when implementing S-ERP, due to the significant changes and challenges brought by digitalization [27], intelligent systems, and data management. The integration of ethics in Industry 4.0 is crucial for several reasons. Firstly, it helps handle risks and uncertainties associated with the increased autonomy and decision-making capabilities of intelligent systems. These systems can make decisions that may not always be beneficial, and ethical considerations help mitigate such risks. Additionally, the vast amount of data generated and used in Industry 4.0 poses risks related to privacy and misuse. Ethical guidelines ensure that data is used responsibly and transparently.

Incorporating ethics into traditional performance metrics transforms the efficiency-effectiveness-relevance triangle into a tetrahedron, ensuring that long-term performance and sustainability are maintained. This approach addresses human concerns such as job displacement, increased pressure, and demotivation due to AI and automation, ensuring that technological advancements do not come at the expense of human well-being. Moreover, ethics in Industry 4.0 aligns with Corporate Social Responsibility (CSR) by promoting fair and equitable financial policies and practices, ensuring that companies operate justly and beneficially for all stakeholders. Ethical frameworks also support compliance with regulations such as the General Data Protection Regulation (GDPR), which protects individual data rights and promotes accountability.

By integrating ethics into their operations, companies can build and maintain trust with stakeholders, including customers, employees, and partners, which is crucial for long-term success and reputation management. Ethical practices ensure transparency in operations, fostering a culture of honesty and integrity within the organization. In conclusion, incorporating ethics into the S-ERP within Industry 4.0 is not only about compliance and risk management but also about fostering a sustainable, fair, and transparent industrial ecosystem that prioritizes human well-being and long-term societal benefits.

Implementing ethics in Industry 4.0, especially when integrating S-ERP, requires a structured and comprehensive methodology. The integration of ethics into the traditional performance model, which historically focuses on efficiency, effectiveness, and relevance, is essential to address the unique challenges and uncertainties posed by Industry 4.0 technologies. This methodology can be structured into a performance tetrahedron, incorporating ethics as the fourth dimension of the existing performance triangle. The methodology begins with a clear description of the situation, focusing on the set objectives, the achieved results, and the utilized means. This step involves collecting data relevant to performance without initially considering ethical aspects. Once the situation is described, the next step is to analyze the consequences of the actions taken, observing both the immediate and potential long-term impacts on all stakeholders, including employees, customers, and society at large.

Following the analysis, it is essential to determine the ethical points of interest. This involves systematically questioning the objectives, means, and results from an ethical perspective. For example, managers should ask whether the set objectives are ethical, if the means used to achieve these objectives are ethical, and whether the results obtained respect ethical standards. Once the ethical points of interest are identified, the next step is to propose ethical objectives and actions. This could involve redefining goals to ensure they align with ethical standards, adjusting operational practices to mitigate any negative impacts on stakeholders, and ensuring transparency and accountability in all processes. Actions might include extending deadlines for achieving certain objectives to reduce pressure on employees, enhancing communication about the ethical use of data, and providing training and support to help staff adapt to new technologies and responsibilities.

For practical application, the performance tetrahedron helps managers integrate ethics into their strategies by ensuring that all actions and objectives are evaluated through an ethical lens. This approach ensures that profitability and performance are achieved without compromising ethical standards, ultimately leading to sustainable and responsible industrial practices in the context of Industry 4.0. The methodology not only addresses immediate ethical concerns but also prepares organizations to manage future risks associated with technological advancements and data management, thus ensuring long-term sustainability and ethical compliance. Based on the methodology, we explain the case that depicts the ethics violation and mitigation.

5.2.4.1 Case: Ethics and Industry 4.0 from S-ERP Perspectives

In the fast-paced realm of Industry 4.0, one leading automotive parts manufacturer, AutoTech Innovations, decided to implement S-ERP to enhance its operational efficiency and product traceability. The new system promised to streamline production processes, optimize resource use, and meet stringent customer requirements for product traceability. Initially, the implementation of S-ERP seemed successful. The system enabled real-time monitoring of production lines and detailed tracking of components from raw material to finished product. However, shortly after deployment, several ethical issues emerged.

The extensive data collection and surveillance capabilities of the S-ERP system created an environment of constant monitoring, which led to increased stress among employees. Workers felt that their every move was being scrutinized, leading to anxiety and reduced morale. Moreover, the system's emphasis on performance metrics fostered unhealthy competition among staff, as they were constantly compared against one another based on productivity data. This competitive atmosphere not only eroded team cohesion but also incentivized short-term performance gains over long-term innovation and improvement.

Recognizing these issues, AutoTech Innovations decided to implement a structured and comprehensive methodology to integrate ethics into their S-ERP system, following a performance tetrahedron model that incorporates ethics as a fourth dimension to the traditional performance triangle of efficiency, effectiveness, and relevance.

Step 1: Description of the Situation

The company began by clearly describing the current situation, focusing on the set objectives (enhanced traceability and operational efficiency), the achieved results (improved production monitoring and data collection), and the utilized means (S-ERP system and performance metrics).

Step 2: Analysis of Consequences

Next, they analyzed the consequences of these actions. The immediate impacts included improved traceability and productivity metrics, but the long-term impacts revealed significant ethical concerns. Employees were experiencing heightened stress and anxiety due to constant monitoring, and the competitive environment was negatively affecting teamwork and overall job satisfaction.

Step 3: Determination of Ethical Points of Interest

The company systematically questioned the objectives, means, and results from an ethical perspective. They asked whether the objectives of enhanced monitoring were ethical given the stress it caused employees, whether the means of constant surveillance were ethical, and whether the results respected the ethical standards of a supportive and fair workplace.

Step 4: Proposal of Ethical Objectives and Actions

Upon identifying these ethical points of interest, AutoTech Innovations proposed several ethical objectives and actions. They redefined their goals to balance efficiency with employee well-being, adjusted operational practices to mitigate negative impacts, and enhanced transparency and accountability. Specific actions included extending deadlines for achieving certain performance targets to reduce pressure on employees, enhancing communication about the ethical use of data, and providing training and support to help staff adapt to the new system.

For practical application, the performance tetrahedron model helped managers integrate ethics into their strategies by ensuring that all actions and objectives were evaluated through an ethical lens. This approach ensured that profitability and performance were achieved without compromising ethical standards, ultimately leading to sustainable and responsible industrial practices in the context of Industry 4.0.

Through this structured methodology, AutoTech Innovations was able to address the immediate ethical concerns and prepare the organization to manage future risks associated with technological advancements and data management, thus ensuring long-term sustainability and ethical compliance. The successful integration of ethics into their S-ERP system not only improved employee morale and teamwork but also reinforced the company's commitment to ethical standards, positioning them as a responsible leader in the Industry 4.0 landscape. The following section explains the ethics of sustainability and the environment concerning Industry 4.0.

5.2.5 Ethics Related to Sustainability and Environment Concerning S-ERP in Industry 4.0

Integrating environmental ethics into S-ERP within Industry 4.0 is essential to address the challenges of industrialization and globalization. Ethical considerations should ensure transparency, fairness, and responsibility in business operations, encompassing international labor standards and human rights. A moral framework for S-ERP aims to mitigate adverse impacts such as economic inequality and exploitation by promoting fair trade and supporting local economies. Industry 4.0 technologies, including IoT, AI, and big data analytics, must align with environmental sustainability goals. S-ERP systems should incorporate practices to minimize environmental footprints, such as reducing energy consumption, minimizing waste, and utilizing renewable resources [31]. Additionally, S-ERP can facilitate environmental performance monitoring and regulatory compliance, contributing to global efforts against climate change and resource preservation.

Globalization necessitates that S-ERP systems manage the complexities of global operations, including cross-border supply chains, diverse regulatory requirements, and international financial transactions. A cosmopolitan approach to business ethics is required, where companies are responsible to local and global communities [30]. S-ERP systems must support global cooperation, cultural sensitivity, and inclusivity, enabling businesses to navigate globalization challenges effectively and foster sustainable development and international justice.

S-ERP systems must incorporate ethical guidelines and ensure transparency, fairness, and responsibility in all operations, aligning with the communitarian Kantianism approach to foster a culture of integrity and trust [36]. They should support sustainable practices by optimizing resource use, reducing waste, and ensuring compliance with environmental regulations, with the integration of triple bottom line reporting helping to monitor and improve sustainability performance. Additionally, S-ERP must facilitate global cooperation, manage cross-border operations efficiently, and adhere to international standards and norms, including supporting CSR initiatives and aligning with global frameworks like the UN SDGs. Furthermore, S-ERP systems should enable robust stakeholder management, ensuring that the needs and expectations of all stakeholders, including the environment, are met, thereby enhancing corporate legitimacy and fostering a collaborative approach to sustainability.

From a political economy perspective [17], adopting S-ERP in Industry 4.0 reflects broader socio-economic dynamics and regulatory frameworks prioritizing environmental sustainability. Governments and policymakers increasingly enforce regulations and incentives to encourage sustainable practices, influencing corporate behavior and investment in green technologies. S-ERP systems help companies comply with these regulations by providing transparent, efficient, and sustainable operational processes. Moreover, the political economy highlights the role of power dynamics and economic interests in shaping environmental policies. Corporations adopting S-ERP can gain a competitive advantage through cost savings and enhanced brand reputation, driving market shifts toward sustainability. Additionally, integrating environmental ethics into S-ERP aligns corporate strategies with global sustainability goals, fostering a more equitable and sustainable economic system.

From a Hypermodern Experience Economy ethics [31] perspective, the relationship between S-ERP in Industry 4.0 and environmental ethics is multifaceted, reflecting the evolving nature of consumer values and business practices in a hypermodern society. The Hypermodern Experience Economy emphasizes the creation of meaningful and immersive experiences for consumers who seek authenticity, sustainability, and ethical engagement in their interactions with businesses. In this context, S-ERP systems play a crucial role by enabling companies to integrate environmental ethics into their operations, thus aligning with the values of hypermodern consumers. S-ERP systems facilitate transparency and traceability, allowing businesses to provide verifiable information about their sustainability practices. This transparency meets the ethical expectations of hypermodern consumers who demand accountability and authenticity from the brands they support. Companies can use S-ERP to track their environmental impact, implement eco-friendly practices, and communicate their sustainability efforts effectively, thereby creating a positive and ethical consumer experience. The following subsection discusses ethical aspects of computer vision in S-ERP for Industry 4.0.

5.2.6 Ethics of Computer Vision in S-ERP for Industry 4.0

Although computer vision is considered a subbranch of AI, there is a slight difference between AI and computer vision ethics. The ethics of computer vision differ from general AI ethics in several vital aspects [31]. One significant difference lies in the type of data involved. Computer vision ethics primarily deals with visual data, such as images and videos, while AI ethics encompasses a broader range of data types, including text, audio, and numerical data. The nature of visual data introduces unique challenges, particularly concerning privacy issues related to identifiable imagery.

Another difference is in the applications of these technologies. Computer vision is often applied in specific areas like surveillance, healthcare, and autonomous vehicles, each presenting its own set of ethical challenges. In contrast, AI ethics covers a wider range of applications, including those where computer vision may not be involved, such as natural language processing and predictive analytics. The impact on individuals also differs between the two. Computer vision technologies tend to have a more direct and visible effect on individuals, such as being recognized by a facial recognition system, whereas the impact of other AI systems may be less obvious or more indirect, such as algorithmic decision-making in credit scoring.

Finally, societal implications vary between computer vision and AI in general. While both fields consider societal impacts, computer vision ethics must address specific concerns, such as the potential normalization of surveillance and the societal acceptance of visual data collection. These concerns may differ from those arising in broader AI applications.

In S-ERP for Industry 4.0, the integration of AI technologies is critical for optimizing processes, enhancing decision-making, and improving overall efficiency. However, as these technologies become more embedded in industrial operations, the ethical considerations associated with their use must be carefully examined. While AI ethics provides a broad framework for addressing issues such as fairness, accountability, and transparency in decision-making processes, computer vision ethics introduces additional layers of complexity that are equally vital in this context.

In S-ERP systems, computer vision technologies are increasingly used for tasks such as quality control, predictive maintenance, inventory management, and security surveillance. These applications involve the processing of visual data, which raises specific ethical concerns distinct from those addressed by general AI ethics. For example, the use of computer vision in surveillance within an industrial setting might lead to privacy infringements, where workers' movements and actions are constantly monitored without their explicit consent. This could create a work environment where individuals feel under constant scrutiny, potentially leading to ethical dilemmas around autonomy and consent.

Moreover, the reliance on computer vision for tasks such as facial recognition or employee monitoring in S-ERP systems can introduce biases and discriminatory practices if the underlying algorithms are not designed and tested carefully. These biases could result in unfair treatment of certain groups of employees or inaccurate assessments of performance, leading to broader ethical and legal implications. In addition, the societal implications of deploying computer vision technologies in Industry 4.0 environments extend beyond the immediate workplace. The normalization of surveillance practices and the extensive collection of visual data could contribute to broader societal shifts, where privacy expectations are diminished, and the acceptance of constant monitoring becomes more prevalent. This shift could have long-term consequences, not only within industrial settings but also in society at large. Therefore, in the context of S-ERP for Industry 4.0, it is not sufficient to address AI ethics alone; computer vision ethics must also be rigorously considered. Ensuring that both AI and computer vision technologies are implemented in a manner that respects privacy, avoids bias, and acknowledges the broader societal implications is essential for creating sustainable and ethically responsible industrial practices. By integrating these ethical considerations, S-ERP systems can better align with the principles of sustainability and social responsibility that underpin Industry 4.0, ensuring that technological advancements contribute positively to both industry and society.

Next, we plan to develop a framework for ethical computer vision that needs to be used in S-ERP of Industry 4.0. The proposed framework is shown in Table 5.1.

At this point, we have discussed all the ethical aspects related to S-ERP in Industry 4.0, as discussed in Fig. 5.1. Next, we aim to discuss the cybersecurity aspects of S-ERP in Industry 4.0.

Table 5.1 Framework for ethics in S-ERP pertaining to Industry 4.0

Name of the Principle	Principle	Actionable steps in S-ERP for Industry 4.0
Data Traceability	Ensure that all visual data used in computer vision applications within S-ERP is traceable throughout its lifecycle	– Implement robust documentation practices that record the origin of all visual data – Maintain a clear chain of custody for datasets, ensuring that data subjects can trace how and where their images are used
Consent and Privacy	Prioritize obtaining informed consent from individuals whose images are used in computer vision datasets	– Establish protocols for gaining consent before collecting or using visual data in S-ERP systems – Ensure that data subjects are aware of how their data will be used and provide mechanisms for them to withdraw consent if desired
Bias and Fairness	Address and mitigate biases in computer vision algorithms to ensure fair treatment of all individuals	– Regularly audit computer vision models for biases related to race, gender, age, and other characteristics – Include diverse data in training datasets to minimize bias and ensure the model's fairness
Transparency and Accountability	Ensure that the use of computer vision technologies in S-ERP is transparent and that accountability mechanisms are in place	– Provide clear documentation on how computer vision models are developed, trained, and deployed within S-ERP systems – Establish accountability frameworks that outline who is responsible for the ethical use of computer vision technologies
Security and Protection	Safeguard the data used in computer vision systems to prevent unauthorized access and misuse	– Implement strong security measures to protect datasets containing visual data – Regularly review and update security protocols to address emerging threats

(continued)

Table 5.1 (continued)

Name of the principle	Principle	Actionable steps in S-ERP for Industry 4.0
Legal Compliance and Ethical Standards	Ensure that all computer vision activities in S-ERP comply with relevant legal and ethical standards	– Stay updated on the latest legal requirements and ethical guidelines related to computer vision and data protection – Incorporate ethical considerations into every stage of the computer vision development process, from data collection to deployment
Ongoing Monitoring and Evaluation	Continuously monitor and evaluate the ethical implications of computer vision technologies in S-ERP	– Conduct regular ethical reviews of computer vision systems to identify and address potential issues – Engage with stakeholders, including employees and external experts, to assess the ethical impact of these technologies
Societal Implications	Consider the broader societal impact of deploying computer vision technologies in Industry 4.0	– Assess how the use of computer vision in S-ERP might influence societal norms, particularly regarding privacy and surveillance – Ensure that the deployment of these technologies contributes positively to society, avoiding any potential harm

5.3 Cybersecurity of S-ERP in Industry 4.0

In the rapidly evolving landscape of Industry 4.0, S-ERP systems play a crucial role in integrating various operational processes, enhancing efficiency, and driving sustainability. However, this interconnectedness also brings forth significant cybersecurity challenges. As S-ERP systems collect, store, and process vast amounts of sensitive data, they become prime targets for cyberattacks. The need for robust cybersecurity measures is paramount to protect critical business information, maintain the integrity of industrial operations, and prevent disruptions that could lead to significant financial losses or reputational damage. In an environment where cyber threats are increasingly sophisticated, ensuring the security of S-ERP systems is not just a technical necessity but a strategic imperative that supports the long-term sustainability and resilience of Industry 4.0 enterprises.

As shown in Fig. 5.2, this section will delve into several key aspects of cybersecurity essential for safeguarding S-ERP systems in Industry 4.0. We will begin by exploring data security, highlighting the importance of protecting sensitive information. Next, we will address network security, emphasizing the need to secure communication channels, followed by examining hardware and software (sensory aspects) to ensure the integrity of physical and digital components. The impact of cybersecurity breaches will be analyzed, illustrating potential consequences on business operations. We will then discuss managing risk through proactive measures and outline best practices for cybersecurity in S-ERP. The section will also explore the balance between shared resources, usability, and security and conclude with a discussion on the cryptography logic underpinning modern security protocols. These topics collectively provide a comprehensive understanding of the cybersecurity challenges and solutions pertinent to S-ERP in Industry 4.0. The following section explores the data and network security for S-ERP in Industry 4.0

Fig. 5.2 Cybersecurity aspects of S-ERP in Industry 4.0 that we are going to discuss in this section

5.3.1 Data and Network Security for S-ERP in Industry 4.0

In the context of S-ERP systems within Industry 4.0, data security is critical to safeguard digital information from unauthorized access, alteration, corruption, or theft throughout its entire lifecycle. Central to data security are the three fundamental principles: confidentiality, integrity, and availability. These principles are maintained by implementing comprehensive policies, controls, and technologies that ensure data is securely protected and accessed only when necessary and with appropriate authorization.

Properly executed data security measures are vital for shielding an organization's information assets from cyberattacks as well as internal threats [39], such as malicious actions by employees or inadvertent data exposure. Additionally, in light of the increasing number of regulatory compliance requirements, robust data security protocols help minimize data exposure and enforce the principle of least privilege, ensuring that access to information is granted only after careful verification of the request. This not only enhances the overall security posture of S-ERP systems but also helps the organization meet legal and regulatory obligations.

Although S-ERP applications provide certain security controls, they often lack the necessary granularity required to fully meet stringent data protection standards. Regulatory frameworks such as the General Data Protection Regulation (GDPR) [40] and the California Privacy Rights Act (CPRA) [12, 14] impose rigorous data privacy requirements, necessitating multiple layers of internal controls at various levels, including field, transaction, and master data. Implementing these precise controls is often challenging, if not unfeasible, using only the native security features inherent to most S-ERP systems. Consequently, organizations must consider additional measures or advanced solutions to achieve compliance and ensure the comprehensive protection of sensitive data within S-ERP environments in Industry 4.0.

In many instances, S-ERP data security is relegated to a secondary priority, frequently overlooked by organizations that regard it as non-essential, irrespective of the prevailing economic conditions [1]. This perception often stems from a belief that existing security measures are "sufficient," leading companies to rely on outdated legacy technologies as they navigate increasingly complex and unpredictable environments, especially in Industry 4.0. Such an approach, however, leaves critical data vulnerable to theft, fraud, and other potential threats. Neglecting to proactively strengthen S-ERP data security not only jeopardizes the integrity of the system but also poses a significant risk to the organization's future stability and success. Failure to adequately prepare for evolving security challenges may ultimately lead to severe consequences.

The capability for remote access to critical data in S-ERP, which is essential for enabling flexibility and ensuring continuous operations, further complicates the security landscape. These practices have been increased after COVID [32]. Employees accessing sensitive information from remote locations, often over public or less secure networks, increase the risk of data interception, unauthorized access, and data leakage [2]. This scenario underscores the urgent need for stringent data security measures within S-ERP systems. These measures must encompass advanced encryption protocols, robust authentication mechanisms, and continuous monitoring to safeguard against potential security breaches.

For S-ERP in Industry 4.0, many types of attacks may affect data security, such as accidental exposure, social engineering attacks, ransomware, and cloud data loss. In S-ERP for Industry 4.0, accidental exposure occurs when employees gain access to sensitive data without proper authorization or need, often due to oversight or lack of awareness about security policies. This risk can be minimized by enhancing access controls and providing thorough employee training. Phishing and other social engineering attacks manipulate employees into divulging sensitive information [10]. Phishing emails, appearing as legitimate communications, trick recipients into sharing credentials or clicking malicious links, compromising the organization's network and systems. Ransomware is malware that encrypts corporate data, rendering it inaccessible unless a ransom is paid for the decryption key [16]. It spreads quickly across networked devices, and organizations can only recover without paying the ransom by relying on backup servers. With the shift to cloud computing, data loss becomes a concern when access and distribution controls are weak [15]. Without proper security measures, unauthorized parties could gain access to cloud-stored data, posing significant risks to the organization. The same issues also exist in Service-Oriented Architecture (SOA) or client-server architecture [7].

One effective approach to enhancing data security within S-ERP systems in Industry 4.0 is to achieve granular visibility into the specifics of ERP data access, including the identification of who accessed the data, when, and from where [34]. Implementing a comprehensive Change Log Module facilitates detailed analysis of data access patterns, enabling the organization to meticulously track modifications to both transaction and master data [5]. This includes identifying the source of any change and the user responsible for initiating the change, as well as capturing the before and after values of the data, including records that have been deleted. Such detailed monitoring capabilities are essential for maintaining data integrity, ensuring accountability, and mitigating potential security risks associated with unauthorized or inadvertent data alterations. Further, to mitigate insider threats, situational awareness is critical for data security [8].

Achieving network security for S-ERP systems in Industry 4.0 requires a multi-layered approach that addresses both external and internal threats. First, robust encryption protocols must be employed to secure data in transit, ensuring that sensitive information remains protected as it moves across the network. This includes using technologies like SSL/TLS for secure communication and VPNs for remote access. Next, implementing strong authentication mechanisms, such as multi-factor authentication (MFA), ensures that only authorized users can access the network. Additionally, network segmentation can be utilized to isolate critical systems and data, reducing the potential impact of a breach by limiting lateral movement within the network. Regular monitoring and intrusion detection systems (IDS) are essential for identifying and responding to suspicious activities in real time. Furthermore, maintaining up-to-date firewalls and regularly patching software vulnerabilities is crucial to defending against potential exploits. Lastly, continuous employee training on cybersecurity best practices helps to mitigate risks associated with social engineering attacks, ensuring that network security is not compromised by human error. These measures, when integrated into the S-ERP framework, create a secure network environment that supports the reliability and integrity of Industry 4.0 operations. The following subsection discusses hardware and software security for S-ERP in Industry 4.0.

5.3.2 Hardware and Software Security of S-ERP in Industry 4.0

In the context of S-ERP in Industry 4.0, hardware and software security are pivotal for ensuring the integrity and reliability of the entire system. With the proliferation of IoT sensors and devices within industrial environments, the security landscape has become increasingly complex. Hardware security involves safeguarding physical devices such as sensors, actuators, and network nodes from tampering, unauthorized access, and other forms of physical compromise. As highlighted in the attached document, the integration of IoT sensors in industrial settings facilitates real-time monitoring and control, but it also introduces vulnerabilities that could be exploited by malicious actors [29].

To mitigate these risks, robust authentication and encryption protocols must be implemented to ensure that only authorized devices can interact with the network. Additionally, continuous monitoring for signs of tampering or unauthorized access is essential for maintaining hardware security. Software security, on the other hand, focuses on protecting the data and communication processes managed by these devices. This includes securing the software that controls IoT devices, as well as the data transmitted between them. The document suggests that blockchain technology can play a significant role in enhancing software security by providing a transparent and immutable record of all transactions and interactions within the network. This not only ensures data integrity but also helps in detecting and preventing unauthorized changes or breaches within the system. By integrating such security measures, S-ERP systems in Industry 4.0 can maintain the confidentiality, integrity, and availability of critical industrial processes.

Blockchain technology can significantly enhance both hardware and software security in S-ERP for Industry 4.0 by providing a robust, decentralized framework that ensures data integrity, transparency, and trust across the system. For hardware security, blockchain can be instrumental in securely registering and authenticating IoT devices, such as sensors and actuators, that are integral to S-ERP systems. Each IoT device in an industrial setting can be registered on a blockchain network, where its identity and activity can be continuously monitored and verified. This registration process ensures that only legitimate devices are granted access to the network, preventing unauthorized or compromised devices from infiltrating the system. By leveraging blockchain's immutable ledger, any tampering or unauthorized modification of device data can be detected and traced back to its source, enhancing the overall security of the hardware infrastructure.

In terms of software security, blockchain provides a secure and transparent way to manage and protect the data generated and transmitted by these loT devices. The document discusses how blockchain can store the data generated by sensors in a decentralized manner, making it resistant to tampering or unauthorized alterations. This ensures that the data remains accurate and trustworthy throughout its lifecycle, which is crucial for making informed decisions in an S-ERP environment. Additionally, blockchain can facilitate secure communication between devices by verifying the authenticity of messages and transactions, thus preventing malicious actors from intercepting or altering critical information. Overall, blockchain's decentralized and transparent nature makes it a powerful tool for bolstering both hardware and software security in S-ERP systems, ensuring that the integrity, confidentiality, and availability of industrial data and devices are maintained in the increasingly connected landscape of Industry 4.0.

To formally represent how blockchain can be used to achieve hardware and software security in S-ERP for Industry 4.0, we can break down the process into a series of equations that encapsulate the core cryptographic operations involved. These equations will highlight how blockchain ensures data integrity, authenticity, and tamper-proofing for hardware and software components. We present seven steps for the effective use of blockchain for S-ERP in Industry 4.0.

1. Hashing Function for Data Integrity

For any data D (which could be sensor data or hardware status information), a cryptographic hash function $H(x)$ is used to generate a fixed-size hash value:

$$H_D = H(D)$$

where:

- H_D is the hash of the data D.
- $H(x)$ is a secure cryptographic hash function (e.g., SHA-256).

2. Digital Signature for Authenticity

To ensure that the data D comes from a legitimate source, the data is signed using a private key k_{priv} associated with the secure element or the hardware device:

$$\text{Sig}_D = \text{Sign}\left(H_D, k_{\text{priv}}\right)$$

where:

- Sig_D is the digital signature of the hash H_D.
- and $\mathrm{Sign}\left(x, k_{\mathrm{priv}}\right)$ represents the signing operation using the private key.

3. Verification of Signature for Tamper-Proofing

The signature is verified at the receiving end (e.g., the cloud server) using the corresponding public key k_{pub}:

$$\text{Verify}\left(H_D, \ \text{Sig}_D, k_{\mathrm{pub}}\right) = \begin{cases} \text{True} & \text{if valid} \\ \text{False} & \text{if invalid} \end{cases}$$

where:

- $\text{Verify}\left(x, y, k_{\mathrm{pub}}\right)$ checks if the signature y was generated using the private key corresponding to k_{pub} and matches the hash x.

4. Block Formation for Immutable Records

The data, along with its signature, is stored in a block. Each block contains the hash of the previous block to ensure immutability:

$$\begin{aligned} \text{Block}_n &= \left\{D_n, \text{Sig}_{D_n}, H_{\text{prev}}\right\} \\ H_{\text{prev}} &= H\left(\text{Block}_{n-1}\right) \end{aligned}$$

where:

- Block_n is the current block containing data D_n, its signature Sig_{D_n}, and the hash of the previous block H_{prev}.

5. Consensus Mechanism for Network Security

A consensus mechanism (e.g., Proof of Work PoW or Proof of Stake PoS) ensures that only valid blocks are added to the blockchain:

$$\text{Consensus}\left(\text{Block}_n\right) = \begin{cases} \text{Add to chain} & \text{if valid} \\ \text{Reject block} & \text{if invalid} \end{cases}$$

6. Continuous Monitoring and Anomaly Detection

The secure element periodically emits a secure pulse P that includes the current status S of sensors or hardware:

$$P_t = \{S_t, \mathrm{Sig}_{S_t}, T_t\}$$

where:

- S_t is the status at time t.
- Sig_{S_t} is the signature of the status.
- T_t is the timestamp.

This pulse is verified and added to the blockchain, ensuring continuous monitoring and anomaly detection.

7. Anomaly Detection Logic

To detect anomalies, compare the current status S_t with the expected status S_{expected}:

$$\text{Anomaly} = \begin{cases} \text{True} & \text{if } S_t \neq S_{\text{expected}} \\ \text{False} & \text{if } S_t = S_{\text{expected}} \end{cases}$$

If an anomaly is detected, it triggers an alert and is logged on the blockchain for audit purposes.

These steps formalize how blockchain technology is applied to ensure the security of both hardware and software in S-ERP systems for Industry 4.0. They emphasize data integrity through hashing, authenticity via digital signatures, immutability with block chaining, and continuous security monitoring through secure pulse emission and verification. The following section discusses the impact of cyberattacks in S-ERP for Industry 4.0.

5.3.3 Impact of Cybersecurity Violations in S-ERP for Industry 4.0

In the rapidly evolving landscape of Industry 4.0, where S-ERP systems are integral to the seamless integration of advanced technologies like AI, IoT, and big data analytics, the importance of cybersecurity cannot be overstated. Cybersecurity violations within S-ERP systems pose significant risks, not just to the integrity and confidentiality of data but also to the operational continuity of the entire enterprise. These violations can lead to substantial financial losses, damage to a company's reputation, and potential legal consequences. Moreover, as S-ERP systems are increasingly interconnected with other industrial systems and supply chains, a breach in one area can cascade, causing widespread disruption across multiple sectors. Therefore, addressing cybersecurity vulnerabilities in S-ERP is not merely a technical necessity but a strategic imperative for safeguarding the future of Industry 4.0.

A cybersecurity breach in an S-ERP system within Industry 4.0 can have profound implications across the triple bottom line-economic, environmental, and social. Economically, cybercriminals may gain access to and exploit sensitive operational and financial data, potentially leading to significant business disruptions, loss of competitive advantage, and regulatory fines for failing to protect customer and proprietary information. Socially, such breaches can erode stakeholder trust, damage the organization's reputation, and affect employee morale, particularly if personal or personnel data is compromised. From an environmental perspective, a security breach that disrupts S-ERP operations could lead to inefficiencies in resource management, resulting in increased waste and energy consumption, thus undermining the sustainability goals central to Industry 4.0. Consequently, the impact of cybersecurity violations extends far beyond immediate financial costs, affecting the broader organizational ecosystem and its commitment to sustainable practices.

A cybersecurity breach in an S-ERP system within Industry 4.0 can severely disrupt a company's operations, leading to significant downtime and loss of productivity. In critical situations, businesses may be forced to halt operations entirely until the breach is resolved, which is particularly detrimental for those dependent on real-time data and automated processes. This disruption not only affects operational efficiency but can also lead to increased energy consumption and waste generation as systems fail to function optimally. The resulting operational inefficiencies may strain resources, contributing to environmental degradation and undermining the sustainability objectives central to Industry 4.0 practices.

A cybersecurity violation in an S-ERP system can have far-reaching consequences on trust, brand reputation, and customer expectations, especially among those committed to purchasing eco-friendly products. Consumers who prioritize sustainability expect the companies they support to uphold the highest standards of integrity, including robust data security. When a breach occurs, it not only jeopardizes sensitive customer information but also raises concerns about the company's ability to maintain the transparency and ethical standards promised by its brand. This loss of trust can be particularly damaging for businesses whose core value proposition is based on environmental responsibility. Customers may begin to question the authenticity of the company's sustainability claims, fearing that a lack of security oversight might extend to other areas of the business, including environmental practices. The brand's reputation as a leader in eco-friendly products could be significantly tarnished, leading to a loss of customer loyalty and a decline in market share. In an era where consumer trust is closely tied to brand identity, especially in the eco-conscious market, a cybersecurity violation can undermine years of effort in building a credible and trustworthy brand, ultimately affecting customer retention and long-term business success.

A cybersecurity violation in an S-ERP system within the context of Industry 4.0 can lead to severe regulatory fines and legal repercussions. S-ERP systems are integral to managing critical data, including sensitive customer information, operational processes, and intellectual property. A breach that compromises this data can result in violations of stringent data protection regulations such as the General Data Protection Regulation (GDPR) in the European Union or the California Consumer Privacy Act (CCPA) in the United States. These regulations impose hefty fines for non-compliance, especially when companies fail to implement adequate security measures to protect personal and sensitive data. Beyond financial penalties, businesses may also face legal consequences, including lawsuits from affected parties, which can further strain resources and damage the company's reputation. Additionally, industry-specific regulations may impose additional fines and sanctions, particularly in healthcare, finance, or energy sectors where data security is paramount. The legal ramifications of a cybersecurity breach can extend beyond immediate financial penalties, potentially leading to long-term operational restrictions, mandatory audits, and increased scrutiny from regulators. This underscores the critical importance of robust cybersecurity measures in S-ERP systems to avoid costly legal entanglements and protect the integrity of the enterprise. The following subsection discusses managing risk in S-ERP for Industry 4.0.

5.3.4 Managing Risk in S-ERP for Industry 4.0

Managing risk in S-ERP for Industry 4.0 involves identifying, assessing, and mitigating potential threats to digital systems, including data breaches, network vulnerabilities, and hardware failures. It requires a proactive approach to cybersecurity, integrating best practices like regular audits, encryption, and access control. Additionally, balancing usability with security is crucial to ensure systems remain functional while safeguarding critical information. Effective risk management in S-ERP also includes contingency planning, such as disaster recovery protocols, to minimize impact and ensure continuity in case of security incidents or system failures.

Data governance is the first step required to manage risk in S-ERP for Industry 4.0. Data governance in S-ERP for Industry 4.0 is critical for managing risk and ensuring data integrity, security, and compliance. It involves the systematic management of data availability, usability, integrity, and security within an organization. Effective data governance frameworks define roles, responsibilities, and processes to ensure that data is managed as a valuable asset.

Formal Equation for Data Quality Measurement:

Let DQ represent data quality, where:

$$DQ = \frac{\sum_{i-1}^{n} w_i \cdot q_i}{\sum_{i=1}^{n} w_i}$$

Here, q_i represents the quality score of data attribute i, and w_i is the weight assigned to the importance of that attribute. This equation helps in assessing the overall data quality.

Equation for Risk Assessment in Data Management:

Risk R can be expressed as:

$$R = P \times I$$

where P is the probability of a data-related incident (e.g., data breach), and I is the impact or consequence of that incident.

In Industry 4.0, data governance also ensures compliance with regulatory requirements, minimizing legal and financial risks. It integrates risk management practices such as data encryption, access controls, and continuous monitoring to protect sensitive information, thus securing the S-ERP ecosystem against potential cyber threats. This holistic approach to data governance is crucial for maintaining the reliability and security of S-ERP systems in the dynamic environment of Industry 4.0.

The second step to managing risk in S-ERP for Industry 4.0 is access control and shared responsibility. Access control and shared responsibility are key components of managing risk in S-ERP for Industry 4.0. Access control ensures that only authorized users have access to critical data and systems, minimizing the risk of unauthorized access and data breaches. This is typically enforced using Role-Based Access Control (RBAC), where roles R_i are assigned to users U_i, and each role has permissions P_j associated with it.

Formal Equation for RBAC:

$$A\left(U_i\right) = \sum_{j=1}^{m} P_j \cdot f\left(R_i, P_j\right)$$

where $A\left(U_i\right)$ represents the access level of user U_i, and $f\left(R_i, P_j\right)$ is a function that equals 1 if role R_i has permission P_j, and 0 otherwise.

Shared responsibility in S-ERP involves collaboration between different stakeholders such as IT, operations, and third-party vendors to ensure that security practices are consistently applied across the system. This shared responsibility model is crucial in Industry 4.0, where interconnected systems and data exchanges between multiple entities are common. Similarly, the equation for risk distribution is as follows:

The equation for Risk Distribution:

$$R_{\text{total}} = \sum_{k=1}^{n} R_k \cdot S_k$$

where R_{total} is the total risk, R_k is the risk associated with a specific entity, and S_k is the responsibility share of that entity. By combining access control with a shared responsibility model, S-ERP systems in Industry 4.0 can effectively mitigate risks and ensure secure, efficient operations.

The third step in managing risk in S-ERP for Industry 4.0 is the implementation strategy and security training. Implementing an effective strategy and security training are vital for managing risk in S-ERP in Industry 4.0. The implementation strategy should include risk assessment, identification of critical assets, and the deployment of security controls tailored to the Industry 4.0 environment. A layered defense approach, combining physical, technical, and administrative controls, ensures comprehensive protection. Formal Equation for Risk Mitigation:

$$R_{\text{residual}} = R_{\text{initial}} - \sum_{i=1}^{n} C_i \cdot E_i$$

where R_{residual} is the residual risk after implementing controls, R_{initial} is the initial risk, C_i represents the control effectiveness, and E_i is the extent to which the control is applied.

Security training focuses on raising awareness among employees and stakeholders about potential threats, proper use of the S-ERP system, and incident response procedures [6]. Regular training sessions should be integrated into the organization's security strategy to ensure that all personnel are equipped to recognize and respond to security incidents.

The equation for Training Effectiveness:

$$T_{eff} = \frac{\sum_{j-1}^{m} S_j \cdot P_j}{m}$$

where T_{eff} is the training effectiveness, S_j is the security awareness score for participant j, and P_j is the participant's proficiency level post-training. Organizations can significantly reduce the risks associated with S-ERP systems in Industry 4.0 by strategically implementing security measures and providing continuous training.

The fourth and final step to managing risk in S-ERP for Industry 4.0 is regulatory compliance and vulnerability assessment. Regulatory compliance and vulnerability assessment are essential for managing risk in S-ERP in Industry 4.0. Regulatory compliance ensures that the S-ERP system adheres to relevant laws, standards, and guidelines, such as GDPR for data protection or ISO/IEC 27001 for information security management. Compliance reduces legal and financial risks, ensuring that the organization operates within the framework of established norms.

Formal Equation for Compliance Score:

$$C_s = \frac{\sum_{i-1}^{n} r_i \cdot c_i}{n}$$

where C_s is the compliance score, r_i represents the requirement i, and c_i is the level of compliance with that requirement.

Vulnerability assessment involves systematically identifying, classifying, and mitigating security weaknesses in the S-ERP system. This process includes scanning for potential vulnerabilities, evaluating their impact, and prioritizing remediation efforts. Regular assessments help in identifying new vulnerabilities as they emerge, ensuring the system remains secure.

The equation for Risk from Vulnerabilities:

$$R_v = \sum_{j=1}^{m} V_j \cdot I_j$$

where R_v is the risk of vulnerabilities, V_j is the severity of vulnerability j, and I_j is its impact on the system.

By ensuring regulatory compliance and conducting regular vulnerability assessments, organizations can proactively manage risks, protect sensitive data, and maintain the integrity of their S-ERP systems in the dynamic environment of Industry 4.0. Next, the following subsection discusses the best cybersecurity practices in S-ERP for Industry 4.0.

5.3.5 Best Cybersecurity Practices in S-ERP for Industry 4.0

In S-ERP for Industry 4.0, robust cybersecurity practices are essential to safeguard sensitive data and ensure the continuity of operations. The best practices include a multi-layered security approach, beginning with network security through firewalls, intrusion detection systems, and encryption protocols to protect data in transit. Regular software updates and patch management are crucial to protecting against known vulnerabilities. Additionally, conducting regular vulnerability assessments and penetration testing helps identify and mitigate potential threats before they can be exploited. Data encryption, both at rest and in transit, is vital to protect sensitive information from unauthorized access. Backup and disaster recovery plans ensure that data can be restored in the event of a cyberattack, minimizing downtime and data loss.

The first best practice is recognizing assets and continuously assessing risk for S-ERP in Industry 4.0. Recognizing assets and continuously assessing risk is the cornerstone of cybersecurity practices in S-ERP for Industry 4.0. The process begins by identifying all critical assets within the S-ERP system, including hardware, software, data, and network components. Each asset is then classified based on its importance and the potential impact of its compromise on the organization's operations.

Formal Equation for Asset Valuation:

$$A_v = \sum_{i=1}^{n} V_i \cdot I_i$$

where A_v represents the asset value, V_i is the value of asset i, and I_i is its impact on the organization.

Continuous risk assessment involves evaluating the likelihood and potential impact of threats to these assets. The risk associated with each asset can be calculated using the formula:

Risk Assessment Equation:

$$R = P \times I$$

where R represents the risk, P is the probability of a threat occurring, and I is the impact of the threat.

By regularly updating these assessments, organizations can prioritize security measures to address the most significant risks. This proactive approach ensures that the S-ERP system remains resilient against evolving threats, enabling businesses to maintain operational continuity and protect sensitive information within the Industry 4.0 framework. Continuous monitoring and assessment help dynamically adjust security protocols and keep the system secure in an ever-changing technological landscape.

The second-best practice is to implement zero trust architecture. Implementing Zero Trust Architecture (ZTA) is a critical cybersecurity practice for S-ERP in Industry 4.0. Unlike traditional security models that assume trust within the network perimeter, Zero Trust operates on the principle of "never trust, always verify." This means that no entity, whether inside or outside the network, is trusted by default. Every access request to S-ERP systems is authenticated, authorized, and encrypted regardless of its origin. The core components of ZTA include strict identity verification, micro-segmentation, and continuous monitoring.

Formal Equation for Access Verification:

$$A_v = I + M + C$$

where A_v is the access verification score, I represents identity verification, M is the monitoring factor, and C is the context-based authentication.

Micro-segmentation divides the network into smaller zones, allowing for granular control over access to each segment. Continuous monitoring ensures that any anomalies or suspicious activities are detected and addressed in real time.

The equation for Threat Detection:

$$T_d = \sum_{i=1}^{n} (A_i \cdot P_i)$$

where T_d represents the threat detection score, A_i is the anomaly score of activity i, and P_i is the probability of it being a threat.

Implementing Zero Trust Architecture in S-ERP for Industry 4.0 significantly reduces the risk of insider threats and lateral movement by continuously verifying access and isolating critical assets. This approach enhances the overall security posture of the organization, ensuring that sensitive data and operations remain protected in the dynamic environment of Industry 4.0.

The third practice that organizations should consider as the best practice for cybersecurity is to automate user provisioning. Automating user provisioning is a crucial cybersecurity practice in S-ERP for Industry 4.0. This practice involves automatically managing user access rights, roles, and permissions across the S-ERP system. Automation ensures that new employees are granted appropriate access quickly, while former employees or those changing roles have their permissions adjusted or revoked immediately, reducing the risk of unauthorized access. It also minimizes human error, ensuring consistency and compliance with security policies. Automated provisioning integrates with identity and access management (IAM) systems, enabling the seamless application of security policies across various modules of S-ERP. It also supports the principle of least privilege, ensuring users only have access to the data and systems necessary for their roles. This approach not only enhances security but also streamlines operations, making the management of user credentials more efficient and less prone to errors. Additionally, automated provisioning provides detailed logs and audit trails, enabling organizations to track who accessed what, when, and why. These logs are invaluable for compliance audits and forensic investigations, ensuring that any security breaches or policy violations can be quickly identified and addressed.

The fourth best cybersecurity practice for organizations implementing S-ERP in Industry 4.0 is code security. Code security is an essential cybersecurity practice in S-ERP for Industry 4.0, where custom code and scripts are often developed to extend system functionality. Ensuring that code is secure from the outset is crucial to protecting the integrity of the S-ERP system. This involves implementing secure coding practices, such as input validation, error handling, and secure data storage, to prevent vulnerabilities like SQL injection, cross-site scripting (XSS), and buffer overflows. Developers should adhere to coding standards and guidelines that prioritize security and use static and dynamic code analysis tools to identify and remediate vulnerabilities during the development process. Additionally, conducting regular code reviews and penetration testing helps uncover potential security flaws before they can be exploited. Automating code security checks as part of the continuous integration/continuous deployment (CI/CD) pipeline ensures that security is integrated into every stage of the software development lifecycle. This proactive approach reduces the risk of introducing vulnerabilities into the S-ERP system and ensures that updates and patches do not inadvertently compromise security. By prioritizing code security, organizations can protect their S-ERP systems from internal and external threats, ensuring that the underlying codebase remains robust, secure, and resilient against cyberattacks.

The final best practice related to cybersecurity for organizations implementing S-ERP in Industry 4.0 is to audit third-party security. Auditing third-party security is a critical cybersecurity practice in S-ERP for Industry 4.0, where systems often rely on external vendors and partners for various services, from cloud hosting to software modules. These third-party entities can introduce significant risks if their security practices are not up to par. Regular security audits of third-party providers ensure that they comply with the organization's security standards and industry regulations. The audit process involves evaluating the third party's security controls, such as data encryption, access controls, incident response procedures, and compliance with standards like ISO/IEC 27001. It also includes reviewing their vulnerability management practices and history of data breaches or security incidents. Formal contracts and Service Level Agreements (SLAs) should clearly define the security expectations, responsibilities, and consequences for non-compliance. Additionally, organizations should require third parties to provide regular security reports and allow for independent security assessments. Conducting thorough and regular audits ensures that third-party vendors do not become weak links in the S-ERP ecosystem. This practice helps organizations maintain a strong security posture, even when relying on external partners, ensuring that the integrity, confidentiality, and availability of their S-ERP systems are not compromised. The following subsection discusses shared system resources and related cybersecurity aspects of S-ERP in Industry 4.0.

5.3.6 Shared System Resources and Related Cybersecurity Aspects of S-ERP in Industry 4.0

Shared system resources in S-ERP for Industry 4.0 can introduce security issues primarily due to the interdependent and interconnected nature of these resources. In systems where multiple users or processes share a resource, the time taken by a shared resource (such as a server or network component) to process a task can vary depending on the workload [9]. An attacker could exploit this variability to infer sensitive information, such as the number of active users or the type of operations being performed. For instance, if a resource takes longer to process a particular job, it might reveal that a significant or sensitive operation is being carried out, compromising security.

The risk of information leakage due to timing analysis can be quantified using Shannon's entropy. The equation for this is:

$$E_t = \lim_{k \to \infty} \frac{H\left(x_k \mid t_k, t_k'\right)}{H\left(x_k \mid t_k\right)}$$

where:

- E_t is the quantified risk due to timing analysis.
- x_k is the job or task or token arrival pattern at time t_k.
- t_k' is the token departure time.
- $H\left(x_k \mid t_k\right)$ is the intrinsic uncertainty of the token arrival pattern.
- $H\left(x_k \mid t_k, t_k'\right)$ is the entropy considering both arrival and departure times.

In systems where multiple users or processes share a resource, the time taken by a shared resource (such as a server or network component) to process a task can vary depending on the workload. An attacker could exploit this variability to infer sensitive information, such as the number of active users or the type of operations being performed. For instance, if a resource takes longer to process a particular job, it might reveal that a significant or sensitive operation is being carried out, compromising security. This is why researchers are actively working on the air-gapped network [37].

The risk of information leakage due to processing sensitive and non-sensitive information together can be expressed as:

$$E_i = H\left(M_o\right) + H\left(M_i\right) - H\left(M_{(o,i)}\right)$$

where:

- E_i is the quantified risk due to information processing.
- $H\left(M_i\right)$ is the marginal entropy of the information associated with the task (token) that goes into the shared system resource.
- $H\left(M_o\right)$ is the marginal entropy of the information associated with the task (token) that comes out of the shared system resource.
- $H\left(M_{(o,i)}\right)$ is the joint marginal entropy of the input and output information.

Shared resources often consume varying amounts of power based on the tasks they perform. By monitoring the power consumption patterns of these resources, an attacker might deduce information about the tasks being performed or the data being processed. For example, if a shared resource is carrying out a cryptographic operation, variations in power consumption could reveal cryptographic keys or other sensitive data.

The risk of information leakage due to power analysis can be captured using the following equation:

$$E_p = H\left(P_o\right) - H\left(P_{S|S'}\right)$$

where:

- E_p is the quantified risk due to power analysis.
- P_o is the power-related probability function of the information that comes out with the token after being processed.
- $H\left(P_o\right)$ is the entropy of the power-related information.
- $H\left(P_{S|S'}\right)$ is the conditional entropy considering the actual secret content S and the guessed content S' by the attacker.

In the context of S-ERP for Industry 4.0, these security issues could manifest in the following ways:

- **Shared Industrial Equipment:** Equipment like manufacturing robots, conveyor belts, or digital control systems might be shared across different processes or even between different plants. An attacker with access to power consumption data, for example, might infer production schedules or the type of products being manufactured.

- **Shared IT Infrastructure:** The use of cloud-based services or centralized databases in an S-ERP system means that critical business data might be processed alongside less sensitive information. Without proper isolation, this could lead to leakage of sensitive data.
- **Network Resources:** In interconnected Industry 4.0 environments, shared network resources could be susceptible to timing attacks, where delays in data transmission could provide clues about network congestion, the presence of critical operations, or the type of data being transmitted.

To mitigate these risks, organizations implementing S-ERP systems need to adopt robust security measures, such as isolating sensitive processes, implementing strict access controls, and monitoring power consumption and timing anomalies. The upper bound of the combined security risk associated with a shared system resource can be expressed as:

$$E_{\text{total}} = E_t + E_i + E_p$$

where:

- E_{total} is the combined quantified security risk of the shared system resource.
- E_t, E_i, and E_p are the risks associated with timing, information processing, and power analysis, respectively.

For a shared system resource-based architecture with multiple resources n, the overall security risk can be summed as:

$$E_{\text{architecture}} = \sum_{i=1}^{n} E_{\text{total},i}$$

where:

- $E_{\text{architecture}}$ is the overall quantified security risk of the entire S-ERP architecture.
- $E_{\text{total},\,i}$ is the combined security risk for each individual shared system resource i.

These formal equations provide a structured approach to quantify the cybersecurity risks associated with shared system resources in S-ERP for Industry 4.0, accounting for potential vulnerabilities in timing, information processing, and power consumption. The following subsection discusses the logic of cryptography in S-ERP for Industry 4.0.

5.3.7 Logic of Cryptography in S-ERP for Industry 4.0

Cryptography is essential in S-ERP for Industry 4.0, as it ensures the confidentiality, integrity, and authenticity of data within the increasingly complex and interconnected systems that characterize this industrial era [38]. With the integration of advanced technologies like IoT, AI, and big data analytics, sensitive information is continuously exchanged across various platforms, devices, and networks. Cryptography protects this information from unauthorized access, tampering, and cyberattacks, which are increasingly prevalent in Industry 4.0 environments.

One of the key roles of cryptography is maintaining data confidentiality. Encryption algorithms, such as AES (Advanced Encryption Standard), convert readable data into an unreadable format, ensuring that only authorized entities can access sensitive information. Additionally, cryptographic methods help preserve data integrity, which is crucial in Industry 4.0, where any alteration in data can lead to incorrect decisions and operational disruptions. Cryptographic hash functions like SHA-256 provide a unique fingerprint of the data, ensuring it has not been altered during transmission or storage.

Authentication and non-repudiation are also critical aspects supported by cryptography in S-ERP. Digital signatures, based on asymmetric cryptography, allow for the verification of the authenticity of data and the identity of entities involved in transactions. This capability is vital in S-ERP systems, where secure and verifiable transactions are necessary to maintain operational trust. Moreover, cryptography aids organizations in complying with various data protection regulations, such as GDPR, by providing tools to protect personal and sensitive data, thereby helping to avoid legal penalties and maintain customer trust.

Finally, as Industry 4.0 introduces new vulnerabilities due to the interconnected nature of systems, cryptography serves as a robust defense mechanism. It makes it challenging for attackers to breach the system or access sensitive information, thereby enhancing the overall security posture of S-ERP systems. In essence, cryptography is indispensable for safeguarding data in S-ERP systems within Industry 4.0, enabling secure operations, ensuring regulatory compliance, and providing protection against the evolving landscape of cyber threats.

Elliptic Curve Cryptography (ECC) is increasingly recognized as a pivotal component in securing S-ERP systems within the framework of Industry 4.0. ECC is a form of public-key cryptography based on the algebraic structure of elliptic curves over finite fields, offering significant advantages in terms of security and efficiency. As S-ERP systems in Industry 4.0 involve the processing and transmission of vast amounts of data across diverse and interconnected devices, the need for robust and efficient cryptographic methods becomes paramount. ECC meets this need by providing a higher level of security with smaller key sizes compared to traditional cryptographic methods like RSA, making it particularly well-suited for the resource-constrained environments of Industry 4.0.

One of the key advantages of ECC in S-ERP for Industry 4.0 is its efficiency. ECC can achieve the same level of security as RSA but with much smaller key sizes. For example, a 256-bit key in ECC provides comparable security to a 3,072-bit key in RSA. This smaller key size translates to faster computations, reduced storage requirements, and lower power consumption, which are critical factors in the resource-constrained environments typical of Industry 4.0, where IoT devices and embedded systems are prevalent. The efficiency of ECC allows S-ERP systems to operate more smoothly, with faster encryption and decryption processes, enhancing overall system performance without compromising security.

Moreover, the enhanced security provided by ECC is vital for protecting the vast and sensitive data flows within S-ERP systems. In Industry 4.0, where data integrity and confidentiality are critical, ECC's strong security properties help safeguard against various cyber threats, including those that exploit the vulnerabilities of traditional cryptographic methods. The difficulty of solving the Elliptic Curve Discrete Logarithm Problem (ECDLP), on which ECC's security is based, ensures that even with smaller key sizes, the encryption remains robust against sophisticated attacks.

Another important aspect of ECC in S-ERP for Industry 4.0 is its adaptability to the increasing complexity of industrial networks. As Industry 4.0 continues to evolve, with more devices and systems becoming interconnected, the scalability of ECC becomes a significant asset. ECC can be seamlessly integrated into existing cryptographic infrastructures, providing a flexible solution that can scale with the growing needs of S-ERP systems. This adaptability is crucial for maintaining security across a broad range of devices and communication channels within Industry 4.0, ensuring that all components of the system remain secure as the network expands.

In addition, the adoption of ECC aligns with the growing emphasis on sustainability in Industry 4.0. By reducing the computational load and energy consumption associated with cryptographic processes, ECC contributes to the overall energy efficiency of S-ERP systems. This is particularly important in the context of Industry 4.0, where the environmental impact of industrial operations is a key consideration. By enabling secure and efficient operations, ECC supports the broader goals of sustainability in Industry 4.0, making it an essential technology for modern S-ERP systems.

Thus, ECC is a critical enabler of secure, efficient, and scalable S-ERP systems in Industry 4.0. Its ability to provide robust security with minimal computational overhead makes it an ideal choice for the complex and interconnected environments of modern industrial operations, where both security and sustainability are of paramount importance.

Quantum cryptography is emerging as a transformative technology for securing S-ERP systems in Industry 4.0, offering unprecedented levels of security through the principles of quantum mechanics. As the industrial landscape evolves with the integration of advanced technologies like IoT, AI, and big data, the need for robust and future-proof cryptographic methods becomes critical. Quantum cryptography addresses this need by providing security mechanisms that are fundamentally resistant to the computational power of classical and even quantum computers, making it a vital component for safeguarding S-ERP systems against emerging threats in Industry 4.0.

One of the most significant contributions of quantum cryptography to S-ERP is its ability to provide unbreakable encryption through Quantum Key Distribution (QKD). QKD allows two parties to securely share encryption keys by transmitting quantum bits (qubits) over a communication channel. The unique properties of qubits, such as superposition and entanglement, ensure that any attempt to eavesdrop on the key exchange will be detected, as the act of measurement alters the state of the qubits. This feature guarantees that any interception of the communication will be immediately known, allowing the parties to discard the compromised key and generate a new one. As a result, QKD provides a level of security that is theoretically unbreakable, which is crucial for protecting the sensitive data and communications within S-ERP systems in Industry 4.0.

Another critical advantage of quantum cryptography is its resilience against future threats posed by quantum computers. While traditional cryptographic methods, such as RSA and ECC, rely on mathematical problems that are difficult for classical computers to solve, these problems could potentially be solved quickly by quantum computers using algorithms like Shor's algorithm. This would render traditional encryption methods vulnerable to attack. Quantum cryptography, however, is immune to such threats, as its security is based on the principles of quantum mechanics rather than computational complexity. This makes it a future-proof solution for securing S-ERP systems, ensuring that the data remains protected even as quantum computing technology advances.

Furthermore, quantum cryptography supports the scalability and adaptability of S-ERP systems in Industry 4.0. As industrial networks grow more complex and interconnected, the ability to maintain secure communication across all devices and platforms becomes increasingly challenging. Quantum cryptography can be integrated into existing infrastructures, providing a scalable solution that can evolve alongside the expanding needs of S-ERP systems. This adaptability is essential for maintaining security in the dynamic and rapidly changing environment of Industry 4.0.

In addition to its security benefits, quantum cryptography aligns with the broader goals of Industry 4.0, including sustainability and efficiency. By leveraging the properties of quantum mechanics, quantum cryptography can enhance the energy efficiency of cryptographic processes, contributing to the overall sustainability of S-ERP systems. As Industry 4.0 emphasizes the need for environmentally responsible and resource-efficient operations, the adoption of quantum cryptography supports these objectives while ensuring that security is not compromised.

Thus, quantum cryptography is a crucial technology for securing S-ERP systems in Industry 4.0, offering unparalleled security through quantum mechanics, resilience against quantum computing threats, and the ability to scale with the evolving industrial landscape. Its adoption will be essential for ensuring the long-term security and sustainability of S-ERP systems in the face of emerging challenges and technological advancements. The following section relates security and ethics for S-ERP in Industry 4.0.

5.4 Relating Cybersecurity and Ethics for S-ERP in Industry 4.0

In the context of S-ERP for Industry 4.0, cybersecurity and ethics are deeply interconnected. Industry 4.0, characterized by interconnected systems, data-driven decision-making, and advanced technologies like Al, requires a robust ethical framework to ensure that cybersecurity measures are not just technically sound but also ethically responsible. Cybersecurity in S-ERP involves protecting sensitive data, ensuring system integrity, and preventing unauthorized access. These aspects are crucial as S-ERP systems often handle vast amounts of data, including personal, financial, and operational information. Any breach can lead to significant financial loss, reputational damage, and legal consequences.

On the other hand, ethics in S-ERP requires organizations to consider the broader impact of their cybersecurity practices on stakeholders. This includes ensuring privacy, preventing misuse of data, and maintaining transparency in how data is collected, stored, and protected. Ethical cybersecurity practices in S-ERP should also address issues of fairness, equity, and accountability, ensuring that all stakeholders are treated with respect and that the system's security measures do not disproportionately disadvantage any group. In Industry 4.0, where S-ERP systems are integral to business operations, the ethical implications of cybersecurity extend to the societal level. For instance, decisions about data access and control can have far-reaching consequences, such as reinforcing power imbalances or contributing to surveillance practices. Therefore, a strong ethical foundation is necessary to guide cybersecurity decisions, ensuring that they align with values like fairness, respect for privacy, and social responsibility.

Thus, in S-ERP for Industry 4.0, cybersecurity and ethics are not separate concerns but must be integrated. Ethical considerations should guide the development and implementation of cybersecurity practices to ensure that they protect not only the technical integrity of systems but also the rights and well-being of individuals and society.

Relating cybersecurity and ethics in S-ERP for Industry 4.0 from a cybernetics perspective involves understanding how systems govern themselves, process information, and maintain control. Cybernetics, the study of communication and control in systems, provides a useful framework for examining the interplay between cybersecurity and ethics in S-ERP. In cybernetics, a system's control is maintained through feedback loops that adjust the system's behavior based on deviations from desired outcomes. For S-ERP in Industry 4.0, cybersecurity functions as a control mechanism, ensuring the integrity, confidentiality, and availability of information within the system. The ethical dimension of this control involves ensuring that these cybersecurity measures do not infringe on stakeholders' rights, such as privacy or fairness. The ethical perspective also demands that the feedback mechanisms in place are transparent, accountable, and inclusive.

Consider an S-ERP system as a dynamic system $S(t)$, where t represents time. The state of the system at any time t is governed by a set of control inputs $u(t)$, disturbances $d(t)$, and feedback $f(t)$. The system's behavior can be represented by the following differential equation:

$$\dot{S}(t) = AS(t) + Bu(t) + Cd(t) + Df(t)$$

where:

- $\dot{S}(t)$ represents the rate of change of the system state.
- A, B, C, D are matrices that represent system parameters.
- $u(t)$ represents control inputs (e.g., cybersecurity measures).
- $d(t)$ represents disturbances (e.g., cyberattacks).
- $f(t)$ represents feedback mechanisms (e.g., intrusion detection systems).

From an ethical perspective, the control inputs $u(t)$ must be designed to respect ethical norms, such as privacy and fairness. This can be represented by introducing an ethical constraint E that modifies the control input:

$$u(t) = u_{\text{cyber}}(t) \cdot E(t)$$

where:

- $u_{\text{cyber}}(t)$ represents the cybersecurity control measures.
- $E(t)$ is an ethical function that ensures the control inputs align with ethical standards.

The ethical function $E(t)$ can be defined as:

$$E(t) = \frac{1}{1 + \alpha \cdot \text{ Ethical Violation } (t)}$$

where:

- α is a weighting factor.
- Ethical Violation (t) is a function that $a' \cdot$ notifies the ethical violations at time t.

The system's objective is to minimize the impact of disturbances $d(t)$ while ensuring that the ethical constraints are respected. This can be represented by an optimization problem:

$$\min_{u(t)} \int_0^T \left(\|d(t)\|^2 + \beta \cdot \text{ Ethical Violation } (t) \right) dt$$

where:

- T is the time horizon.
- and β is a weighting factor for the ethical considerations.

By using a cybernetics approach, we can see that cybersecurity in S-ERP for Industry 4.0 is not just about technical control but also about ethical governance. The system's behavior must be regulated not only by cybersecurity measures but also by ethical considerations, ensuring that the feedback loops respect stakeholders' rights and societal values. The formal equations provided illustrate how these principles can be mathematically modeled and integrated into the system's control strategy. In summary, the chapter answers the following research questions.

Research Questions Addressed

The chapter titled "Ethical and Security Aspects of Sustainable Enterprise Resource Planning (S-ERP) in Industry 4.0" addresses the following key research questions:

1. **What ethical considerations are involved in deploying S-ERP systems within Industry 4.0?**

- The chapter explores various ethical frameworks, such as utilitarianism, virtue ethics, deontological ethics, and others, as they apply to integrating S-ERP with advanced technologies like loT, AI, and big data within Industry 4.0.

2. **How do security concerns in S-ERP systems differ from those in traditional ERP systems within Industry 4.0?**

- The chapter discusses the enhanced security challenges of Industry 4.0 technologies, such as IoT and AI, and how these require more sophisticated cybersecurity measures than traditional ERP systems.

3. **What are the potential cybersecurity risks associated with implementing S-ERP in Industry 4.0, and how can these risks be managed?**

- The chapter identifies cybersecurity threats like data breaches, network vulnerabilities, and hardware/software security challenges. It suggests strategies for managing these risks, including using blockchain technology to secure loT devices and data.

4. **How can ethical principles be integrated into designing and implementing S-ERP systems in Industry 4.0?**

- The chapter proposes steps for making S-ERP algorithms ethical, such as integrating ethical principles, human-in-the-loop systems, transparency, cultural and social sensitivity, and continuous ethical training.

5. **How do cybersecurity violations impact S-ERP systems in Industry 4.0?**

- The chapter examines the broader implications of cybersecurity breaches, including their economic, social, and environmental impacts, and stresses the importance of robust security measures to protect S-ERP systems.

6. **How can organizations manage the risks associated with using S-ERP systems in Industry 4.0?**

- The chapter outlines a comprehensive approach to risk management, including data governance, access control, shared responsibility, security training, regulatory compliance, and vulnerability assessment.

7. **What are the best practices for ensuring cybersecurity in S-ERP systems within Industry 4.0?**

- The chapter provides best practices such as recognizing assets and continuously assessing risk, implementing zero trust architecture, automating user provisioning, ensuring code security, and auditing third-party security.

These research questions guide the chapter's exploration of both ethical and security aspects of S-ERP systems in the context of Industry 4.0.

References

1. Al-Amin Md, Tanjim Hossain Md, Jahidul Islam Md, Biwas SK (2023) History, features, challenges, and critical success factors of enterprise resource planning (ERP) in the era of industry 4.0. https://papers.ssrn.com/abstract=4376127
2. Almorsy M, Grundy J, Müller I (2016) An analysis of the cloud computing security problem, September 2016. http://arxiv.org/abs/1609.01107. arXiv:1609.01107 [cs]
3. Anjaria K (2021) A framework for ethical artificial intelligence - from social theories to cybernetics-based implementation. Int J Soc Humanist Comput 4(1):1–28. ISSN 1752-6124. https://doi.org/10.1504/IJSHC.2021.116870. https://www.inderscienceonline.com/doi/abs/10.1504/IJSHC.2021.116870. Publisher: Inderscience Publishers
4. Anjaria K (2024) Enhancing sustainability integration in Sustainable Enterprise Resource Planning (S-ERP) system: application of transaction cost theory and case study analysis. Int J Inf Manage Data Insights 4(2):100243. ISSN 2667-0968. https://doi.org/10.1016/j.jjimei.2024.100243. https://www.sciencedirect.com/science/article/pii/S2667096824000326
5. Anjaria K, Mishra A (2017) Information leakage analysis of software: how to make it useful to IT industries? Future Comput Inf J 2(1), 10–18. ISSN 2314–7288. https://doi.org/10.1016/j.fcij.2017.04.002. https://www.sciencedirect.com/science/article/pii/S2314728816300654
6. Anjaria K, Mishra A (2017b) Neuroevolution for effective information security training. In: 2017 international conference on soft computing and its engineering applications (icSoftComp), pp 1–6. https://doi.org/10.1109/ICSOFTCOMP.2017.8280078. https://ieeexplore.ieee.org/abstract/document/8280078
7. Anjaria K, Mishra A (2017c) Quantitative analysis of information leakage in service-oriented architecture-based Web services. Kybernetes 46(3):479–500. ISSN 0368-492X. https://doi.org/10.1108/K-07-2016-0178. https://doi.org/10.1108/K-07-2016-0178. Publisher: Emerald Publishing Limited

8. Anjaria K, Mishra A (2017d) Relating Wiener's cybernetics aspects and a situation awareness model implementation for information security risk management. Kybernetes 47(1):58–79. ISSN 0368-492X. https://doi.org/10.1108/K-06-2017-0226. https://doi.org/10.1108/K-06-2017-0226. Publisher: Emerald Publishing Limited
9. Anjaria K, Mishra A (2018) Information processing and security analysis of shared system resource based architectures. Int J Cooperat Inf Syst 27(04):1850009. ISSN 0218-8430. https://doi.org/10.1142/S0218843018500090. https://www.worldscientific.com/doi/abs/10.1142/S0218843018500090. Publisher: World Scientific Publishing Co
10. Astakhova LV, Medvedev IA (2021) An information tool for increasing the resistance of employees of an organization to social engineering attacks. Sci Tech Inf Process 48(1):15–20. ISSN 1934-8118. https://doi.org/10.3103/S0147688221010020. https://doi.org/10.3103/S0147688221010020
11. Chetan B (2023) Liberation and purity: race. new religious movements and the ethics of post-modernity. Routledge, London
12. Bushey KN (2023) One size does not fit all: how the california privacy rights act will not improve employee data collection and privacy rights. Catholic Univ J Law Technol 32:171
13. Bytniewski A, Matouk K, Rot A, Hernes M, Kozina A (2020) Towards industry 4.0: functional and technological basis for ERP 4.0 systems. In: Hernes M, Rot A, Jelonek D (eds) Towards industry 4.0 – current challenges in information systems. Springer International Publishing, Cham, pp 3–19
14. Carmi L, Zohar M, Riva GM (2023) The European General Data Protection Regulation (GDPR) in mHealth: theoretical and practical aspects for practitioners–use. Med, Sci Law 63(1):61–68. ISSN 0025-8024. https://doi.org/10.1177/00258024221118411. https://doi.org/10.1177/00258024221118411. Publisher: SAGE Publications
15. Chang BR, Tsai H-F, Chen C-Y, Tsai Y-C (2015) Assessment of in-cloud enterprise resource planning system performed in a virtual cluster. Math Prob Engin 2015(1):213461. ISSN 1563-5147. https://doi.org/10.1155/2015/213461. https://onlinelibrary.wiley.com/doi/abs/10.1155/2015/213461. _eprint: https://onlinelibrary.wiley.com/doi/pdf/10.1155/2015/213461
16. Chen P-H, Bodak R, Gandhi NS (2021) Ransomware recovery and imaging operations: lessons learned and planning considerations. J Digit Imaging 34(3):731–740. ISSN 1618-727X. https://doi.org/10.1007/s10278-021-00466-x. https://doi.org/10.1007/s10278-021-00466-x
17. Cohn Theodore H, Anil H (2020) Global political economy: theory and practice, 8th edn. Routledge, New York
18. Dong S, Li C, Yang S, An B, Li W, Gao Y (2024) Egoism, utilitarianism and egalitarianism in multi-agent reinforcement learning. Neural Netw 178:106544. ISSN 0893-6080. https://doi.org/10.1016/j.neunet.2024.106544. https://www.sciencedirect.com/science/article/pii/S0893608024004684
19. Furlan TJ (2024) Principlism, uncodifiability, and the problem of specification. In: Cambridge quarterly of healthcare ethics, pp 1–22. ISSN 0963-1801, 1469-2147. https://doi.org/10.1017/S0963180123000622. https://www.cambridge.org/core/journals/cambridge-quarterly-of-healthcare-ethics/article/abs/principlism-uncodifiability-and-the-problem-of-specification/89BCCFF317A1447BB6B8DF234BE42326
20. Ghai B, Mueller K (2023) D-BIAS: a causality-based human-in-the-loop system for tackling algorithmic bias. IEEE Trans Visualizat Comput Graph 29(1):473–482. ISSN 1941-0506. https://doi.org/10.1109/TVCG.2022.3209484. https://ieeexplore.ieee.org/abstract/document/9903601. Conference Name: IEEE Transactions on Visualization and Computer Graphics
21. Hampson T, McKinley J (2023) Problems posing as solutions: criticising pragmatism as a paradigm for mixed research. Res Educat 116(1):124–138. ISSN 0034-5237. https://doi.org/10.1177/00345237231160085. https://doi.org/10.1177/00345237231160085. Publisher: SAGE Publications Ltd STM
22. Lagerkvist A, Tudor M, Smolicki J, Ess CM, Lundström JE, Rogg M (2024) Body stakes: an existential ethics of care in living with biometrics and AI. AI & Society 39(1):169–181. ISSN 1435-5655. https://doi.org/10.1007/s00146-022-01550-8. https://doi.org/10.1007/s00146-022-01550-8

23. Megías A, de Sousa L, Jiménez-Sánchez F (2023) Deontological and consequentialist ethics and attitudes towards corruption: a survey data analysis. Soc Indicat Res 170(2):507–541. ISSN 1573-0921. https://doi.org/10.1007/s11205-023-03199-2. https://doi.org/10.1007/s11205-023-03199-2
24. Mittelstadt BD, Allo P, Taddeo M, Wachter S, Floridi L (2016) The ethics of algorithms: mapping the debate. Big Data Soc 3(2):2053951716679679. ISSN 2053-9517. https://doi.org/10.1177/2053951716679679. https://doi.org/10.1177/2053951716679679. Publisher: SAGE Publications Ltd
25. Munger M, Vanberg G (2023) Contractarianism, constitutionalism, and the status quo. Public Choice 195(3):323–339, June 2023. ISSN 1573-7101. https://doi.org/10.1007/s11127-021-00878-x. https://doi.org/10.1007/s11127-021-00878-x
26. Maayan P, Niva E-K (2015) Accountability in algorithmic copyright enforcement. Stanford Technol Law Rev 19:473
27. Pevkur A (2022) Ethical challenges in digital transformation for entrepreneurship. In Digital transformation for entrepreneurship volume 5 of Digital transformation: accelerating organizational intelligence. World Scientific, pp 17–30. ISBN 9789811270161
28. Queloz M, van Ackeren M (2024) Virtue ethics and the morality system. Topoi 43(2):413–424. ISSN 1572-8749. https://doi.org/10.1007/s11245-023-09964-9. https://doi.org/10.1007/s11245-023-09964-9
29. Rathee G, Balasaraswathi M, Prabhu Chandran K, Gupta SD, Boopathi CS (2021) A secure IoT sensors communication in industry 4.0 using blockchain technology. J Ambient Intell Human Comput 12(1):533–545. ISSN 1868-5145. https://doi.org/10.1007/s12652-020-02017-8. https://doi.org/10.1007/s12652-020-02017-8
30. Rendtorff JD (2019) References. In Philosophy of management and sustainability: rethinking business ethics and social responsibility in sustainable development. Emerald Publishing Limited, pp 221–236. ISBN 978-1-78973-453-9 978-1-78973-454-6. https://doi.org/10.1108/978-1-78973-453-920191022. https://doi.org/10.1108/978-1-78973-453-920191022
31. Rendtorff JD, Rendtorff JD (2019) Philosophy of management in the hypermodern experience economy. In Philosophy of management and sustainability: rethinking business ethics and social responsibility in sustainable development. Emerald Publishing Limited, pp 187–201. ISBN 978-1-78973-453-9 978-1-78973-454-6. https://doi.org/10.1108/978-1-78973-453-920191015. https://doi.org/10.1108/978-1-78973-453-920191015
32. Sahut JM, Lissillour R (2023) The adoption of remote work platforms after the Covid-19 lockdown: new approach, new evidence. J Busin Res 154:113345. ISSN 0148-2963. https://doi.org/10.1016/j.jbusres.2022.113345. https://www.sciencedirect.com/science/article/pii/S0148296322008104
33. Schuck AM, Rabe-Hemp C, Harris C (2023) A feminist ethics of care approach to de-escalation in policing. In: Staller Mario S, Swen K, Benni Z (eds) Police conflict management, vol I. challenges and opportunities in the 21st century. Springer International Publishing, Cham, pp 169–193
34. Panah AS, Yavari A, van Schyndel R, Georgakopoulos D, Yi X (2019) Context-driven granular disclosure control for internet of things applications. IEEE Trans Big Data 5(3):408–422. ISSN 2332-7790. https://doi.org/10.1109/TBDATA.2017.2737463. https://ieeexplore.ieee.org/abstract/document/8007332. Conference Name: IEEE Transactions on Big Data
35. Soltysik-Piorunkiewicz A, Krysiak M (2020) The cyber threats analysis for web applications security in industry 4.0. In: Hernes M, Rot A, Jelonek D (eds) Towards industry 4.0 – current challenges in information systems. Springer International Publishing, Cham, pp 127–141
36. Steffen Jason R (2021) A Kantian & communicative approach to criminal adjudication. Syracuse Law Rev 71:1333
37. Khan SM, Ali AAN, Laouini G, Alshammari FS (2023) Quantitative analysis of worm transmission and insider risks in air-gapped networking using a novel machine learning approach. IEEE Access 11:111034–111052. ISSN 2169-3536. https://doi.org/10.1109/ACCESS.2023.3322924. https://ieeexplore.ieee.org/abstract/document/10274075. Conference Name: IEEE Access

38. Pillai SE, Polimetla K (2024) Analyzing the impact of quantum cryptography on network security. In: 2024 international conference on integrated circuits and communication systems (ICICACS), pp 1–6, February 2024. https://doi.org/10.1109/ICICACS60521.2024.10498417. https://ieeexplore.ieee.org/abstract/document/10498417
39. Veerabadran V, Goldman J, Shankar S, Cheung B, Papernot N, Kurakin A, Goodfellow I, Shlens J, Sohl-Dickstein J, Mozer MC, Elsayed GF (2023) Subtle adversarial image manipulations influence both human and machine perception. Nat Commun 14(1):4933. ISSN 2041-1723. https://doi.org/10.1038/s41467-023-40499-0. https://www.nature.com/articles/s41467-023-40499-0. Publisher: Nature Publishing Group
40. Warkentin M, Willison R (2009) Behavioral and policy issues in information systems security: the insider threat. Eur J Inf Syst 18(2):101–105. ISSN 0960-085X. https://doi.org/10.1057/ejis.2009.12. Publisher: Taylor & Francis _ eprint: https://doi.org/10.1057/ejis.2009.12
41. Wong DB (2023) Moral relativism and pluralism. In: Elements in ethics. https://doi.org/10.1017/9781009043496. https://www.cambridge.org/core/elements/moral-relativism-and-pluralism/3DA6C896B00ACDFE5A64E5228E797BC3. ISBN: 9781009043496 9781009044301 Publisher: Cambridge University Press

Chapter 6
Circular Economy in S-ERP and Its Integration with Business Functions in Industry 4.0

6.1 Introduction

The circular economy concept represents a transformative approach to economic development, emphasizing sustainability, resource efficiency, and environmental responsibility. Unlike the traditional linear economy, which follows a "take-make-dispose" model, the circular economy seeks to minimize waste and extend the life-cycle of products through practices such as reuse, recycling, and regeneration [15]. In a circular economy, resources are kept in use for as long as possible, extracting the maximum value from them before recovering and regenerating materials at the end of their service life. This model not only reduces the environmental impact of production and consumption but also fosters innovation, creates new economic opportunities, and promotes resilience against resource scarcity. By decoupling economic growth from the consumption of finite resources, the circular economy offers a sustainable alternative that aligns with global efforts to address climate change, reduce pollution, and preserve natural ecosystems for future generations.

The adoption of a circular economy is crucial in the modern-day business world due to its potential to address the growing challenges of resource scarcity, environmental degradation, and economic sustainability. As businesses face increasing pressure to reduce their environmental footprint and comply with stringent regulatory requirements, the circular economy provides a strategic framework that aligns economic activities with sustainable development goals. By embracing circular principles such as resource efficiency, product lifecycle extension, and waste minimization, businesses can reduce their dependence on finite resources, mitigate risks associated with supply chain disruptions, and enhance their competitiveness in a rapidly changing global market. Moreover, the circular economy is a hotbed for innovation, encouraging the development of new business models, such as product-as-a-service, that prioritize value creation over mere product ownership. This shift not only drives economic growth but also strengthens the resilience of businesses by promoting long-term sustainability and stakeholder trust. In an era where consumers and investors

K. Anjaria, *Sustainable Enterprise Resource Planning (S-ERP) for Industry 4.0*,
Contributions to Management Science, https://doi.org/10.1007/978-981-96-7734-4_6

increasingly prioritize environmental and social responsibility, the circular economy emerges as a vital paradigm for businesses striving to achieve economic success while contributing to the broader goal of sustainable development.

The circular economy has a profound impact on several core business functions, necessitating a paradigm shift in how organizations operate and create value. Firstly, product design and development are significantly affected, as companies must prioritize eco-design principles that enhance product longevity, facilitate repair, and enable easy disassembly for recycling or remanufacturing. This shift requires close collaboration between designers, engineers, and sustainability experts to ensure that products are developed with the end-of-life stage in mind. Supply chain management is also deeply influenced, with an emphasis on sourcing sustainable materials, optimizing resource use, and developing closed-loop supply chains that recover and reintroduce materials back into the production cycle. This necessitates the adoption of innovative logistics strategies and stronger partnerships with suppliers and recycling facilities.

In marketing and sales, the circular economy drives the need for new business models, such as product-as-a-service or leasing arrangements, where the ownership of products remains with the company, incentivizing durability and the return of products at the end of their use. This requires rethinking customer relationships and value propositions, with a focus on offering sustainable solutions that meet consumer demand for environmentally responsible products. Furthermore, operations and manufacturing processes must be reconfigured to minimize waste, reduce energy consumption, and integrate advanced technologies for resource efficiency. This includes implementing lean manufacturing techniques, adopting renewable energy sources, and utilizing Industry 4.0 technologies to monitor and optimize production in real time.

Finally, finance and accounting functions are impacted by the circular economy as companies must develop new metrics and reporting frameworks to assess the long-term value of circular initiatives, account for the costs and benefits of sustainable practices, and communicate these efforts to stakeholders. Overall, the circular economy demands a holistic approach that integrates sustainability into every aspect of business operations, driving innovation, reducing environmental impact, and enhancing long-term profitability.

The integration of circular economy principles within the Industry 4.0 environment is crucial as it leverages advanced technologies to create more sustainable and resilient industrial systems. Industry 4.0, characterized by the convergence of digital technologies such as the Internet of Things (IoT), artificial intelligence, and big data analytics, offers unprecedented opportunities to enhance resource efficiency and reduce waste through real-time monitoring, predictive maintenance, and optimized production processes. By embedding circular economy strategies into the fabric of Industry 4.0, businesses can achieve a more efficient use of resources, enabling the recovery, recycling, and repurposing of materials at a scale previously unattainable. This synergy not only reduces the environmental impact of industrial activities but also drives innovation in product design, manufacturing, and supply chain management, resulting in products and systems that are inherently more sustainable.

Furthermore, the circular economy in Industry 4.0 supports the transition toward a regenerative industrial model, where waste is designed out of the system, and economic growth is decoupled from resource consumption. As industries seek to navigate the complexities of global markets and the pressures of environmental regulation, the integration of circular economy principles within Industry 4.0 is essential for achieving sustainable growth, enhancing competitiveness, and contributing to the broader goals of climate action and resource conservation.

However, literature suggests that it is not easy to implement circular economy aspects in business. Several barriers hinder the adoption of circular economy (CE) business models, including a lack of information on product lifecycles, limited technological expertise in product design and integration, and significant institutional and regulatory challenges. Additionally, businesses often face socio-cultural obstacles, such as risk aversion among managers and consumer resistance to departing from traditional products [8]. However, the advent of Industry 4.0 technologies presents opportunities to overcome these barriers. Innovations such as robotic sorting, artificial intelligence, digitalization, and additive manufacturing are increasingly enabling businesses to implement CE principles effectively, particularly in areas like electronic waste recycling and solid waste management. These implementations not only contribute to environmental sustainability but also offer significant cost savings and potential revenue streams, thereby enhancing the long-term profitability of businesses.

With this motivation, the present chapter plans to integrate circular economy in S-ERP for Industry 4.0. Before developing the model, it is critical to understand where we are in implementing a circular economy with Industry 4.0 or S-ERP. The following section discusses where we are.

6.2 Implementing a Circular Economy with Industry 4.0 and S-ERP: Where We Are at Present

The present literature has attempted this complex task of integrating Industry 4.0 and circular economy using various methods. Some studies were just impact studies described in [3], some were limited to specific functions or products, as explained in [19], and some were specific to industries, as indicated in [14]. Our present work aims to discuss the generalized models independent of industry, functions, and product types.

Readers might question the significance of this issue, considering it a straightforward task to simply collect the data provided by Industry 4.0 technologies, as discussed in previous chapters. However, developing a generalized data-driven model for Industry 4.0 is far from simple. Challenges such as data unavailability, inconsistencies, security concerns, equipment malfunctions, and intermittent violations create significant obstacles to the effective utilization and integration of Industry 4.0 data [5]. Still, many researchers have attempted to develop the methodology for

Table 6.1 Literary models to integrate the circular economy and Industry 4.0

Method name	Explanation	References
Roadmapping Circular Economy and Industry 4.0	We have already discussed Roadmapping extensively in the previous chapter. It creates a holistic plan to integrate the circular economy and Industry 4.0	[13]
Business and product lifecycle assessment	In this approach, the entire value chain is assessed and analyzed to implement a circular economy with Industry 4.0	[6]
Aligning KPIs with Circular Economy	The authors develop a measurement a platform for the circular economy utilizing cloud computing and IoT technologies to enable real-time monitoring of economic indicators, early warning thresholds, and critical values	[10]
Digitization of business function to implement Circular Economy	The study proposes an integrative framework aimed at deepening the understanding of the relationship between the circular economy (CE) and digital technologies by digitizing data- driven operations	[11]
Integrative framework with business model	Drawing on the theoretical foundations of the circular economy and Industry 4.0, along with resource-based and dynamic capabilities perspectives, the authors propose an integrative framework for the circular economy	[4]

Industry 4.0 and circular economy integration. We have tabulated these attempts in Table 6.1.

The models presented in Table 6.1 demonstrate the integration of the circular economy with Industry 4.0, but they do not incorporate the aspect of S-ERP. Recognizing the potential of S-ERP, we introduce models in this study that leverage AI, data analytics, visualization, integration, and sensors to implement the circular economy. By doing so, we aim to present frameworks that, while existing within the realm of digitization, have yet to be explored in the context of the circular economy.

In developing our model, we adhere to the framework proposed by Porter and Heppelmann [16], which asserts that models designed for Industry 4.0 should encompass four key aspects: monitoring, optimization, control, and autonomy. Based on this aspect, we first formally attempt to understand the circular economy.

6.2.1 Formal Modeling of Circular Economy

A formal circular economy model can be conceptualized by defining the key variables and relationships that govern the flow of resources, energy, and waste within an economic system. Below is a simplified mathematical representation:

Variables and Parameters

- Resource input at time t: $R(t)$
- Production output at time t: $P(t)$
- Waste generated at time t: $W(t)$
- Energy consumption at time t: $E(t)$
- Recycled resource at time t: $R_c(t)$
- Resource usage efficiency at time t: $R_u(t)$
- Cost of production at time t: $C(t)$
- Value generated at time t: $V(t)$
- Efficiency of the recycling process at time t: $\eta(t)$

1. Resource Input:

$$R(t) = R_u(t) \cdot P(t) - \eta(t) \cdot R_c(t)$$

This equation represents the total resource input required at time t, considering the efficiency of resource usage and the amount of recycled resources.

2. Waste Generation:

$$W(t) = P(t) - R_c(t)$$

Waste generation is defined as the difference between production output and recycled resources.

3. Energy Consumption:

$$E(t) = \alpha \cdot P(t) + \beta \cdot R_c(t)$$

Here, α and β represent the energy intensities of production and recycling processes, respectively.

4. Cost of Production:

$$C(t) = c_r \cdot R(t) + c_e \cdot E(t) + c_w \cdot W(t)$$

The cost of production is determined by the cost of resources (c_r), energy consumption (c_e), and waste management (c_w).

5. Value Generated:

$$V(t) = v_p \cdot P(t) - C(t)$$

The value generated by the system is the difference between the value of production output (v_p) and the cost of production.

6. Circular Economy Efficiency:

$$\xi(t) = \frac{R_c(t)}{R(t)}$$

This equation represents the circular economy efficiency, defined as the ratio of recycled resources to total resource input.

Interpretation

- Maximizing Circular Economy Efficiency ($\xi(t)$) requires increasing $R_c(t)$ (the recycled resource) and reducing $R(t)$ (the total resource input).
- Optimizing Waste Generation involves reducing $W(t)$ by enhancing $\eta(t)$ (the efficiency of recycling processes).
- Balancing Costs and Value is critical, where the goal is to maximize $V(t)$ by reducing $C(t)$ through efficient resource and energy use and minimizing waste.

This formal model captures the essence of the circular economy by focusing on resource efficiency, recycling, waste minimization, and the balance between costs and value generation. It provides a mathematical framework that can be adapted and expanded based on specific industry contexts and additional factors such as environmental impact and regulatory constraints. In the following subsection, we move step-wise and see the formal model of circular economy when it is implemented with Industry 4.0.

6.2.2 *Formal Modeling of Circular Economy in Industry 4.0*

In the context of Industry 4.0, the circular economy model would be enhanced by incorporating advanced technologies and data-driven processes, leading to more dynamic and real-time optimization of resource flows, energy use, and waste management. Here are the key changes and additions to the formal model:

Additional Variables and Parameters

- $D(t)$: Data-driven insights at time t
- $\text{IoT}\ (t)$: Internet of Things (IoT) data from sensors at time t
- $AI(t)$: Artificial Intelligence optimization factor at time t
- $\mu(t)$: Machine learning model accuracy at time t
- $S(t)$: Smart manufacturing output at time t

1. Resource Input with IoT and AI Optimization:

$$R(t) = R_u(t) \cdot S(t) - \eta(t) \cdot R_c(t) - AI(t) \cdot D(t)$$

Here, $AI(t)$ and $D(t)$ represent the impact of real-time data and Al-driven optimization, reducing the resource input by enhancing efficiency through predictive analytics and smart decision-making.

2. Smart Manufacturing Output:

$$S(t) = P(t) + \mu(t) \cdot \text{IoT}(t)$$

The smart manufacturing output $S(t)$ is influenced by traditional production $P(t)$ and adjustments based on IoT data, fine-tuned by the accuracy of machine learning models $\mu(t)$.

3. Energy Consumption with Smart Grid Integration:

$$E(t) = \alpha \cdot S(t) + \beta \cdot R_c(t) - \gamma \cdot AI(t)$$

Energy consumption now considers smart grid integration, where γ reflects the reduction in energy use due to AI-driven optimizations in production and recycling processes

4. Waste Generation with Predictive Maintenance:

$$W(t) = S(t) - R_c(t) - \delta \cdot AI(t)$$

Predictive maintenance, enabled by AI and IoT, helps reduce waste generation ($\delta. AI(t)$) by preventing inefficiencies and equipment failures.

5. Cost of Production with Digital Twin Simulation:

$$C(t) = c_r \cdot R(t) + c_e \cdot E(t) + c_w \cdot W(t) - \lambda \cdot D(t)$$

Digital twin technology, represented by $\lambda \cdot D(t)$, contributes to cost reduction by simulating production scenarios and optimizing resource allocation before implementation.

6. Value Generated with Industry 4.0 Enhancements:

$$V(t) = v_p \cdot S(t) - C(t) + \epsilon \cdot AI(t)$$

The value generated now includes an additional term $\epsilon \cdot AI(t)$, reflecting the increased value from AI-driven process optimizations and efficiencies.

7. Circular Economy Efficiency with Real-Time Data:

$$\xi(t) = \frac{R_c(t) + \lambda \cdot D(t)}{R(t)}$$

Circular economy efficiency is now enhanced by incorporating real-time data and simulations, which improve recycling rates and resource input efficiency.

Interpretation in Industry 4.0 Context

- **Enhanced Efficiency and Waste Reduction:** The inclusion of AI, IoT, and machine learning allows for real-time monitoring and predictive maintenance, leading to significant reductions in resource input, energy use, and waste generation.
- **Smart Manufacturing:** Industry 4.0 technologies facilitate adaptive production processes through IoT and digital twins, optimizing output and reducing inefficiencies.
- **Data-driven Decision Making:** Real-time data from IoT devices and insights from Al models provide dynamic adjustments to resource flows, minimizing costs and maximizing value.
- **Advanced Resource and Waste Management:** The model now integrates smart grid technologies and predictive analytics, contributing to more sustainable energy use and effective recycling.

In the Industry 4.0 environment, the circular economy model becomes more dynamic, responsive, and optimized, leveraging the latest technologies to achieve greater sustainability, efficiency, and value creation across all product lifecycle stages. The following subsection discusses the formal modeling of circular economy in S-ERP pertaining to Industry 4.0.

6.2.3 Formal Modeling of Circular Economy in S-ERP Pertaining to Industry 4.0

When adapting the circular economy model to the context of S-ERP within Industry 4.0, additional considerations related to sustainability, integrated resource planning, and cross-functional collaboration within an enterprise need to be incorporated. The following are the key changes and enhancements to the previous model:

Additional Variables and Parameters

- $S\_ERP(t)$: Sustainable Enterprise Resource Planning integration factor at time t
- $S_c(t)$: Sustainable supply chain efficiency at time t
- $C_S(t)$: Carbon footprint or environmental cost at time t
- $G(t)$: Governance and regulatory compliance factor at time t
- $H(t)$: Human capital or workforce sustainability index at time t

1. Resource Input with S-ERP Optimization:

$$R(t) = R_u(t) \cdot S(t) - \eta(t) \cdot R_c(t) - AI(t) \cdot D(t) - S\_ERP(t) \cdot S_c(t)$$

In this equation, $S\_ERP(t)$ represents the optimization provided by S-ERP systems, which enhances resource efficiency by integrating sustainable supply chain practices ($S_c(t)$) and real-time data from Industry 4.0 technologies.

2. Sustainable Manufacturing Output:

$$S(t) = P(t) + \mu(t) \cdot \mathrm{IoT}(t) + S\_ERP(t) \cdot H(t)$$

Sustainable manufacturing output is now influenced not only by traditional production and IoT data but also by S-ERP systems, which incorporate human capital sustainability $(H(t))$ into the production process.

3. Energy Consumption with S-ERP and Environmental Costs:

$$E(t) = \alpha \cdot S(t) + \beta \cdot R_c(t) - \gamma \cdot AI(t) - S\_ERP(t) \cdot C_s(t)$$

The energy consumption equation now includes an adjustment for environmental costs $(C_s(t))$ managed through S-ERP, which integrates sustainability metrics into energy use and production processes.

4. Waste Generation with S-ERP Compliance:

$$W(t) = S(t) - R_c(t) - \delta \cdot AI(t) - G(t) \cdot S\_ERP(t)$$

Waste generation is further reduced by ensuring compliance with governance and regulatory standards $(G(t))$, as enforced through S-ERP systems, in addition to AI-driven predictive maintenance.

5. Cost of Production with S-ERP Integration:

$$C(t) = c_T \cdot R(t) + c_e \cdot E(t) + c_w \cdot W(t) - \lambda \cdot D(t) - \phi \cdot S\_ERP(t)$$

The cost of production now considers the efficiency and cost savings provided by S-ERP ($\phi.S\_ERP(t)$), which optimizes resource allocation, reduces environmental costs, and integrates real-time data for better decision-making.

6. Value Generated with S-ERP and Sustainability:

$$V(t) = v_p \cdot S(t) - C(t) + \epsilon \cdot AI(t) + \zeta \cdot S\_ERP(t)$$

The value generated is now enhanced by both AI optimizations and S-ERP contributions ($\zeta . S\_ERP(t)$), reflecting the added value from integrating sustainability into enterprise resource planning.

7. Circular Economy Efficiency with S-ERP and Governance:

$$\xi(t) = \frac{R_c(t) + \lambda \cdot D(t) + \zeta \cdot S\_ERP(t)}{R(t)}$$

Circular economy efficiency is further improved by incorporating the positive impacts of S-ERP systems and governance factors, which enhance recycling rates, reduce resource input, and ensure regulatory compliance.

Interpretation in S-ERP and Industry 4.0 Context

- **Enhanced Sustainability and Compliance:** S-ERP integration ensures that sustainability is embedded across all business processes, from supply chain management to production, in alignment with environmental regulations and corporate governance.
- **Comprehensive Resource Planning:** S-ERP facilitates holistic resource management by integrating real-time data, AI, and sustainability metrics, leading to more efficient use of resources and energy.
- **Sustainable Human Capital Management:** S-ERP systems consider the sustainability of human capital, ensuring that workforce factors are optimized alongside production processes, contributing to overall enterprise sustainability.
- **Integrated Environmental Costing:** S-ERP allows for the inclusion of environmental costs in energy consumption and production decisions, leading to more sustainable and cost-effective operations.
- **Governance and Regulatory Compliance:** S-ERP systems play a crucial role in ensuring that all processes comply with relevant regulations, further enhancing the circular economy model's effectiveness in the Industry 4.0 environment.

Incorporating S-ERP into the circular economy model within Industry 4.0 results in a more comprehensive and sustainable approach to enterprise management. It integrates advanced technologies with sustainability practices and governance, enabling businesses to achieve greater efficiency, compliance, and long-term value creation.

The above model states that implementing a circular economy is mathematically a multi-objective, multi-criteria-driven problem, where some elements need to be optimized, some need to be maximized, and some need to be balanced. Further, most of the equations were timebound. So, keeping all these complexities in mind, we propose models to integrate circular economy in S-ERP for Industry 4.0 in the following section.

6.3 Models to Integrate Circular Economy in S-ERP for Industry 4.0

Based on the above discussion, we propose six models to implement circular economy in S-ERP for Industry 4.0. The critical thing in these models is they would be related to time and hold significant importance from the system perspective as they are related to S-ERP. Figure 6.1 shows the models we will discuss in this section.

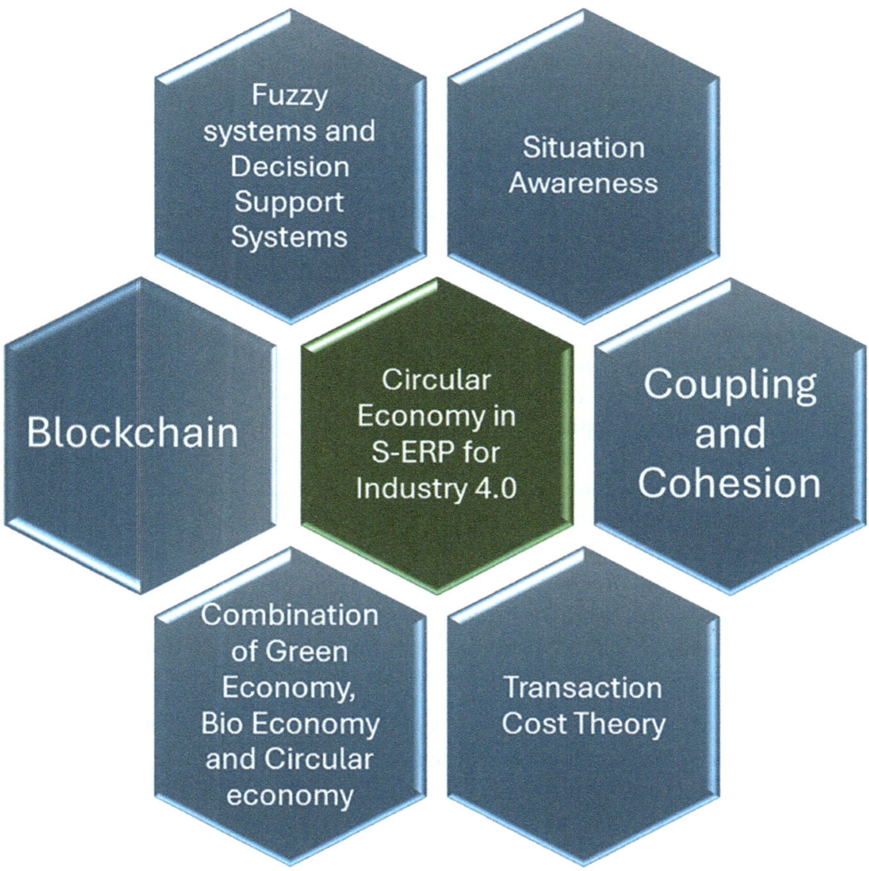

Fig. 6.1 Models to implement circular economy in S-ERP for Industry 4.0

The first model we will implement to integrate circular economy in S-ERP for Industry 4.0 is the situation awareness model in the following subsection.

6.3.1 Situation Awareness to Implement Circular Economy in S-ERP for Industry 4.0

Situation Awareness (SA) is a critical concept in the field of human factors, particularly in environments where people operate complex and dynamic systems, such as aviation, nuclear power, and advanced manufacturing. Situation awareness is formally defined as

> The perception of the elements in the environment within a volume of time and space, the comprehension of their meaning, and the projection of their status in the near future (Endsley, 2021) [7].

SA involves three levels:

- **Level 1 SA - Perception:** This is the recognition of key elements in the environment. For example, in aviation, this might include awareness of critical flight parameters, the state of onboard systems, and the location of other aircraft.
- **Level 2 SA - Comprehension:** This is the ability to understand the significance of those elements, particularly when they are integrated with one another. For instance, a pilot might interpret a pattern of flight parameters as indicating that the aircraft is near a stall point.
- **Level 3 SA - Projection:** The highest level of SA, where the operator is able to predict the future status of the system based on their understanding. In aviation, this might involve predicting the movement or behavior of other aircraft based on their current formation.

SA is crucial for effective decision-making and performance in dynamic systems, as it allows operators to anticipate and respond to changes in their environment before they occur. SA is often impaired in automated systems, leading to the "out-of-the-loop" performance problem where operators, who are more passive observers rather than active controllers, may have reduced awareness of the system's state and dynamics. This can result in slower detection of issues, misinterpretation of system data, and delayed or inappropriate responses to system failures.

Integrating SA into S-ERP for Industry 4.0 is a highly effective approach for implementing a circular economy due to the dynamic and complex nature of both environments. SA enhances decision-making by providing real-time awareness of critical factors like material flows, energy consumption, and waste generation, allowing for informed decisions that align with circular economy principles. It improves system comprehension and future projection, enabling stakeholders to understand the interconnections between circular processes and predict potential disruptions or bottlenecks. This predictive capability is crucial for optimizing resource loops and ensuring continuous improvement.

By leveraging Industry 4.0 technologies like IoT and AI, SA allows for dynamic adaptation and real-time monitoring of feedback loops, ensuring the system adapts to changes and improves its sustainability performance. It also supports broader Environmental, Social, and Governance (ESG) objectives by aligning enterprise operations with sustainability goals. Furthermore, maintaining SA minimizes the out-of-the-loop issues common in automated systems, reducing human error and enhancing system responsiveness. Overall, SA is essential for driving the successful adoption of circular economy principles in S-ERP for Industry 4.0, ensuring a sustainable and resilient approach to resource management.

To implement SA as a method to integrate the circular economy into S-ERP for Industry 4.0, a detailed approach is required, which can be structured as shown in Fig. 6.2.

Step 1 Developing a Comprehensive SA Framework:

This method requires organizations to develop three levels of SA discussed above

- Perception (SA Level 1): Start by designing systems that enable operators to perceive critical data related to the circular economy, such as resource usage, waste generation, and recycling rates. This involves integrating loT sensors and data collection tools that provide real-time information on these parameters.
- Comprehension (SA Level 2): Enhance the ability of operators to comprehend the implications of the data collected. This can be achieved by developing algorithms that analyze and interpret data in the context of circular economy goals, such as reducing waste or increasing resource efficiency. Visualization tools should be employed to display the relationships between different data points, helping operators understand how their actions impact the circular economy.
- Projection (SA Level 3): Equip the system with predictive analytics capabilities that allow operators to foresee the future state of the system based on current

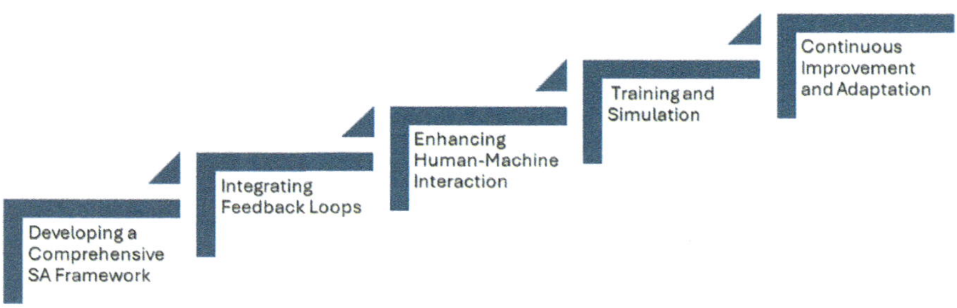

Fig. 6.2 Structure of a method to integrate the circular economy into S-ERP for Industry 4.0

data. For example, by projecting the impact of current resource consumption on future availability, operators can make decisions that align with circular economy principles.

Step 2 Integrating Feedback Loops:

In this step, organizations should implement real-time feedback loops that continuously monitor and report on the effectiveness of circular economy initiatives within the S-ERP system. This ensures that operators remain aware of how well the system performs in relation to circular economy goals and allows for immediate adjustments when necessary.

Step 3 Enhancing Human-Machine Interaction:

This step requires S-ERP to have adaptive automation and better UI/UX to implement a circular economy.

- Adaptive Automation: Implement adaptive automation strategies that keep human operators actively involved in decision-making processes. This helps maintain a high level of SA by ensuring that operators remain engaged and aware of the system's state, even as automation takes over routine tasks. Adaptive automation can be particularly useful in managing the complexities of a circular economy, where dynamic decision-making is essential. Dynamic decision-making is crucial in a circular economy because it involves continuously adjusting strategies based on real-time data about resource availability, waste generation, and market demands. In such a dynamic environment, operators need to make quick decisions to optimize resource use, respond to unexpected changes, and ensure that sustainability goals are met. Adaptive automation helps by allowing the system to handle routine tasks while keeping operators engaged in these critical, ever-changing decision-making processes.
- User Interface Design: Develop user interfaces that present data in a clear, intuitive manner, facilitating quick comprehension and decision-making. The interface should provide operators with easy access to information about resource flows, production cycles, and waste streams, allowing them to quickly assess the current state and take appropriate actions.

Step 4 Training and Simulation:

- Scenario-Based Training: Use scenario-based training programs to improve operators' SA in relation to the circular economy. By simulating different scenarios, such as disruptions in resource supply or changes in market demand, operators can

practice making decisions that support circular economy objectives under varying conditions.
- SA Assessment Tools: Implement tools like the Situational Awareness Global Assessment Technique (SAGAT) to evaluate and improve the SA of operators in real time. These tools can help identify gaps in SA and provide targeted training to address these gaps.

Step 5 Continuous Improvement and Adaptation:

- Data-Driven Adjustments: Regularly analyze system performance data to identify areas where the circular economy initiatives can be improved. This might involve adjusting production processes, changing resource procurement strategies, or enhancing recycling efforts.
- Feedback from Operators: Encourage operators to provide feedback on the system's performance and their experience with the SA tools. This feedback can be used to refine the SA framework and ensure it remains effective in supporting circular economy goals.

By systematically implementing SA within S-ERP for Industry 4.0, enterprises can better manage the complexities of integrating circular economy principles, leading to more sustainable and resilient operations. For the implementation of SA to integrate circular economy in S-ERP for Industry 4.0, we formalize layers of SA, adaptive automation, feedback, and implementation strategy as follows:

Creation of Perception Layer (SA Level 1)

The first step involves collecting real-time data from various sensors and systems within the S-ERP, such as IoT devices that monitor resource usage, production cycles, and waste generation. This can be represented as:

$$D_t = \{d_1(t), d_2(t), \ldots, d_n(t)\}$$

where D_t is the set of all data points collected at the time t, and $d_i(t)$ represents the data from the sensor i at time t. Later, the perception of the current state can be captured using a state vector S_t, which is a function of the data collected:

$$S_t = f(D_t)$$

where f is the function that maps the collected data to the current system state.

Creation of Comprehension Layer (SA Level 2)

Comprehension involves interpreting the state vector S_t to understand the current situation. This requires applying domain-specific knowledge, which can be modeled using a comprehension function C :

$$C_t = g(S_t)$$

where C_t is the comprehension of the current state at time t, and g is the function that applies domain-specific rules to interpret S_t. The comprehension function can be extended to assess risks or opportunities related to the circular economy, such as resource shortages or potential waste reduction:

$$R_t = h(C_t)$$

where R_t is the risk or opportunity assessment at time t, and h is the function that evaluates the comprehension in the context of circular economy goals.

Projection Layer (SA Level 3)

The projection layer involves predicting future states based on the current comprehension. This can be modeled using a prediction function P :

$$S_{t+\Delta t} = P(C_t)$$

where $S_{t+\Delta t}$ is the predicted state at a future time $t + \Delta t$, and P is the function that forecasts future states based on the current comprehension. The projection function aids in decision-making by identifying potential actions and their outcomes:

$$A_t = \arg\max_{a \in A} U(P(C_t), a)$$

where A_t is the optimal action at time t, A is the set of possible actions, and U is the utility function that evaluates the outcomes of actions based on the predicted future state.

Adaptive Automation

The system should dynamically adjust automation levels based on the operator's situation awareness. The level of automation (LoA) can be modeled as a function of SA levels:

$$\text{LoA}_t = \phi(S_t, C_t, R_t, A_t)$$

where ϕ is a function that determines the appropriate level of automation based on the current perception, comprehension, risk assessment, and selected action. To ensure that operators remain actively involved, the system should present key information and decision points in a user-friendly manner, enabling operators to override or adjust automation as needed.

Feedback Loops

Implement feedback loops that continuously monitor the effectiveness of decisions and automation levels. The effectiveness E_t at time t can be represented as:

$$E_t = \psi\,(S_t, C_t, R_t, A_t, \mathrm{LoA}_t)$$

where ψ is a function that evaluates the effectiveness of the overall system, ensuring that it aligns with circular economy goals. The system should learn from past actions and outcomes to improve future decision-making. This can be modeled using a learning function L that updates the decision-making model:

$$M_{t+1} = L\,(M_t, E_t)$$

where M_t is the decision-making model at time t, and L updates the model based on the effectiveness evaluation.

Implementation Strategy

Use simulation tools to model different scenarios and train operators on how to maintain situation awareness in dynamic environments. This includes practicing decision-making under various levels of automation and circular economy constraints. Implement tools like the Situational Awareness Global Assessment Technique (SAGAT) to measure and improve operators' SA during real-time operations, ensuring that they can effectively manage the complexities of a circular economy.

By formalizing these equations and methods, the S-ERP system can be effectively integrated with situation awareness to support the circular economy, enhancing both operational efficiency and sustainability. Next, we relate the above formal model with the formal model discussed in Sects. 6.2.1, 6.2.2 and 6.2.3. To effectively integrate Situation Awareness (SA) into the formalized circular economy model for S-ERP in Industry 4.0, we can align the SA levels with the variables and equations of the circular economy model to enhance decision-making and system efficiency. Here's how SA can be systematically incorporated:

1. Perception Layer (SA Level 1) Integration with Circular Economy

- Real-Time Data Collection: The perception layer of SA corresponds to gathering real-time data from various sources like loT devices, sensors, and other Industry 4.0 technologies. This aligns with variables such as $D(t)$, $IoT(t)$, and $S(t)$, which are crucial for understanding the current state of resource usage $R(t)$, production $P(t)$, and energy consumption $E(t)$.
- State Vector Formation: The perception function $S_t = f(D_t)$ can be integrated into the circular economy model by forming a state vector that includes resource inputs, production outputs, energy consumption, and waste generation. This state vector directly informs the values of $R(t)$, $P(t)$, $W(t)$, and $E(t)$, thereby setting the foundation for higher-level SA processes.

2. Comprehension Layer (SA Level 2) Integration

- System Understanding and Risk Assessment: The comprehension layer involves interpreting the data to assess the efficiency and sustainability of the circular economy processes. For instance, by evaluating $R(t)$ and $R_c(t)$, operators can comprehend how effectively resources are being recycled and what the current waste generation $W(t)$ is. The risk assessment $R_t = h(C_t)$ then helps in identifying potential inefficiencies or risks, such as low recycling efficiency $\eta(t)$ or high-energy consumption $E(t)$.
- Equations Integration: The comprehension function can be directly related to the key equations of the circular economy model:

$$C_t = \{R(t), P(t), W(t), E(t), R_c(t), \eta(t)\}$$

This set of comprehension factors allows operators to understand and manage the interdependencies among resource input, waste generation, and energy consumption in realtime, optimizing the circular economy's efficiency $\xi(t)$.

3. Projection Layer (SA Level 3) Integration

- Predictive Analytics and Future State Projection: In the projection layer, the model uses AI and machine learning (represented by $AI(t)$ and $\mu(t)$) to predict future states of the system, such as future resource needs $R(t + \Delta t)$, potential waste reductions $W(t + \Delta t)$, and energy consumption $E(t + \Delta t)$.
- Decision Support for Optimization: The projection function $P(C_t)$ assists in making informed decisions that optimize circular economy efficiency. For instance, decisions about adjusting resource input $R(t)$ can be informed by projecting the efficiency of recycling processes $\eta(t)$ and the impact of smart manufacturing outputs $S(t)$.

4. Adaptive Automation and Dynamic Decision-Making

- Incorporating Adaptive Automation: The system dynamically adjusts automation levels based on the situation awareness levels to ensure optimal resource use, waste management, and energy efficiency. The level of automation function $\mathrm{LoA}_t = \phi(S_t, C_t, R_t, A_t)$ is influenced by data-driven insights $D(t)$ and AI optimizations $AI(t)$, enhancing the model's ability to maintain high circular economy efficiency $\xi(t)$.
- Formalized Dynamic Adjustments: The circular economy model can be expanded to include:

$$R(t) = R_u(t) \cdot S(t) - \eta(t) \cdot R_c(t) - AI(t) \cdot D(t) - \mathrm{LoA}_t \cdot S_{ERP}(t)$$

This equation reflects how real-time situation awareness and automation dynamically reduce resource input and improve recycling efficiency.

5. Feedback Loops and Continuous Improvement

- Real-Time Monitoring and Feedback: Continuous monitoring of the system's performance is crucial for feedback loops. This can be modeled as:

$$E_t = \psi(S_t, C_t, R_t, A_t, \mathrm{LoA}_t)$$

where ψ evaluates the system's effectiveness, contributing to ongoing adjustments in S-ERP practices.

- Learning and Adaptation: Incorporating learning functions $L(M_t, E_t)$ into the S-ERP system ensures that the circular economy model evolves based on past data, improving resource efficiency, waste reduction, and overall system sustainability.

6. Integration with Formal Circular Economy Model

- Resource Input Optimization: Aligning the SA-driven adjustments with the equation $R(t) = R_u(t) \cdot S(t) - \eta(t) \cdot R_c(t) - AI(t) \cdot D(t)$ optimizes the resource input by integrating situation awareness into real-time decision-making, minimizing resource usage, and maximizing recycling.
- Energy and Waste Management: Energy consumption $E(t)$ and waste generation $W(t)$ can be optimized using predictive analytics and SA, ensuring that energy use is minimized through AI optimizations and that waste generation is kept low by enhancing recycling and predictive maintenance.

- Circular Economy Efficiency: The circular economy efficiency $\xi(t)$ is enhanced by integrating real-time data, AI, and situation awareness, ensuring that the S-ERP system remains responsive and adaptive, thereby maximizing sustainability and economic value.

By incorporating SA into the formal circular economy model for S-ERP in Industry 4.0, the system can dynamically optimize resource usage, energy consumption, and waste management, leading to a more sustainable and efficient circular economy.

One of the ways to implement a circular economy in S-ERP for Industry 4.0 is to use granular computing [18]. To implement a circular economy in S-ERP for Industry 4.0 using granular computing, the key is to break down complex systems and processes into smaller, more manageable granules of information, which can be processed at different levels of granularity to enhance decision-making, resource optimization, and waste reduction. Granular computing enables the representation and processing of information in a structured way that aligns with the multi-layered and dynamic nature of Industry 4.0 environments. Here's how granular computing facilitates the implementation of a circular economy:

1. Information Granulation: Granular computing divides data into granules, which are collections of entities grouped by similarity, proximity, or function. In S-ERP, this means organizing data on resources, energy, waste, and production into meaningful subsets that can be managed more effectively.
2. Hierarchical Processing: The granules are processed at different levels of granularity, allowing the system to handle complex information more efficiently. For instance, at a high level of granularity, the system might manage overall resource inputs and outputs, while at a finer granularity, it could optimize specific processes like recycling or energy use.
3. Dynamic Decision-Making: Granular computing supports adaptive decision-making by allowing the system to adjust the level of detail or granularity based on the current situation. This is crucial in a circular economy, where decisions must be made quickly to respond to changes in resource availability or waste generation.
4. Multi-Agent Systems: By using multi-agent systems within the granular computing framework, S-ERP can distribute tasks and processes across different agents, each focusing on a specific aspect of the circular economy. These agents can collaborate and share information, ensuring that the system as a whole operates efficiently and sustainably.
5. Improved Situation Awareness: Granular computing enhances situation awareness by organizing and processing data in a way that highlights critical information and patterns, enabling better comprehension and projection of future states. This supports the continuous improvement and optimization of circular economy processes within S-ERP.

Thus, granular computing provides a systematic approach to managing and optimizing the complex, interrelated processes involved in implementing a circular economy in S-ERP for Industry 4.0, making it a powerful tool for achieving sustainability and efficiency. After the complete framework of SA to integrate circular economy in S-ERP for Industry 4.0, the next subsection discusses coupling and cohesion to integrate circular economy in S-ERP for Industry 4.0.

6.3.2 Coupling and Cohesion to Integrate Circular Economy in S-ERP for Industry 4.0

Coupling refers to the degree of interdependence between software modules, indicating how closely connected they are. Lower coupling is generally preferred as it allows for greater modularity and easier maintenance. Cohesion measures the degree to which the elements within a single module belong together. Higher cohesion within a module is desirable as it signifies that the module is well-focused and functions effectively as a single unit.

In the context of S-ERP for Industry 4.0, coupling and cohesion are critical for ensuring system efficiency and scalability. Low coupling between S-ERP modules enhances flexibility, enabling easier updates and integration with new technologies without disrupting the entire system. High cohesion within each module ensures that functions are tightly aligned with specific sustainability goals, leading to more effective resource management and streamlined processes. Together, optimized coupling and cohesion contribute to the robustness and adaptability of S-ERP systems in the dynamic environment of Industry 4.0.

Coupling and cohesion are pivotal in integrating circular economy principles into S-ERP for Industry 4.0. Low coupling among S-ERP modules facilitates the seamless integration of circular economy practices by allowing independent modules to interact with minimal dependency. This flexibility is crucial for incorporating new circular processes, such as closed-loop supply chains or waste management systems, without extensive reconfiguration of the existing S-ERP infrastructure.

Conversely, high cohesion within each S-ERP module ensures that all components within a module are tightly aligned with the circular economy's objectives, such as resource efficiency, waste reduction, and product lifecycle management. This internal consistency enables each module to effectively support specific aspects of the circular economy, leading to more efficient resource utilization and sustainable operations. Thus, the strategic management of coupling and cohesion is essential for successfully embedding circular economy practices within S-ERP systems, aligning them with the broader sustainability goals of Industry 4.0. Next, we present the methodology for integrating circular economy in S-ERP for Industry 4.0 using coupling and cohesion.

Methodology for Integrating Circular Economy in S-ERP for Industry 4.0 Using Coupling and Cohesion

1. System Analysis and Module Design: The integration of circular economy principles into S-ERP for Industry 4.0 begins with a comprehensive analysis of the existing system architecture. This phase involves identifying key business

processes that are critical to circular economy practices, such as resource management, waste reduction, and product lifecycle optimization. Based on this analysis, modules within the S-ERP system should be designed or restructured to ensure high cohesion, where each module is focused on a specific aspect of the circular economy. For instance, one module might be dedicated to recycling processes, while another focuses on remanufacturing and product recovery.

2. Coupling Reduction and Interface Standardization: To achieve low coupling, it is essential to standardize the interfaces between S-ERP modules. This can be accomplished by adopting open standards and protocols that allow for modular interaction with minimal dependencies. By reducing coupling, each module can be modified or replaced without necessitating changes to the entire system. This flexibility is particularly important for integrating new circular economy practices as they evolve. Interface standardization also facilitates the integration of third-party applications that specialize in circular economy solutions, thereby enhancing the overall capability of the S-ERP system.
3. Cohesion Enhancement Through Focused Functionalities: Each S-ERP module should be designed with a specific focus on circular economy practices, ensuring high cohesion. This involves aligning all functionalities within a module to support a unified goal. For example, a module focused on sustainable procurement should encompass all related functions, such as supplier evaluation, green product selection, and inventory management for reusable materials. High cohesion ensures that each module operates effectively and efficiently, contributing directly to the circular economy's objectives.
4. Iterative Development and Testing: The integration process should be iterative, with ongoing testing and refinement to ensure that the principles of low coupling and high cohesion are maintained. During each iteration, modules should be tested for their ability to operate independently (low coupling) while delivering focused and consistent outputs (high cohesion). Feedback from these tests should be used to make necessary adjustments, ensuring that the S-ERP system remains adaptable to the dynamic requirements of the circular economy.
5. Continuous Monitoring and Adaptation: Once the S-ERP system is operational, continuous monitoring is required to ensure that coupling remains low and cohesion remains high. This involves regular audits of module interactions and performance assessments of individual modules in relation to their circular economy goals. As new circular economy practices and technologies emerge, the S-ERP system should be adaptable enough to incorporate these changes with minimal disruption. Continuous adaptation ensures that the S-ERP system remains aligned with the evolving sustainability goals of Industry 4.0.
6. Documentation and Knowledge Management: Comprehensive documentation should accompany the integration process, detailing the design choices, interface standards, and specific functionalities of each module. This documentation serves as a reference for future modifications and helps maintain the low coupling and high cohesion of the system. Additionally, a knowledge management

system should be established to capture lessons learned and best practices, ensuring that the integration of circular economy principles into S-ERP systems can be replicated and improved over time.

7. Stakeholder Engagement and Training: Effective integration of circular economy principles requires the involvement of all relevant stakeholders, including business leaders, IT professionals, and sustainability experts. Regular training sessions should be conducted to familiarize stakeholders with the newly integrated S-ERP modules, emphasizing the importance of maintaining low coupling and high cohesion. Stakeholder engagement ensures that the system is used effectively and that the principles of the circular economy are embedded in the organizational culture.
8. Evaluation and Continuous Improvement: Finally, the success of the integration process should be evaluated based on key performance indicators (KPIs) related to both system performance and sustainability outcomes. Metrics such as resource efficiency, waste reduction, and product lifecycle management should be tracked to assess the impact of the S-ERP system on circular economy goals. Continuous improvement processes should be implemented to refine the system based on these evaluations, ensuring that it remains responsive to the evolving demands of Industry 4.0.

This methodology provides a structured approach to integrating circular economy principles into S-ERP systems for Industry 4.0, leveraging the concepts of coupling and cohesion to achieve a sustainable, flexible, and efficient enterprise resource planning system. Next, we relate this methodology with the formal model discussed in Sects. 6.2.1, 6.2.2 and 6.2.3. The methodology for integrating circular economy in S-ERP for Industry 4.0 using coupling and cohesion significantly enhances the formal modeling of the circular economy in several ways:

1. Enhanced Modularity and Flexibility (Impact on $R(t)$, $S(t)$, $E(t)$):

The reduction of coupling among S-ERP modules allows for independent optimization of each module, such as resource input ($R(t)$) and energy consumption ($E(t)$). This modular approach enables the efficient integration of real-time data (D(t)) and AI -driven insights (Al(t)), leading to more accurate and dynamic resource planning (R(t)) and smart manufacturing outputs (S(t)). Consequently, the system becomes more responsive to changes in resource availability, production demands, and environmental conditions, directly influencing the effectiveness of the circular economy model.

2. Focused Functionalities and High Cohesion (Impact on $\eta(t)$, $W(t)$, $C(t)$):

High cohesion within S-ERP modules ensures that each module is tightly aligned with specific circular economy goals, such as improving recycling efficiency ($n(t)$) and minimizing waste generation (W(t)). By ensuring that all functions within a module are concentrated on a particular objective, the system enhances the accuracy and efficiency of processes like waste reduction and cost management (C(t)). This focused approach enables better resource utilization and supports the overall sustainability objectives of the circular economy.

3. Optimization of Circular Economy Efficiency (ξ(t)):

The integration of circular economy principles within S-ERP systems, guided by the principles of coupling and cohesion, directly impacts the circular economy efficiency metric ($\xi(t)$). By enhancing the system's ability to manage resources efficiently through low coupling and high cohesion, the recycling rate (R_c(t)) is increased, and total resource input ($R(t)$) is reduced. This optimization leads to a higher $\xi(t)$, indicating a more effective circular economy implementation within the S-ERP system.

4. Adaptive and Scalable Integration (Impact on S_ERP(t), $S\_c(t)$, $C\_s(t)$):

The methodology's emphasis on modularity (low coupling) allows S-ERP systems to be easily adapted and scaled to incorporate new sustainability practices, such as those driven by Industry 4.0 technologies. The incorporation of variables like S_ERP (t) and S_c (t) in the formal model reflects this adaptive capacity, ensuring that the system remains aligned with evolving environmental regulations and sustainability standards. This adaptability is crucial for maintaining an optimal balance between resource input, waste generation, and energy consumption in a dynamic industry environment.

5. Governance and Compliance Integration (Impact on G(t),H(t)):

High cohesion within S-ERP modules ensures that governance and regulatory compliance factors ($G(t)$) are consistently enforced across all processes. This integration not only reduces waste (W(t)) and energy consumption (E(t)) but also aligns the system with broader sustainability goals, including human capital sustainability ($H(t)$). By ensuring that all modules operate in concert toward these objectives, the system supports a more holistic and sustainable enterprise management approach.

6. Long-term Value Creation (Impact on $V(t)$):

The combined effects of low coupling and high cohesion within the S-ERP system enhance the overall value generated ($V(t)$) by the enterprise. By optimizing resource use, reducing costs, and maximizing efficiency, the system creates sustainable value that aligns with both economic and environmental objectives. This holistic approach ensures that the circular economy model not only reduces waste and improves efficiency but also contributes to long-term profitability and sustainability.

The above-described formal model helps us understand the importance of circular economy integration in S-ERP for Industry 4.0 with the help of coupling and cohesion. They have existed in software engineering research for the last few years, but they have not been used extensively for sustainability. Hence, we propose their use here. The following section uses Transaction Cost Theory to integrate circular economy in S-ERP for Industry 4.0.

6.3.3 *Transaction Cost Theory to Integrate Circular Economy in S-ERP for Industry 4.0*

Transaction Cost Theory (TCT) is an economic theory that explains and predicts the structures and behaviors of firms by analyzing the costs associated with economic exchanges [21]. These costs, known as transaction costs, include search and information costs, bargaining and decision costs, and monitoring and enforcement costs. According to TCT, firms will organize themselves in ways that minimize these costs, either by performing transactions internally (within the firm) or through the market (external transactions). The theory, developed by Ronald Coase and further expanded by Oliver Williamson, provides a framework for understanding why firms exist, how they are structured, and why they might choose to outsource or internalize certain activities.

TCT is significant in S-ERP for Industry 4.0 as it provides a framework for optimizing resource allocation and decision-making processes. Minimizing transaction costs such as those related to information exchange, coordination, and compliance-S-ERP systems can enhance operational efficiency and sustainability, as discussed in detail by Anjaria [2].

TCT helps firms determine whether to internalize processes or collaborate with external partners, ensuring that sustainability goals are met without incurring excessive costs. This alignment of economic efficiency with environmental and social objectives is crucial for the successful implementation of S-ERP in the context of Industry 4.0.

TCT is pivotal in integrating the circular economy into S-ERP within Industry 4.0 by guiding the efficient management of economic exchanges and resource flows. In a circular economy, where resources are reused, recycled, and regenerated, TCT helps firms minimize the costs of these complex transactions, such as sourcing secondary materials, coordinating with multiple stakeholders, and enforcing sustainable practices across the supply chain. Applying TCT allows S-ERP systems to optimize decision-making processes, ensuring that adopting circular economy principles does not lead to prohibitive transaction costs. This integration supports achieving sustainability objectives while maintaining economic efficiency, making TCT an essential component in the transition to a circular economy within the framework of Industry 4.0.

To effectively integrate the principles of a circular economy within S-ERP for Industry 4.0, a structured approach grounded in TCT is essential. TCT offers a lens through which the economic exchanges and organizational structures necessary for circular economy implementation can be optimized. With seven steps, the following methodology outlines the framework to integrate circular economy practices within S-ERP systems, leveraging TCT to minimize transaction costs and promote sustainability.

Step 1 Identifying Key Transaction Costs:

The first step is to identify the key transaction costs associated with the shift from a linear to a circular economy within the S-ERP context. These include information cost, coordination cost, and monitoring and enforcement cost. Information costs are the costs related to gathering, processing, and distributing information necessary for circular practices. Coordination costs are the expenses incurred in aligning activities and stakeholders across the supply chain. Finally, monitoring and enforcement costs are the costs involved in ensuring compliance with circular economy principles and sustainability standards across the product lifecycle.

Step 2 Designing the S-ERP Framework:

Develop an S-ERP framework that incorporates the principles of circular economy. This involves Modular System Design (MSD) [12] and Lifecycle Tracking (LT) [20]. MSD creates an adaptable S-ERP system that can integrate with various circular economy activities such as recycling, remanufacturing, and resource recovery. On the other hand, (LT) implements tools within the S-ERP to track products throughout their lifecycle, enabling the system to manage and optimize resource use, reduce waste, and facilitate reuse.

Step 3 Minimizing Information Asymmetry:

The S-ERP should enhance transparency and standardize the data to reduce information asymmetry. Enhanced transparency ensures real-time visibility of data across all stages of the product lifecycle, from raw material sourcing to post-consumer recycling. The standardize data adopt standardized reporting and data management practices within the S-ERP to ensure consistent and comparable information across the supply chain.

Step 4 Eco-Opportunism Mitigation:

To mitigate the risks of eco-opportunism, where organizations might exploit sustainability claims without genuine compliance, the S-ERP must include rigorous monitoring mechanisms (RMMs) and contractual safeguards (CSs). RMMs integrate automated monitoring systems within the S-ERP to detect and report any discrepancies or non-compliance with circular economy practices. CSs design contracts with suppliers and partners that include specific clauses for sustainability performance, backed by penalties for non-compliance.

Step 5 Reducing Coordination and Bargaining Costs:

Effective coordination across multiple stakeholders is critical in a circular economy. The S-ERP should facilitate this through Collaborative Platforms and Integrated Decision-Making Tools. Collaborative Platforms introduce platforms within the S-ERP for stakeholders to collaborate, share information, and coordinate activities seamlessly. On the other hand, Integrated Decision-Making Tools provide decision-support tools within the S-ERP that optimize resource allocation, scheduling, and supply chain coordination to reduce the need for extensive bargaining and negotiation.

Step 6 Monitoring and Enforcement Through S-ERP:

The S-ERP system must include robust monitoring and enforcement mechanisms to ensure adherence to circular economy principles, such as Automated Compliance Checks and Performance Matrix. Automated Compliance Checks implement automated compliance checks within the S-ERP that continuously monitor adherence to environmental regulations and circular economy practices. Performance Metrics define and track specific sustainability performance metrics within the S-ERP to ensure that circular economy goals are met efficiently.

Step 7 Continuous Improvement and Adaptation:

Finally, the integration of circular economy in S-ERP for Industry 4.0 should be an ongoing process with Feedback Loops and Scalability and Flexibility. Feedback Loops incorporate feedback loops within the S-ERP to capture data on the effectiveness of circular economy practices and use this data to continuously refine and improve the system. The Scalability and Flexibility design of the S-ERP is scalable and flexible, allowing it to adapt to evolving circular economy practices and Industry 4.0 technologies.

Integrating the circular economy within S-ERP for Industry 4.0 using Transaction Cost Theory ensures that sustainability efforts are economically viable and operationally efficient. By focusing on minimizing transaction costs and enhancing organizational efficiency, this methodology provides a pathway to achieving sustainable enterprise operations within the framework of Industry 4.0. The methodology outlined for integrating the circular economy into S-ERP for Industry 4.0 using Transaction Cost Theory (TCT) directly impacts the formal model of the circular economy by providing a structured approach to optimizing the variables and equations defined in the model. Here's how:

1. Resource Input Optimization:

The inclusion of TCT within S-ERP minimizes resource input costs by improving resource usage efficiency ($R\_u(t)$) and enhancing recycling efficiency ($\eta(t)$). This aligns with the equation for resource input ($R(t)$), where reduced transaction costs lead to more efficient use of resources, decreasing the total resource input needed.

2. Waste Generation Reduction:

By incorporating compliance and monitoring mechanisms, as suggested in the methodology, waste generation (W(t)) is minimized. The S-ERP system's ability to enforce regulations and optimize recycling processes reduces waste, directly impacting the waste generation equation by lowering W(t).

3. Energy Consumption Management:

The methodology's focus on integrating sustainability into energy use through S-ERP impacts the energy consumption equation ($E(t)$). By reducing transaction costs related to energy management and incorporating Al-driven optimizations, energy use becomes more efficient, decreasing $E(t)$ over time.

4. Cost of Production:

The cost of production ($C(t)$) is impacted by the methodology's emphasis on reducing coordination, monitoring, and enforcement costs. The integration of real-time data and predictive analytics within S-ERP reduces costs associated with resources, energy, and waste management, thereby optimizing $C(t)$.

5. Value Generation:

The methodology enhances value generation (V(t)) by increasing efficiency and reducing costs through the optimized S-ERP system. The focus on minimizing transaction costs and improving sustainability directly contributes to maximizing V(t), as the value generated is the difference between production value and costs.

6. Circular Economy Efficiency:

The formal model's circular economy efficiency ($\xi(t)$) is directly influenced by the methodology's aim to enhance recycling rates and reduce resource input. By integrating TCT with S-ERP, the ratio of recycled resources to total resource input is improved, thus increasing $\xi(\mathrm{t})$.

Thus, the methodology impacts the formal model by optimizing resource input, reducing waste and energy consumption, lowering production costs, and maximizing value generation. It enhances circular economy efficiency through systematic transaction cost management, making the entire economic system more sustainable and aligned with the principles of Industry 4.0. The following subsection discusses a combination of a green economy, bio-economy, and circular economy for S-ERP in Industry 4.0.

6.3.4 A Combination of a Green Economy, Bio-Economy, and Circular Economy for S-ERP in Industry 4.0

A combination of a green economy, bio-economy, and circular economy represents an integrated approach to sustainable development [9]. The green economy emphasizes reducing environmental risks and ecological scarcities while promoting sustainable development without degrading the environment. The bio-economy focuses on the sustainable use of biological resources, processes, and innovations to produce food, materials, and energy. Meanwhile, the circular economy aims to minimize waste and make the most of resources by closing the loop in product lifecycles through recycling, reuse, and regeneration. Together, these paradigms foster an economy that is environmentally restorative, resource-efficient, and rooted in biological and renewable resources, aiming for long-term sustainability and resilience.

The integration of a green economy, bio-economy, and circular economy into S-ERP for Industry 4.0 is crucial for achieving long-term sustainability and competitive advantage. This combination ensures that S-ERP systems not only optimize resource efficiency but also align with environmental and social goals. By incorporating green economy principles, S-ERP minimizes environmental impact, while the bio-economy promotes the use of renewable biological resources, fostering innovation in sustainable production processes. The circular economy further enhances S-ERP by enabling closed-loop systems that reduce waste and extend product lifecycles. Together, these approaches ensure that Industry 4.0 enterprises can achieve sustainable growth, enhance resource efficiency, and meet the evolving demands of a responsible, resilient economy.

The combination of a green economy, bio-economy, and circular economy plays a pivotal role in integrating circular economy principles into S-ERP for Industry 4.0. This integration facilitates the development of a holistic framework that prioritizes environmental sustainability, resource efficiency, and innovation. The green economy's emphasis on reducing environmental impacts ensures that S-ERP systems incorporate practices that minimize ecological footprints. The bio-economy contributes by leveraging renewable biological resources and promoting sustainable production methods, thereby enhancing the sustainability of the supply chain within Industry 4.0.

Most critically, the circular economy provides the operational blueprint for S-ERP, enabling the transition from linear to circular processes by embedding waste reduction, resource reuse, and product lifecycle extension into enterprise planning and operations. This integration ensures that Industry 4.0 systems are not only technologically advanced but also aligned with sustainable development goals, leading to a more resilient, resource-efficient, and environmentally conscious industrial ecosystem.

The method to implement this strategy in the business is a five-step process. These steps are (1) Framework development, (2) Process Reengineering, (3) Technological Integration, (4) Stakeholder Engagement and Training, and (5) Continuous Monitoring and Improvement. The detailed implementation of the above steps is as follows:

1. Framework Development

- **Green Economy Alignment:** Begin by embedding green economy principles into the S-ERP framework. This involves identifying key environmental metrics such as carbon footprint, energy consumption, and waste generation and integrating them into the resource planning modules of the S-ERP. Define sustainability goals aligned with these metrics to guide decision-making processes across the enterprise.

- **Bio-Economy Integration:** Incorporate bio-economy principles by mapping out the biological resources utilized in the production processes. This includes identifying renewable resources, optimizing their use, and minimizing reliance on non-renewable inputs. Integrate bio-based innovations, such as bioplastics and bioenergy, into the S-ERP supply chain management to ensure that all production activities align with sustainable biological practices.
- **Circular Economy Mapping:** Develop a circular economy roadmap within the S-ERP by identifying key areas where waste can be minimized and resources can be reused or recycled. This involves mapping out the entire product lifecycle within the S-ERP, from design and production to consumption and disposal. Integrate lifecycle assessment tools into the S-ERP to continuously evaluate the environmental impact and resource efficiency at each stage.

2. Process Reengineering

- **Design for Sustainability:** Reengineer product design processes to prioritize sustainability. Use the principles of eco-design and biomimicry, which are central to the green and bio-economy, to create products that are durable, repairable, and recyclable. Integrate these principles into the S-ERP product design modules to ensure that sustainability is considered from the outset.
- **Supply Chain Optimization:** Redesign the supply chain to support circular economy practices. This includes implementing closed-loop supply chains within the S-ERP, where materials are continuously cycled back into production. Use bio-economy principles to source materials sustainably and green economy principles to minimize transportation emissions. The S-ERP should be programmed to prioritize suppliers and partners that adhere to circular and sustainable practices.
- **Production Process Modification:** Modify production processes to reduce waste and energy consumption. Integrate bio-economy innovations such as bio-based materials and renewable energy sources into the S-ERP's production modules. Implement green manufacturing practices, such as lean production, to enhance resource efficiency. The S-ERP should track and optimize energy and material usage, aiming for zero waste production.

3. Technological Integration

- **IoT and Smart Sensors:** Utilize Industry 4.0 technologies such as the Internet of Things (IoT) and smart sensors to monitor and manage resources in real time. Integrate these technologies into the S-ERP to provide accurate data on resource consumption, waste generation, and environmental impact. This real-time data can be used to make dynamic adjustments in production processes, ensuring that circular economy principles are adhered to at all times.

- **Blockchain for Traceability:** Implement blockchain technology within the S-ERP to enhance transparency and traceability in the supply chain. This ensures that all materials and products are tracked throughout their lifecycle, enabling closed-loop processes and verifying adherence to green economy and bio-economy standards. Blockchain can also facilitate the verification of sustainable practices, ensuring that all stakeholders in the supply chain are contributing to the circular economy.
- **Advanced Analytics and AI:** Utilize advanced analytics and artificial intelligence (AI) within the S-ERP to optimize resource use and predict waste generation. AI algorithms can be trained to identify inefficiencies in the production process and suggest improvements that align with circular economy principles. Analytics can also be used to forecast demand and optimize inventory management, reducing overproduction and waste.

4. Stakeholder Engagement and Training

- **Employee Training:** Develop comprehensive training programs for employees to ensure that they understand the principles of the green economy, bio-economy, and circular economy and how these are integrated into the S-ERP. Training should cover the use of new technologies, the importance of sustainable practices, and the role of each employee in achieving the enterprise's sustainability goals.
- **Supplier and Partner Collaboration:** Engage with suppliers and partners to align their practices with the enterprise's sustainability goals. Use the S-ERP to facilitate collaboration and ensure that all stakeholders are adhering to circular economy principles. This may involve revising contracts, developing shared sustainability targets, and offering incentives for sustainable practices.
- **Customer Engagement:** Integrate customer feedback into the S-ERP to ensure that products meet sustainability expectations. Use the S-ERP to gather data on customer preferences and behavior, which can be used to design products that are more sustainable and align with circular economy principles.

5. Continuous Monitoring and Improvement

- **Performance Metrics:** Develop and integrate a set of sustainability performance metrics within the S-ERP to monitor the effectiveness of circular economy integration. Metrics should include resource efficiency, waste reduction, carbon emissions, and the percentage of recycled or reused materials. Regularly review these metrics to assess progress and identify areas for improvement.
- **Feedback Loops:** Implement feedback loops within the S-ERP to continuously gather data on resource use, waste generation, and environmental impact. Use this data to make ongoing adjustments to production processes, supply chain management, and product design. The S-ERP should facilitate continuous improvement by providing insights and recommendations based on real-time data.

- **Reporting and Compliance:** Ensure that the S-ERP generates regular sustainability reports that align with industry standards and regulatory requirements. These reports should provide a transparent account of the enterprise's sustainability performance, demonstrating how the integration of green economy, bio-economy, and circular economy principles is contributing to Industry 4.0 goals.

This methodology outlines a comprehensive approach to embedding circular economy principles within S-ERP for Industry 4.0, leveraging the synergies between green economy, bio-economy, and circular economy to achieve sustainable industrial practices. The detailed methodology for integrating a circular economy in S-ERP for Industry 4.0, using a combination of green economy, bio-economy, and circular economy, directly impacts the formal modeling of the circular economy (shown in Sects. 6.2.1, 6.2.2 and 6.2.3) by enhancing the efficiency and sustainability of the system. Here's how:

1. **Resource Input Optimization:** The methodology's focus on green economy and bio-economy principles reduces the overall resource input $R(t)$ by improving resource usage efficiency $R_u(t)$ and increasing the recycling rate $R_c(t)$. The integration of AI and real-time data analytics further reduces $R(t)$ by predicting and optimizing resource flows, aligning with the formal equation $R(t) = R_u(t) \cdot P(t) - \eta(t) \cdot R_c(t)$
2. **Waste Reduction:** By embedding circular economy practices, such as waste minimization and resource reuse, the methodology helps reduce waste generation $W(t)$. Predictive maintenance and smart manufacturing techniques lower $W(t)$ as represented in the equation $W(t) = P(t) - R_c(t)$. This leads to an overall decrease in waste and enhances recycling efficiency $\eta(t)$
3. **Energy Consumption Efficiency:** The methodology's integration of green technologies and bio-economy practices, combined with AI-driven smart grids, reduces energy consumption $E(t)$. The equation $E(t) = \alpha \cdot P(t) + \beta \cdot R_c(t) - \gamma \cdot AI(t)$ is directly influenced by these optimizations, leading to more sustainable energy use in production and recycling processes.
4. **Cost of Production:** The methodology's approach to optimizing production costs $C(t)$ through sustainable resource use, energy efficiency, and waste reduction is reflected in the equation $C(t) = c_r \cdot R(t) + c_e \cdot E(t) + c_w \cdot W(t) - \lambda \cdot D(t)$. The S-ERP system integrates these elements, lowering costs by minimizing resource inputs and energy consumption and by incorporating digital twin simulations and AI-driven insights.
5. **Value Generation:** The integration of circular economy practices enhances value generation $V(t)$ by increasing the efficiency of production and resource use. The equation $V(t) = v_p \cdot P(t) - C(t) + \epsilon \cdot AI(t)$ benefits from the added value created through AI optimizations and sustainable practices embedded in the S-ERP system.
6. **Circular Economy Efficiency:** The methodology improves circular economy efficiency $\xi(t)$ by increasing the recycled resource rate $R_c(t)$ and reducing total resource

input $R(t)$. The equation $\xi(t) = \frac{R_c(t)}{R(t)}$ is optimized through the integration of sustainable supply chain practices, AI, and real-time data, leading to higher recycling rates and reduced dependency on virgin resources.

Thus, the methodology enhances each component of the formal circular economy model by embedding sustainable practices, advanced technologies, and real-time data analytics into the S-ERP system. This results in a more efficient, sustainable, and resilient economic model within the context of Industry 4.0, aligning with the mathematical framework outlined in the formal model. The following section discusses blockchain to integrate circular economy in S-ERP for Industry 4.0.

6.3.5 Blockchain to Integrate Circular Economy in S-ERP for Industry 4.0

Blockchain is a decentralized and distributed digital ledger technology that securely records transactions across a network of computers [17]. It operates on a consensus mechanism, ensuring that all participants in the network agree on the validity of transactions without the need for a central authority. Each transaction is grouped into blocks cryptographically linked to the previous block, forming an immutable chain. This structure provides transparency, security, and traceability, making blockchain particularly valuable for applications requiring trust and verification, such as financial services, supply chain management, and digital identity.

Blockchain technology plays a critical role in S-ERP within Industry 4.0 by enhancing transparency, security, and traceability across the supply chain and business processes. By enabling decentralized data management, blockchain ensures that all transactions are recorded in an immutable ledger, reducing the risk of fraud and errors. This system supports real-time auditing and compliance, which is essential for maintaining sustainability standards. Additionally, blockchain's ability to facilitate smart contracts automates and enforces agreements, further streamlining operations and ensuring adherence to sustainable practices in Industry 4.0 environments.

Blockchain plays a pivotal role in integrating the circular economy within S-ERP for Industry 4.0 by ensuring transparency, traceability, and accountability across the entire lifecycle of products. It facilitates the recording of every transaction and material flow in a decentralized and immutable ledger, enabling precise tracking of resources from production to disposal or reuse. This capability is essential for implementing circular economy principles, where materials are continuously cycled back into the production process, minimizing waste and resource consumption. Blockchain also supports the verification of sustainable practices and compliance with environmental standards through smart contracts, which automatically enforce circular economy policies. This integration not only enhances resource efficiency but also fosters greater trust among stakeholders, promoting the adoption of sustainable practices in Industry 4.0.

The integration of circular economy (CE) principles into S-ERP for Industry 4.0 can be significantly enhanced through the use of blockchain technology. Blockchain offers a decentralized, transparent, and secure system that aligns with the core objectives of CE, such as resource optimization, waste reduction, and product lifecycle management. This methodology outlines a structured approach for leveraging blockchain to embed circular economy principles into S-ERP systems within the Industry 4.0 framework. It has four steps described as follows:

1. Assessment of S-ERP System Requirements

- **Identification of Circular Economy Goals:** The first step involves aligning the S-ERP objectives with circular economy principles, such as reducing resource consumption, enhancing recycling processes, and extending product lifecycles. This alignment ensures that the S-ERP system is designed to meet the sustainability goals of the organization.
- **Evaluation of Blockchain Capabilities:** Assess the specific capabilities of blockchain that can support these CE goals. Key blockchain features to be considered include transparency, traceability, decentralization, and smart contract functionality.

2. Design of Blockchain-Enabled S-ERP Architecture

- **Decentralized Ledger Integration:** Incorporate a decentralized ledger into the S-ERP system to record and track every transaction across the supply chain. This integration enables real-time monitoring of resource use, waste generation, and product lifecycle stages, ensuring compliance with CE principles.
- **Smart Contracts for Automation:** Develop and deploy smart contracts within the S-ERP system to automate processes such as material sourcing, product recycling, and compliance verification. These contracts can enforce CE-related rules and agreements without the need for intermediaries, thus reducing transaction costs and enhancing efficiency.
- **Data Transparency and Traceability:** Ensure that the blockchain provides complete transparency and traceability of all materials and products within the S-ERP system. This feature is crucial for verifying the sustainability of supply chains and for providing stakeholders with reliable information on the environmental impact of their operations.

3. Implementation Strategy

- **Pilot Testing:** Implement a pilot blockchain-enabled S-ERP system in a controlled environment. The pilot should focus on a specific segment of the supply chain

where circular economy practices can be most effectively applied, such as material sourcing or waste management. The pilot will help identify potential challenges and areas for improvement before full-scale implementation.
- **Stakeholder Engagement and Training:** Engage with all relevant stakeholders, including suppliers, customers, and employees, to ensure they understand the benefits and operational changes associated with the blockchain-enabled S-ERP system. Provide comprehensive training on the use of blockchain technology, particularly in relation to CE practices.
- **Scalability and Integration:** After successful pilot testing, scale the blockchain-enabled S-ERP system across the entire organization. Ensure that the system is fully integrated with existing S-ERP modules and compatible with other Industry 4.0 technologies, such as IoT and AI, to maximize its effectiveness in supporting circular economy goals.

4. Continuous Monitoring and Improvement

- **Real-Time Data Analytics:** Utilize the real-time data generated by the blockchain to monitor the performance of CE initiatives within the S-ERP system. Analyze this data to identify trends, inefficiencies, and opportunities for further enhancement of circular economy practices.
- **Feedback Loops for Continuous Improvement:** Establish feedback mechanisms that allow stakeholders to provide input on the effectiveness of the blockchain-enabled S-ERP system. Use this feedback to make iterative improvements to the system, ensuring it continues to meet evolving CE objectives and industry standards.
- **Compliance and Reporting:** Ensure that the blockchain provides accurate and verifiable data for reporting on sustainability metrics. This data can be used to demonstrate compliance with environmental regulations and to showcase the organization's commitment to the circular economy.

The integration of blockchain technology into S-ERP systems within Industry 4.0 offers a robust framework for embedding circular economy principles into organizational processes. By enhancing transparency, automating key processes, and providing real-time data analytics, blockchain can significantly reduce the barriers to implementing a circular economy, making S-ERP systems more sustainable and resilient in the face of environmental challenges.

The methodology for integrating the circular economy in S-ERP for Industry 4.0 using blockchain has a direct impact on the formal modeling of the circular economy, as described in formalization available in Sects. 6.2.1, 6.2.2 and 6.2.3. Below is a brief explanation of how the methodology influences the formal model:

1. Resource Input Optimization ($R(t)$):

The integration of blockchain into S-ERP enhances the accuracy and efficiency of resource input by enabling precise tracking and real-time data analysis of materials. This reduces unnecessary resource consumption, which directly affects the equation for resource input. The blockchain-driven transparency and automation provided by smart contracts can also reduce the total resource input by improving the efficiency of recycling processes ($\eta(t)$) and sustainable resource usage (R_u(t)).

2. Waste Generation ($W(t)$):

Blockchain-enabled S-ERP systems improve waste management by providing better traceability and accountability of waste materials. This reduces waste generation (W(t)) by ensuring that more resources are recycled ($R\_c(t)$) and by optimizing production processes to minimize waste through predictive analytics and compliance with sustainability standards.

3. Energy Consumption ($E(t)$):

The blockchain methodology in S-ERP allows for more efficient energy use through the integration of real-time data and predictive maintenance. By optimizing production and recycling processes, blockchain reduces energy consumption ($E(t)$) by improving the energy intensities (α and β) of these processes, leading to more sustainable operations.

4. Cost of Production ($C(t)$):

The use of blockchain within S-ERP facilitates cost reduction by streamlining resource management and energy consumption, thus lowering the overall cost of production ($C(t)$). Blockchain's ability to enhance transparency and optimize resource allocation also reduces costs associated with waste management and environmental compliance.

5. Value Generation (V(t)):

The incorporation of blockchain into S-ERP increases the value generated ($V(t)$) by improving the efficiency and sustainability of production processes. The transparency and automation provided by blockchain reduce costs and enhance the perceived value of sustainably produced goods, contributing to higher overall value generation.

6. Circular Economy Efficiency ($\xi(t)$):

Blockchain significantly enhances circular economy efficiency ($\xi(t)$) by increasing the proportion of recycled resources ($R\_c(t)$) relative to the total resource input ($R(t)$). The methodology's focus on real-time data integration and automation ensures that recycling rates are maximized, thus improving the overall efficiency of the circular economy within the S-ERP system.

Thus, the methodology proposed for integrating blockchain into S-ERP for Industry 4.0 directly impacts the key variables and equations of the formal circular economy model by improving resource efficiency, reducing waste, optimizing energy use, lowering costs, and enhancing value generation. These improvements lead to a more effective and sustainable circular economy within the Industry 4.0 framework. The following subsection discusses the fuzzy logic and decision support systems to integrate circular economy in S-ERP for Industry 4.0.

6.3.6 Fuzzy Logic and Decision Support Systems to Integrate Circular Economy in S-ERP for Industry 4.0

Fuzzy Logic is a form of reasoning that extends classical Boolean logic to handle the concept of partial truth, where truth values range between completely true and completely false. It operates on the principle that the truth of any statement can be represented as a degree of membership in a set, often using linguistic variables to model uncertainty and imprecision [1]. This approach allows for more flexible and humanlike reasoning, especially in complex systems where binary logic is inadequate. Decision Support Systems (DSS) are interactive software-based systems that assist decision-makers in making informed and data-driven decisions. They integrate data, sophisticated analytical models, and user-friendly interfaces to support problem-solving, particularly in complex and uncertain environments. Fuzzy Logic is often incorporated into DSS to enhance their ability to process ambiguous, imprecise, or incomplete information, thereby improving the robustness and adaptability of the decision-making process.

The integration of Fuzzy Logic with DSS in S-ERP for Industry 4.0 is highly significant. This combination enhances the ability of S-ERP systems to handle the inherent uncertainties and complexities of modern industrial processes. Fuzzy Logic allows for more nuanced decision-making by accommodating imprecise data and linguistic variables, which is common in sustainability assessments and resource optimization. When embedded within DSS, Fuzzy Logic enables S-ERP systems to provide more accurate and adaptable recommendations, facilitating informed decisions that align with sustainability goals. This integration supports the dynamic and real-time nature of Industry 4.0, where adaptability, precision, and sustainability are paramount.

Combining Fuzzy Logic and DSS is critical in integrating the circular economy into S-ERP for Industry 4.0. Circular economy principles emphasize resource reduction, reuse, and recycling, requiring complex decision-making processes that account for various uncertainties and incomplete information. Fuzzy Logic, with its ability to handle imprecise and ambiguous data (we have discussed earlier that sometimes the data can be incomplete or inaccurate in Industry 4.0), enhances DSS by providing a more flexible framework for evaluating sustainable practices within an industrial context.

In S-ERP systems, this integration allows for more accurate modeling of resource flows, waste management, and lifecycle assessments, enabling companies to make informed decisions that promote resource efficiency and waste minimization. By facilitating the assessment of multiple criteria and scenarios, the combination of Fuzzy Logic and DSS supports the dynamic and iterative nature of circular economy practices, ensuring that S-ERP systems can adapt to the evolving demands of sustainability in Industry 4.0. This synergy ultimately contributes to more effective and resilient industrial operations that align with circular economy goals.

Fuzzy logic itself is a huge discipline. To cover all the aspects of fuzzy logic would be difficult here. However, we briefly try to attempt its application in the nine-step methodology. The methodology is as follows:

Step 1. Problem Definition and System Analysis

The first step involves defining the specific goals of integrating circular economy principles into S-ERP within the context of Industry 4.0. This includes identifying key areas where circular economy practices can be applied, such as resource management, waste reduction, and product lifecycle extension. A thorough analysis of the existing S-ERP system is conducted to understand its current capabilities, limitations, and the nature of data available, particularly in relation to sustainability metrics.

Step 2. Data Collection and Preprocessing

Data relevant to circular economy practices, such as resource consumption patterns, waste generation, recycling rates, and environmental impact assessments, is collected from various sources within the organization. Given the complexity and variability of this data, preprocessing steps are necessary to handle missing values, normalize datasets, and convert qualitative assessments into quantifiable metrics. This stage also involves the identification of key performance indicators (KPIs) related to circular economy goals.

Step 3. Design and Integration of Fuzzy Logic Models

Fuzzy Logic models are designed to manage the inherent uncertainties and ambiguities in circular economy data. These models use linguistic variables to represent complex concepts such as "resource efficiency," "waste minimization," and "sustainability," which are difficult to quantify with precise values. Membership functions are defined for these variables, allowing for a graded assessment of performance against circular economy objectives. The design of Fuzzy Logic models also includes the development of rule-based systems that map inputs (e.g., resource usage, waste levels) to outputs (e.g., sustainability scores, decision recommendations).

Step 4. Development of Decision Support System (DSS) Framework

A DSS framework is developed to integrate the Fuzzy Logic models with the S-ERP system. This framework includes modules for data input, Fuzzy Logic processing, decision rule evaluation, and output generation. The DSS framework is designed to be interactive, allowing decision-makers to input various scenarios and obtain real-time recommendations on circular economy practices. The framework also includes visualization tools to represent fuzzy outcomes, helping users understand the implications of different decisions.

Step 5. Integration with S-ERP System

The DSS framework, enhanced with Fuzzy Logic, is integrated into the S-ERP system through an interface that allows seamless data exchange between the two. This integration ensures that the DSS can access real-time data from the S-ERP system, perform necessary computations, and feed back actionable insights into the S-ERP's operational workflows. The integration is tested for compatibility, performance, and scalability, ensuring that it can handle large datasets and complex decision-making processes typical in Industry 4.0 environments.

Step 6. Scenario Analysis and Simulation

To validate the integration and assess its effectiveness, various scenarios reflecting different circular economy strategies are simulated within the integrated S-ERP system. These scenarios include changes in resource inputs, variations in waste generation, and shifts in market demands. The Fuzzy Logic-enhanced DSS processes these scenarios and provides decision-makers with insights into the potential outcomes, helping them identify the most sustainable and circular approaches.

Step 7. Evaluation and Refinement

The results from the scenario analysis are evaluated against predefined circular economy KPIs to assess the effectiveness of the integration. This evaluation includes both quantitative measures (e.g., reduction in resource consumption and increase in recycling rates) and qualitative assessments (e.g., alignment with sustainability goals). Based on this evaluation, the Fuzzy Logic models and DSS framework are refined to improve accuracy, adaptability, and user-friendliness. Feedback from decision-makers is also incorporated to ensure the system meets the practical needs of the organization.

Step 8. Implementation and Continuous Improvement

Once refined, the integrated system is implemented across the organization's S-ERP platform. Continuous monitoring and improvement mechanisms are established to ensure the system remains effective as the organization's needs evolve and as new circular economy practices emerge. This step includes training for users, ongoing data collection for further model refinement, and regular updates to the DSS framework to incorporate new decision rules and fuzzy logic criteria.

Step 9. Documentation and Knowledge Sharing

Finally, comprehensive documentation of the methodology, including the design of Fuzzy Logic models, the DSS framework, integration steps, and results from scenario analyses, is prepared. This documentation serves as a reference for further development and a resource for knowledge sharing within the organization. It also provides a foundation for future research and innovation in integrating circular economy practices into S-ERP systems for Industry 4.0.

This methodology ensures a systematic and robust approach to embedding circular economy principles within S-ERP systems, leveraging the strengths of Fuzzy Logic and Decision Support Systems to enhance sustainability in the dynamic environment of Industry 4.0. The methodology for integrating circular economy into S-ERP for Industry 4.0 using Fuzzy Logic and Decision Support Systems (DSS) significantly impacts the formalization of circular economy as presented in Sects. 6.2.1, 6.2.2 and 6.2.3. Here's how:

1. Enhanced Resource Input Optimization

The formal model's equation for resource input $R(t) = R_u(t) \cdot P(t) - \eta(t) \cdot R_c(t)$ is enhanced by the DSS framework integrated with Fuzzy Logic. Fuzzy Logic allows the system to handle uncertainties and ambiguities in resource usage efficiency $R_u(t)$ and recycling efficiency $\eta(t)$, leading to more precise predictions and adjustments in resource inputs. This optimization aligns with the goal of reducing $R(t)$, which is critical in a circular economy.

2. Improved Waste Generation Control

The waste generation equation $W(t) = P(t) - R_c(t)$ benefits from the Fuzzy Logicdriven DSS by incorporating nuanced decision-making regarding production processes and recycling efficiency. The system can adapt to varying conditions, minimizing waste by enhancing $R_c(t)$, thus reducing $W(t)$. The ability to manage and reduce waste dynamically supports the circular economy's focus on waste minimization.

3. Dynamic Energy Consumption Management

The equation $E(t) = \alpha \cdot P(t) + \beta \cdot R_c(t)$ for energy consumption is optimized through the integration of real-time data processed by Fuzzy Logic in the DSS. The DSS can adjust α and β dynamically based on real-time conditions, leading to more efficient energy use. This adaptability is crucial in minimizing energy consumption and aligning with the circular economy's sustainability objectives.
4. Cost of Production Optimization

In the cost equation $C(t) = c_r \cdot R(t) + c_e \cdot E(t) + c_w \cdot W(t)$, the incorporation of Fuzzy Logic enhances the ability to optimize costs by balancing resource inputs, energy use, and waste management dynamically. The DSS framework allows for the real-time adjustment of production costs, leading to more cost-effective and sustainable operations, a key aspect of the circular economy.

5. Maximization of Value Generated

The equation $V(t) = v_p \cdot P(t) - C(t)$ for value generation is positively impacted by the DSS, which, through Fuzzy Logic, ensures that production processes are continuously optimized. By improving resource efficiency, reducing waste, and minimizing costs, the system maximizes $V(t)$, which is central to the circular economy's goal of generating value while minimizing environmental impact.

6. Circular Economy Efficiency

The circular economy efficiency equation $\xi(t) = \frac{R_c(t)}{R(t)}$ is enhanced by the integration of DSS and Fuzzy Logic, which optimizes $R_c(t)$ and reduces $R(t)$ more effectively. This improvement in efficiency is crucial for achieving higher circular economy metrics, supporting the overall sustainability of the enterprise.

7. Adaptation to Industry 4.0 Enhancements

The formal model's adaptation to Industry 4.0 technologies, such as IoT, AI, and smart manufacturing, is further strengthened by the DSS framework. Fuzzy Logic's ability to process and adapt to complex, real-time data from these technologies ensures that the S-ERP system remains responsive and efficient. This responsiveness is critical in managing the dynamic and interconnected environment of Industry 4.0, further supporting the circular economy.

8. Holistic Integration with S-ERP

The incorporation of S-ERP factors, such as sustainable supply chain efficiency $S_c(t)$, environmental costs $C_s(t)$, and governance $G(t)$, into the formal model is significantly enhanced by the DSS framework. Fuzzy Logic enables the system to balance these variables effectively, ensuring that the S-ERP system optimizes sustainability across all dimensions, which is in line with the circular economy's objectives.

Thus, the methodology described for integrating circular economy into S-ERP using Fuzzy Logic and DSS not only aligns with but also enhances the formal model's equations and objectives. It ensures that the system can manage the complexities and uncertainties inherent in the circular economy, leading to more efficient, sustainable, and value-driven operations within the Industry 4.0 framework. We have formally verified all the methodologies discussed in the above section in the present work. In the future, developing case studies using the above methods would be interesting. In summary, the chapter answers the following research questions.

Research Questions Addressed

1. **How can circular economy principles be integrated into Sustainable Enterprise Resource Planning (S-ERP) within Industry 4.0?**

- The chapter explores methodologies and models for embedding circular economy practices into S-ERP systems, focusing on resource optimization, waste reduction, and sustainable production.

2. **What role does situation awareness play in implementing circular economy practices within S-ERP for Industry 4.0?**

- The chapter discusses the importance of situation awareness in managing dynamic and complex systems, particularly in S-ERP and circular economy integration, highlighting its impact on real-time decision-making and sustainability.

3. **How can blockchain technology enhance the implementation of circular economy principles in S-ERP systems within Industry 4.0?**

- The chapter examines the potential of blockchain to provide transparency, traceability, and automation in S-ERP systems, ensuring that circular economy practices are effectively implemented and monitored.

4. **How does using fuzzy logic and decision support systems affect integrating circular economy principles in S-ERP for Industry 4.0?**

- The chapter analyzes the use of fuzzy logic and decision support systems to manage uncertainties and complexities in S-ERP, particularly in achieving circular economy goals.

5. **How do coupling and cohesion concepts contribute to integrating circular economy practices into S-ERP systems within Industry 4.0?**

- The chapter explores the significance of coupling and cohesion in designing and implementing modular S-ERP systems that effectively support circular economy practices, ensuring system flexibility and sustainability.

6. **How can Transaction Cost Theory be applied to integrate circular economy principles into S-ERP systems within Industry 4.0?**

- The chapter investigates the application of Transaction Cost Theory to optimize resource allocation, minimize costs, and enhance the economic efficiency of S-ERP systems, thereby supporting circular economy objectives.

7. **What are the benefits of combining green economy, bio-economy, and circular economy principles in S-ERP for Industry 4.0?**

- The chapter discusses the synergistic effects of integrating green economy, bio-economy, and circular economy principles into S-ERP systems, highlighting their collective impact on sustainability and long-term business resilience.

These research questions guide the chapter's exploration of advanced methodologies and technologies for integrating circular economy practices into S-ERP systems within the framework of Industry 4.0.

References

1. Anjaria K (2022) Dempster-shafer theory and linguistic intuitionistic fuzzy number-based framework for blending knowledge from knowledge repositories: an approach for knowledge management. Expert Syst Appl 199:117142. ISSN 0957-4174. https://doi.org/10.1016/j.eswa.2022.117142. https://www.sciencedirect.com/science/article/pii/S0957417422005383
2. Anjaria K (2024) Enhancing sustainability integration in Sustainable Enterprise Resource Planning (S-ERP) system: application of transaction cost theory and case study analysis. Int J Inf Manage Data Insights 4(2):100243. ISSN 2667-0968. https://doi.org/10.1016/j.jjimei.2024.100243. https://www.sciencedirect.com/science/article/pii/S2667096824000326
3. Bag S, Telukdarie A, Pretorius JHC, Gupta S (2018) Industry 4.0 and supply chain sustainability: framework and future research directions. Benchmark: Int J 28(5):1410–1450. ISSN 1463-5771. https://doi.org/10.1108/BIJ-03-2018-0056. https://doi.org/10.1108/BIJ-03-2018-0056. Publisher: Emerald Publishing Limited
4. Chabowski BR, Gabrielsson P, Hult GTM, Morgeson FV (2023) Sustainable international business model innovations for a globalizing circular economy: a review and synthesis, integrative framework, and opportunities for future research. J Int Business Stud. ISSN 1478–6990. https://doi.org/10.1057/s41267-023-00652-9
5. Chauhan C, Sharma A, Singh A (2019) A SAP-LAP linkages framework for integrating Industry 4.0 and circular economy. Benchmark: Int J 28(5):1638–1664. ISSN 1463-5771. https://doi.org/10.1108/BIJ-10-2018-0310. https://doi.org/10.1108/BIJ-10-2018-0310. Publisher: Emerald Publishing Limited
6. Cholewa M, Minh LHB (2021) PLM solutions in the process of supporting the implementation and maintenance of the circular economy concept in manufacturing companies. Sustainability 13(19):10589. ISSN 2071-1050. https://doi.org/10.3390/su131910589. https://www.mdpi.com/2071-1050/13/19/10589. Number: 19 Publisher: Multidisciplinary Digital Publishing Institute
7. Endsley MR (2021) Situation awareness. In: Handbook of human factors and ergonomics. Wiley, Ltd, pp 434–455. ISBN 978-1-119-63611-3. https://doi.org/10.1002/9781119636113.ch17. https://onlinelibrary.wiley.com/doi/abs/10.1002/9781119636113.ch17. Section: 17 _ eprint: https://onlinelibrary.wiley.com/doi/pdf/10.1002/9781119636113.ch17
8. Grafström J, Aasma S (2021) Breaking circular economy barriers. J Clean Product 292:126002. ISSN 0959-6526. https://doi.org/10.1016/j.jclepro.2021.126002. https://www.sciencedirect.com/science/article/pii/S0959652621002225
9. Jaroenkietkajorn U, Gheewala SH, Mungkung R, Jakrawatana N, Silalertruksa T, Lecksiwilai N, Prasara-A J, Nilsalab P (2024) Challenges and opportunities of bio-circular-green economy for agriculture. Circul Econ Sustain. ISSN 2730–5988. https://doi.org/10.1007/s43615-024-00355-9
10. Kanellou E, Alexakis K, Kapsalis P, Kokkinakos P, Askounis D (2021) The DigiPrime KPIs' framework for a circular economy transition in the automotive industry. Proc Manufact 54:302–307. ISSN 2351-9789. https://doi.org/10.1016/j.promfg.2021.09.003. https://www.sciencedirect.com/science/article/pii/S2351978921001979
11. Kantola JI, Nazir S, Barath T (2018) Advances in human factors, business management and society: proceedings of the AHFE 2018 international conference on human factors, business management and society, July 21-25, 2018, Loews Sapphire Falls Resort at Universal Studios, Orlando, Florida, USA. Springer. ISBN 978-3-319-94709-9. Google-Books-ID: VgdiDwAAQBAJ
12. Li W, Wang Z, Mai R, Ren P, Zhang Q, Zhou Y, Xu N, Zhuang J, Xin B, Gao L, Hao Z, Fan Z (2023) Modular design automation of the morphologies, controllers, and vision systems for intelligent robots: a survey. Vis Intell 1(1):2. ISSN 2731-9008. https://doi.org/10.1007/s44267-023-00006-x. https://doi.org/10.1007/s44267-023-00006-x

13. Lopes de Sousa Jabbour AB, Jabbour CJ, Godinho Filho M, Roubaud D (2018) Industry 4.0 and the circular economy: a proposed research agenda and original roadmap for sustainable operations. Ann Operat Res 270(1):273–286. ISSN 1572-9338. https://doi.org/10.1007/s10479-018-2772-8. https://doi.org/10.1007/s10479-018-2772-8
14. Majeed AA, Rupasinghe TD (2017) Internet of Things (IoT) embedded future supply chains for Industry 4.0: an assessment from an ERP-based fashion apparel and footwear industry. Int J Supply Chain Manage 6(1):25–40. ISSN 2050-7399. https://doi.org/10.59160/ijscm.v6i1.1395. https://ojs.excelingtech.co.uk/index.php/IJSCM/article/view/1395. Number: 1
15. Milios L (2021) Overarching policy framework for product life extension in a circular economy–A bottom-up business perspective. Environ Policy Govern 31(4):330–346. ISSN 1756-9338. https://doi.org/10.1002/eet.1927. https://onlinelibrary.wiley.com/doi/abs/10.1002/eet.1927. _ eprint: https://onlinelibrary.wiley.com/doi/pdf/10.1002/eet.1927
16. Mohelska H, Sokolova M (2016) Smart, connected products change a company's business strategy orientation. Appl Econ 48(47):4502–4509. ISSN 0003-6846. https://doi.org/10.1080/00036846.2016.1158924. https://www.tandfonline.com/doi/abs/10.1080/00036846.2016.1158924. Publisher: Routledge
17. Ressi D, Romanello R, Piazza C, Rossi S (2024) AI-enhanced blockchain technology: a review of advancements and opportunities. J Netw Comput Appl 225:103858. ISSN 1084-8045. https://doi.org/10.1016/j.jnca.2024.103858. https://www.sciencedirect.com/science/article/pii/S1084804524000353
18. Shen Y, Zhao D, Hu X, Pedrycz W, Chen Y, Li J, Xiao Z (2024) Structure identification of missing data: a perspective from granular computing. Soft Comput. ISSN 1433–7479. https://doi.org/10.1007/s00500-023-09523-9
19. Stock T, Seliger G (2016) Opportunities of sustainable manufacturing in industry 4.0. Proc CIRP 40:536–541. ISSN 2212-8271. https://doi.org/10.1016/j.procir.2016.01.129. https://www.sciencedirect.com/science/article/pii/S221282711600144X
20. Tokede O, Globa A (2024) Life cycle sustainability tracker: a dynamic approach. Engin, Construct Architect Manage, ahead-of-print (ahead-of-print), January 2024. ISSN 0969-9988. https://doi.org/10.1108/ECAM-07-2023-0680
21. Yin S, Ying H, Yan F, Zhao X, Chen L, Collaborative supply chain finance based on transaction cost theory: a multiple case study in the POF scenario. Int J Logist Res Appl 0(0):1–25. ISSN 1367-5567. https://doi.org/10.1080/13675567.2024.2311204. Publisher: Taylor & Francis

Chapter 7
Analytics for Harnessing IoT and Industry 4.0: Data Flows in S-ERP

7.1 Introduction

IoT (Internet of Things) analytics refers to the process of analyzing the data generated by interconnected devices within an IoT ecosystem [3]. These devices, equipped with sensors, actuators, and other technologies, continuously collect and transmit vast amounts of data related to their environment, operation, and interactions. The primary goal of IoT analytics is to extract actionable insights from this data to improve decision-making, optimize processes, enhance user experiences, and create new opportunities for innovation across various industries.

In an academic context, IoT analytics encompasses several critical components. Data collection and storage are fundamental, as IoT devices generate large volumes of data, often in real time, requiring robust mechanisms to manage and store the data. The data collected can be structured, semi-structured, or unstructured, depending on the type of device and the nature of the data. Once collected, the data must be processed to extract meaningful information. This processing can involve data cleaning, transformation, and integration, as well as more complex operations like filtering, aggregation, and normalization. Given the often real-time nature of IoT data, processing must be efficient and scalable.

The core of IoT analytics lies in the analysis phase, where various analytical techniques are applied to the processed data. These techniques range from basic descriptive statistics to advanced machine learning and artificial intelligence algorithms. The choice of analytical method depends on the specific objectives, such as predictive maintenance, anomaly detection, or pattern recognition. To make the insights derived from IoT data actionable, they must be presented in an understandable and accessible format. Data visualization tools and techniques play a crucial role in this process, enabling stakeholders to interpret the results and make informed decisions.

K. Anjaria, *Sustainable Enterprise Resource Planning (S-ERP) for Industry 4.0*,
Contributions to Management Science, https://doi.org/10.1007/978-981-96-7734-4_7

IoT analytics finds applications across diverse domains, including smart cities, healthcare, manufacturing, agriculture, and transportation [4]. The insights gained from IoT analytics can lead to significant improvements in operational efficiency, cost reduction, and enhanced customer satisfaction. Moreover, IoT analytics is increasingly seen as a driver of innovation, enabling the development of new products, services, and business models. IoT analytics is a multidisciplinary field that involves the collection, processing, and analysis of data generated by IoT devices to derive insights that can enhance decision-making and drive innovation across various sectors. Its significance continues to grow as IoT technologies become more pervasive, generating ever-increasing amounts of data that hold the potential to transform industries and societies.

IoT analytics is fundamental to the evolution of Industry 4.0, which is characterized by the integration of cyber-physical systems, advanced automation, and digitalization within industrial environments [8]. The ability to analyze data generated by interconnected devices is crucial for creating intelligent, adaptive, and efficient industrial processes. In Industry 4.0, IoT analytics enables predictive maintenance by analyzing real-time data from sensors embedded in industrial machinery. This analysis identifies patterns and anomalies, allowing for the anticipation of equipment failures and the optimization of maintenance schedules, thereby minimizing downtime and extending equipment lifespan.

Additionally, IoT analytics optimizes production processes by providing insights into the production flow, resource allocation, and potential bottlenecks [9]. Real-time data analysis allows for the dynamic adjustment of production parameters, improving productivity, reducing waste, and enhancing quality control. In supply chain management, IoT analytics offers enhanced visibility and transparency, enabling accurate demand forecasting, inventory optimization, and just-in-time production. This leads to cost reduction and improved customer satisfaction. Moreover, IoT analytics underpins smart manufacturing by facilitating the creation of flexible production systems that can autonomously adapt to changing market conditions, supporting mass customization without compromising efficiency.

Furthermore, IoT analytics plays a critical role in quality management by enabling real-time monitoring and analysis of product quality [14]. This ensures consistent quality standards and reduces waste through immediate corrective actions. Thus, IoT analytics is a cornerstone of Industry 4.0, providing the data-driven insights necessary for optimizing industrial processes, predictive maintenance, supply chain management, smart manufacturing, and quality control. Its role is increasingly vital in driving innovation, efficiency, and competitiveness in the industrial sector as Industry 4.0 advances.

IoT analytics is critically aligned with the objectives of S-ERP within the context of Industry 4.0, where the focus on sustainability is paramount. By leveraging IoT-generated data, S-ERP systems can be enhanced to support sustainable practices across industrial operations, contributing to economic, environmental, and social sustainability. In terms of environmental sustainability, IoT analytics enables real-time monitoring and optimization of energy consumption within industrial processes [12]. Sensors embedded in machinery and facilities collect data on energy usage, which is then analyzed to identify inefficiencies and opportunities for reducing energy consumption. For example, IoT analytics can optimize heating, ventilation, and air conditioning (HVAC) systems or machinery operations, minimizing energy waste and lowering the carbon footprint of industrial activities. This aligns with the goals of S-ERP, which are to reduce environmental impact and promote the efficient use of resources [2].

From an economic sustainability perspective, IoT analytics enhances resource management by providing insights into material usage, waste generation, and supply chain efficiency [6]. By analyzing data from IoT-enabled devices, S-ERP systems can track resource utilization in real time, facilitating just-in-time inventory management and reducing excess material waste. This not only leads to cost savings but also contributes to the circular economy by minimizing waste and promoting the reuse and recycling of materials. Social sustainability, which involves the well-being of employees and communities, is also supported by IoT analytics within S-ERP systems. IoT devices can monitor workplace conditions, such as air quality, noise levels, and worker safety. Analyzing this data allows for the identification and mitigation of health and safety risks, ensuring a safer and healthier working environment. Additionally, IoT analytics can be used to monitor compliance with labor and environmental regulations, helping companies maintain socially responsible practices.

Moreover, IoT analytics can drive sustainable innovation within S-ERP by enabling the development of new products and services that meet sustainability criteria [7]. For instance, data collected from IoT devices can inform the design of products with lower environmental impact or services that enhance resource efficiency and reduce waste. This fosters a culture of sustainability-driven innovation within enterprises, aligning with the broader goals of Industry 4.0 to create sustainable, resilient industrial systems.

IoT analytics is integral to successfully implementing S-ERP in Industry 4.0, as it provides the necessary data-driven insights to promote environmental, economic, and social sustainability. By optimizing resource use, reducing environmental impact, ensuring regulatory compliance, and fostering sustainable innovation, IoT analytics helps S-ERP systems support the transition to more sustainable industrial practices, thus contributing to the broader sustainability goals in Industry 4.0.

IoT analytics is crucial for understanding and optimizing data flows in S-ERP systems within Industry 4.0. By analyzing real-time data generated by IoT-enabled devices across manufacturing, supply chains, and customer interactions, IoT analytics offers insights into the movement and efficiency of information throughout the enterprise. This enables dynamic decision-making, enhances operational efficiency, and aligns with sustainability goals by optimizing resource use, reducing waste, and ensuring timely, data-driven responses to changing conditions. Ultimately, IoT analytics underpins the effective management of complex data flows essential to S-ERP and Industry 4.0 objectives.

Following the introduction, the chapter is organized into several key sections that explore different facets of IoT analytics. Firstly, the chapter delves into the IoT Data for IoT Analytics: S-ERP for Industry 4.0 Perspectives, offering insights into the nature and importance of IoT data within smart enterprises. It examines how data collected from IoT devices is crucial for analytics that drive operational efficiency and strategic decision-making in Industry 4.0 environments. The chapter then discusses the lifecycle of IoT data analytics, where the processes of collecting, processing, analyzing, and utilizing IoT data are detailed. This section highlights the importance of managing the entire lifecycle to ensure data integrity and the effectiveness of analytics within S-ERP systems. Moving forward, the chapter explores Applications of IoT Analytics with IoT Architecture: S-ERP for Industry 4.0 Perspectives. This section is subdivided into three critical areas of application: Smart Metering, Smart Transportation, and Smart Logistics; Supply Chain; and Retailing. Each subsection provides detailed examples of how IoT analytics are applied within these specific domains to optimize operations and enhance productivity in Industry 4.0. The chapter further relates IoT Analytics in S-ERP for Industry 4.0 and Quality 4.0, discussing the integration of analytics in maintaining and improving quality standards across industries. This is followed by an exploration of IoT Analytics as a Service and the Movement from IoT Analytics to IIoT Analytics, which covers the evolution and future directions of IoT analytics. Lastly, the chapter concludes by assessing the Impact of 5G and 6G Technologies on IoT Analytics of S-ERP in Industry 4.0, highlighting how these advanced networks will transform IoT analytics, enabling more sophisticated and real-time analytics capabilities. This structure ensures that readers gain a holistic understanding of the role and potential of IoT analytics in the modern industrial landscape. After this rigorous introduction, the next section aims to understand one of the most critical building blocks of IoT analytics, i.e., IoT data.

7.2 IoT Data for IoT Analytics: S-ERP for Industry 4.0 Perspectives

As discussed in the introduction, IoT data is a critical component before IoT analytics. The question is, what exactly is IoT data? Literature has presented many definitions, but Soldatos [1] has answered this question very straightforwardly. According to Soldatos [1], IoT data is Bigdata. Detailed proof of IoT data being big data is available in the work [1]. The essential characteristics of IoT data are four Vs: volume, velocity, variety, and veracity.

- **Volume** refers to the vast data generated by numerous interconnected devices.
- **Velocity** denotes the rapid speed at which this data is produced and needs to be processed.
- **Variety** encompasses diverse data types, including structured, semi-structured, and unstructured formats.
- **Veracity** addresses the quality and reliability of the data, considering potential inconsistencies and uncertainties.

These characteristics highlight the complexity and challenges of managing IoT data within advanced analytics frameworks. Due to these complexities, the McKinsey report states that only 1% of the IoT data is being consumed and analyzed [5]. IoT data's complexity, volume, velocity, variety, and veracity can be effectively managed within S-ERP systems through integrated data processing and analytics capabilities. S-ERP systems streamline data flows, standardize diverse data formats, and enhance data quality through real-time validation and filtering processes. S-ERP systems reduce complexity by providing a unified platform for managing IoT data, enabling more efficient analysis, improved decision-making, and alignment with sustainability objectives in Industry 4.0. Further, data mining, data stream management, and IoT middleware for interoperability can reduce the complexity of IoT data and make it useful for analytics. Data mining extracts valuable patterns and insights from large datasets, refining raw IoT data into actionable information. Data stream management handles continuous data flow, ensuring real-time processing and filtering to maintain relevance and accuracy. IoT middleware ensures interoperability between diverse IoT devices and systems, standardizing data formats and facilitating seamless integration. Together, these technologies improve data quality, relevance, and accessibility, making IoT data more explainable for advanced analytics.

Data explainability in IoT for S-ERP in Industry 4.0 refers to the clarity and transparency of data-driven insights derived from IoT analytics. It ensures stakeholders understand and interpret the input, outcomes, and decisions generated by IoT systems within S-ERP. This is crucial for maintaining trust, ensuring regulation compliance, and enabling informed decision-making. In the context of Industry 4.0, data explainability supports the alignment of IoT analytics with sustainability goals, allowing for the effective communication of how data influences operational efficiency, resource management, and overall enterprise sustainability.

High-quality data is essential for effective feature engineering and selection in IoT analytics using data mining. Accurate and reliable data enhances the ability to identify meaningful patterns and relationships, enabling the extraction and refinement of the most relevant features. These optimized features improve the performance of analytical models, leading to more accurate predictions and insights. Quality data underpins the entire feature engineering process, ensuring that IoT analytics is robust and effective in generating actionable outcomes.

How one collects the data from IoT also impacts the data, data quality, and features of IoT-based systems like Industry 4.0 for IoT data analytics. Data collected from IoT devices directly impacts the data quality and features of IoT-based systems like those in Industry 4.0. The accuracy, consistency, and completeness of this data influence the reliability of analytical models and decision-making processes. Poor data quality can lead to erroneous insights and suboptimal feature selection, undermining system performance. Conversely, high-quality IoT data enhances the precision and relevance of features, thereby improving the overall efficiency and effectiveness of Industry 4.0 systems. Selecting appropriate methods for data collection from IoT devices is crucial for the success of IoT analytics. The decision of when to employ sensor networks, edge computing, data aggregators, APIs, communication protocols, and cloud integration is vital to the effectiveness of IoT data analytics. This choice is influenced by factors such as data volume, processing speed, latency requirements, and the complexity of the required analytics. For example, sensor networks are essential for real-time data capture; edge computing minimizes latency and reduces bandwidth usage, while cloud integration offers scalable analytics solutions. Thus, aligning the data collection method with the specific needs of the IoT application is essential to ensure optimal data quality, efficiency, and analytical accuracy, underscoring its importance in IoT analytics.

The following section discusses the lifecycle of IoT data analytics.

7.3 The Lifecycle of IoT Data Analytics

This section discusses the IoT analytics lifecycle. The lifecycle is shown in Fig. 7.1. As shown in the figure, the lifecycle has six phases, each playing a vital role in transforming raw IoT data into actionable insights and driving operational improvements.

Here is a brief explanation of each phase:

Phase-1 IoT Data Generation: This initial phase involves the creation of data through IoT devices equipped with sensors, actuators, and other data-capturing mechanisms. These devices continuously monitor various parameters and generate data reflecting the state, performance, and environment of the objects they observe.
Phase-2 IoT Data Collection: Once the data is generated, it needs to be collected from the IoT devices. This phase involves gathering the data using various methods such as sensor networks, edge computing, and communication protocols. Effective data collection ensures that the data is accurately captured and transmitted for further processing.
Phase-3 IoT Data Transformation or Unification: In this phase, the collected data, which may come from various sources and be in different formats, is standardized and integrated into a cohesive dataset. Data unification involves cleaning, transforming, and combining the data to ensure consistency, accuracy, and readiness for analysis.
Phase-4 IoT Data Analysis: This phase involves applying analytical techniques to the unified data to extract meaningful insights. Methods such as data mining, machine learning, and statistical analysis are employed to identify patterns, predict outcomes, and generate actionable insights that can inform decision-making.
Phase-5 IoT Analytics Deployment: After analysis, the insights gained need to be deployed within the operational environment. This phase involves implementing the findings into the workflow, whether through automation, optimization strategies, or decision support systems, to enhance operational efficiency and effectiveness.
Phase-6 Operational Steps: This final phase focuses on the practical application of the deployed analytics. It involves the continuous monitoring and adjustment of operations based on the real-time insights provided by IoT analytics. This ensures that the systems remain optimized and can adapt to any changes in the operational environment.

Each of these phases is interconnected, forming a continuous cycle that drives the effective utilization of IoT data to improve operations, drive innovation, and achieve strategic objectives in an Industry 4.0 context. The above lifecycle helps

Fig. 7.1 IoT analytics lifecycle

perform IoT analytics operations such as predictive analysis, prescriptive analysis, descriptive analysis, monitoring and notification, control, optimization, knowledge generation, and planning. The above lifecycle is useful for all types of data and their sources with different responsiveness. The sources and responsiveness are shown in Fig. 7.2.

The responsiveness of IoT components, as illustrated in Fig. 7.2, is critical for the effectiveness of IoT analytics. Servers play a key role in processing and storing the vast amounts of data generated by IoT devices. Their responsiveness is essential for ensuring that data can be quickly accessed, processed, and analyzed, thereby enabling timely insights. Connections refer to the network infrastructure that links

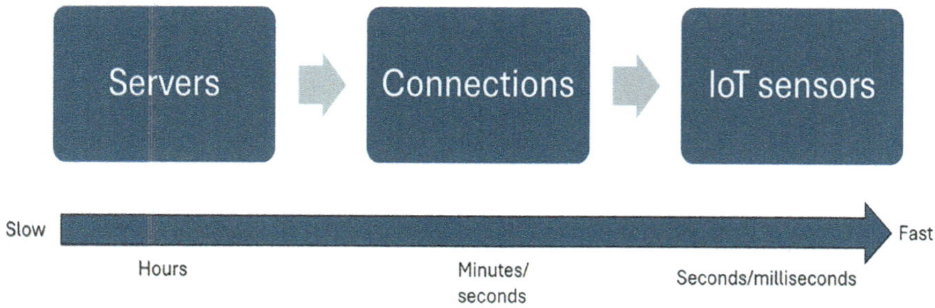

Fig. 7.2 Responsiveness of IoT components for IoT analytics

IoT sensors to servers. The responsiveness of these connections is crucial for minimizing latency, ensuring real-time data transmission, and maintaining the integrity of the data flow. High-speed and reliable connections are vital for supporting the continuous and seamless transfer of data necessary for effective IoT analytics. IoT sensors are responsible for collecting and transmitting data from the physical environment. Their responsiveness is crucial for capturing accurate and timely data, which forms the foundation of IoT analytics. Sensors must operate with minimal delay and high precision to ensure that the data fed into the analytics pipeline is both relevant and accurate. All these components shown in Fig. 7.2 are part of the IoT analytics lifecycle.

Along with the above components, IoT metadata are also part of IoT analytics. In the IoT analytics lifecycle, metadata plays a crucial role in enhancing the understanding, organization, and usability of IoT data across various stages. Metadata, which includes information about the data's origin, structure, format, and context, is generated during the IoT data generation and collection phases. Specifically, as IoT sensors collect data and transmit it through connections to servers, metadata is created to describe the characteristics of the data, such as timestamps, sensor types, and data formats. Throughout the lifecycle, this metadata is used to unify and standardize the data during the unification phase, enabling seamless integration from diverse IoT sources. During the analysis phase, metadata guides the selection of appropriate analytical models and ensures the correct interpretation of the data. Finally, in the deployment and operational steps, metadata supports traceability, reproducibility, and optimization of IoT analytics processes. In essence, metadata is generated at the data generation and collection stages and plays a vital role throughout the IoT analytics lifecycle by ensuring data quality, facilitating integration, and enhancing the accuracy of analysis. The following section discusses the application of IoT analytics with possible architecture.

7.4 Applications of IoT Analytics with IoT Architecture: S-ERP for Industry 4.0 Perspectives

This section relates IoT analytics with the IoT architecture. The applications that we will be discussing are smart metering, smart transportation, and smart logistics; supply chain; and retailing.

7.4.1 Smart Metering in S-ERP for Industry 4.0

Smart Metering is a critical application within the Internet of Things (IoT) ecosystem, especially in the context of Industry 4.0. It involves the use of advanced metering infrastructure to monitor, measure, and analyze energy usage in real time. This application can significantly enhance the efficiency and sustainability of operations within an S-ERP framework, which is integral to Industry 4.0.

Smart meters provide continuous data on energy consumption, allowing for real-time monitoring and analysis. This data can be integrated into S-ERP systems to optimize energy usage, reduce waste, and lower operational costs. By analyzing energy usage patterns, S-ERP systems can suggest operational adjustments to minimize energy consumption during peak times, leading to better demand-side management and more sustainable operations. The data collected from smart meters can be used to predict equipment failures or inefficiencies, allowing for proactive maintenance scheduling. This reduces downtime and extends the lifespan of machinery, aligning with the sustainability goals of Industry 4.0. Smart metering data helps ensure compliance with energy regulations by providing accurate records of energy consumption and emissions. This is crucial for industries aiming to meet environmental standards and certifications. By optimizing energy consumption and improving operational efficiency, smart metering can lead to significant cost savings, which is a core objective of S-ERP systems.

From the architectural perspective, given the nature of smart metering, real-time analytics is essential. Therefore, the real-time architecture is the most suitable. This architecture is designed to analyze large amounts of data generated by sensors, which aligns perfectly with the requirements of smart metering in an Industry 4.0 environment. The tools, such as Greenplum and HANA, are ideal for handling and processing large datasets in real time. These tools are capable of parallel processing and utilizing memory-based computing platforms, making them suitable for managing the continuous data stream from smart meters. The key advantage of this architecture is its ability to perform parallel processing, ensuring that the data from smart meters is processed and analyzed efficiently without any delays. This real-time capability is critical for the timely optimization of energy usage and operational efficiency in S-ERP.

In the context of smart metering, several data analysis methods, from clustering, classification, association rule mining, prediction, and time series analysi, are highly useful. Clustering can be used to group consumers or consumption patterns based on their energy usage profiles. By grouping customers with similar energy usage patterns, utilities can tailor energy-saving recommendations, identify anomalous behaviors (e.g., potential energy theft), and design targeted demand response programs. For example, high-energy consumers might be clustered together for specialized energy-saving programs. Classification methods can be used to categorize energy consumption data into predefined categories, such as "high usage," "medium usage," and "low usage," or to detect specific types of events, like faults or anomalies. Classification helps in identifying patterns in energy consumption that correspond to different user behaviors or operational states. This can be useful for automating the detection of abnormal energy usage, which could indicate faults, inefficiencies, or even tampering with the meter.

Association rule mining in smart metering can identify relationships between different energy consumption patterns or between energy usage and other factors, such as weather conditions or time of day. This method is useful for uncovering hidden patterns in energy consumption that can inform strategies for reducing consumption or shifting usage to off-peak times. For example, an association rule might reveal that high-energy usage is often associated with certain times of day or specific weather conditions. Predictive models can be developed to forecast future energy consumption based on historical data. Prediction is critical for load forecasting, which helps utilities manage supply and demand efficiently. Accurate predictions can also enable proactive maintenance by forecasting potential issues based on trends in energy consumption. Time series analysis is used to analyze the temporal patterns in energy consumption data, such as daily or seasonal trends. Since energy consumption data is inherently time-dependent, time series analysis is essential for understanding consumption patterns over time. This method can help identify peak usage periods, seasonal variations, and long-term trends, which are crucial for demand forecasting and optimizing energy distribution. The following subsection discusses smart transportation.

7.4.2 Smart Transportation in S-ERP for Industry 4.0

Smart Transportation is a vital application within the Internet of Things (IoT) ecosystem, particularly within the framework of Industry 4.0. It involves the integration of sensors, data analytics, and connectivity to enhance the efficiency, safety, and sustainability of transportation systems. In the context of S-ERP for Industry 4.0, smart transportation can play a crucial role in optimizing logistics, reducing environmental impact, and improving operational efficiency.

Smart transportation systems provide real-time data on vehicle locations, traffic conditions, and route efficiency. This data can be integrated into S-ERP systems to optimize logistics and supply chain management, reducing fuel consumption and delivery times. By monitoring and analyzing transportation data, smart transportation systems can suggest more efficient routes and transportation methods, leading to reduced emissions and a lower carbon footprint. This aligns with the sustainability goals of S-ERP systems in Industry 4.0. Real-time monitoring of vehicle conditions and driver behavior helps ensure compliance with safety regulations. This data can be used within an S-ERP system to enforce maintenance schedules, monitor driver performance, and improve overall safety standards. Data from smart transportation systems can be used to predict vehicle maintenance needs, prevent breakdowns, and extend the lifespan of transportation assets. S-ERP systems can integrate this data to plan maintenance schedules and reduce downtime. By analyzing historical and real-time transportation data, smart transportation systems can provide insights that support strategic decision-making within S-ERP systems, such as route planning, fleet management, and resource allocation.

Given the real-time nature and data intensity of smart transportation, the real-time architecture is most suitable for supporting this application in S-ERP for Industry 4.0. This architecture is designed to analyze large amounts of data generated by sensors in real-time, which is essential for smart transportation systems that need to process data continuously from vehicles, traffic systems, and environmental sensors. The tools, such as Greenplum and HANA, are well-equipped to handle large-scale, real-time data processing. These tools support parallel processing and are optimized for memory-based computing platforms, making them ideal for processing the vast amount of data generated by smart transportation systems. The key advantage of this architecture is its ability to provide immediate insights and responses, which are crucial for real-time decision-making in smart transportation. Whether rerouting vehicles to avoid traffic or responding to a maintenance alert, the real-time processing capability ensures that S-ERP systems can act swiftly and efficiently.

In the context of smart transportation, several data analysis methods, from clustering, classification, association rule mining, prediction, and time series analysis, are particularly useful. Clustering can be used to group vehicles, routes, or travel patterns based on similarities in usage, behavior, or geographical location. By clustering transportation data, it's possible to identify common traffic patterns, optimize route planning, and segment vehicles or drivers based on usage patterns. For example, clustering can help distinguish between commercial and private vehicle usage, allowing for tailored traffic management strategies or optimized fleet management. Classification methods can be employed to categorize transportation events, such as identifying different types of traffic incidents and vehicle types or predicting the likelihood of accidents. Classification is essential for automated detection and categorization of various transportation-related events. For instance, classifying road conditions, identifying high-risk zones for accidents, or classifying vehicles based on emissions levels can help in making real-time decisions that enhance safety and efficiency. Association rule mining can identify relationships between different transportation events or factors, such as the correlation between traffic congestion and specific times of the day or weather conditions. This method is useful for uncovering patterns and associations within transportation data that can inform traffic management strategies. For instance, discovering that heavy traffic is often associated with certain weather conditions can lead to proactive measures like deploying additional traffic management resources during those times.

Predictive models in smart transportation can be developed to forecast traffic conditions, vehicle breakdowns, fuel consumption, or travel times based on historical and real-time data. Prediction is critical for anticipating and mitigating potential issues in smart transportation. For example, predicting traffic congestion allows for dynamic route adjustments, while predicting vehicle breakdowns enables proactive maintenance scheduling. This helps improve efficiency, safety, and reliability in transportation systems. Time series analysis is used to analyze temporal patterns in transportation data, such as daily traffic trends, seasonal variations in vehicle usage, or periodic maintenance needs. Time series analysis is indispensable for understanding how transportation metrics evolve over time. This is crucial for planning purposes, such as anticipating peak traffic periods, scheduling maintenance during off-peak times, and managing long-term transportation infrastructure needs. The following subsection discusses smart logistics, supply chain, and retailing in S-ERP for Industry 4.0.

7.4.3 Smart Logistics, Supply Chain, and Retailing in S-ERP for Industry 4.0

Smart Logistics, Supply Chain, and Retailing are critical components of Industry 4.0, enhanced by the Internet of Things (IoT). These applications involve using IoT devices, sensors, and data analytics to optimize the movement, storage, and sale of goods throughout the supply chain. Integrating these smart systems into S-ERP allows organizations to improve efficiency, reduce waste, and increase responsiveness to market demands.

Smart logistics and supply chain systems provide real-time tracking of goods, enabling better visibility and traceability throughout the supply chain. This ensures that inventory levels are optimized and the movement of goods can be monitored and managed efficiently. IoT-enabled smart retailing systems provide accurate, real-time data on stock levels, helping retailers maintain optimal inventory levels. This reduces the risk of overstocking or stockouts and allows for better demand forecasting. Smart systems enable supply chains to be more agile, responding quickly to changes in demand or disruptions. Real-time data from sensors and IoT devices allows for dynamic rerouting of shipments, proactive maintenance of logistics infrastructure, and rapid adaptation to market changes. By optimizing routes, reducing idle times, and improving load management, smart logistics systems help reduce fuel consumption and emissions. This supports the sustainability goals of S-ERP in Industry 4.0, aligning with broader environmental objectives.

In retailing, IoT-enabled systems provide better customer insights and personalized shopping experiences. Real-time data on customer preferences and behaviors allows for targeted marketing and more responsive service, enhancing customer satisfaction. Smart logistics and supply chain systems use IoT data to predict when maintenance is required for vehicles, machinery, or other assets, reducing downtime and extending the lifespan of critical infrastructure.

Given the real-time data requirements and the large scale of operations in smart logistics, supply chain, and retailing, real-time architecture is ideal, but a combination with the Business Intelligence Level could also be necessary depending on the scale and nature of the operations. Real-time architecture is designed to analyze large amounts of data generated by sensors, which is crucial for real-time tracking, monitoring, and decision-making in smart logistics and supply chains. Tools such as Greenplum and HANA are capable of handling real-time data processing, enabling parallel processing and memory-based computing. This ensures that data from across the supply chain is analyzed and acted upon immediately. The ability to process data in real time is critical for dynamic rerouting, demand forecasting, and inventory management. Real-time architecture ensures that the supply chain remains agile and responsive to changes in demand or disruptions.

Business intelligence level can be used when the data scale surpasses the memory level, providing both offline and online analysis capabilities. This is particularly useful for large-scale supply chains and retail operations where historical data analysis and planning are as important as real-time data processing. Tools supporting this architecture are capable of handling large datasets and providing valuable insights for strategic decision-making. This architecture allows for the integration of real-time data with historical data to create comprehensive business intelligence. Combining real-time processing with business intelligence ensures that organizations can make data-driven decisions both in real-time and in long-term strategic planning.

In the context of smart logistics, supply chain, and retailing, various data analysis methods such as clustering, classification, association rule mining, prediction, and time series analysis, are highly useful. Clustering can be used to group products, suppliers, or customers based on similarities in demand patterns, purchasing behavior, or geographical location. Clustering helps segment the supply chain or retail operations into manageable groups. For example, clustering customers based on purchasing behavior allows for more targeted marketing strategies, while clustering suppliers based on delivery performance can help optimize supply chain management. Clustering also aids in inventory management by grouping similar products together for more efficient stock control. Classification can be used to categorize orders, products, or customers into predefined categories, such as "high-priority," "medium-priority," and "low-priority," or to classify risk levels in the supply chain. Classification is essential for decision-making processes that require the identification of specific characteristics within the supply chain or retail operations. For instance, classifying products based on demand patterns can help in prioritizing stock replenishment. Similarly, classifying customers based on their lifetime value can guide loyalty programs and marketing efforts. It can also be used to classify risks, such as potential delays or disruptions, enabling proactive management.

Association rule mining can uncover relationships between different products, suppliers, or customer behaviors, such as identifying which products are frequently purchased together. This method is particularly useful in retailing, where understanding the relationships between products can drive cross-selling and upselling strategies. For example, if customers frequently buy certain products together, these items can be strategically placed together in stores or recommended online. In supply chain management, association rules can identify dependencies between suppliers and logistics partners, helping to streamline operations and reduce bottlenecks. Predictive models can be used to forecast demand, stock levels, delivery times, or sales trends based on historical data. Prediction is critical for efficient supply chain and retail management. Accurate demand forecasting allows companies to maintain optimal inventory levels, reducing both overstocking and stockouts. Predicting delivery times helps manage customer expectations and optimize logistics routes. Predictive analytics can also be used to anticipate market trends, enabling more informed decision-making in procurement and sales strategies. Time series analysis is used to analyze temporal patterns in logistics, supply chain, and retail data, such as seasonal demand fluctuations, delivery time trends, or sales cycles. Since logistics and retail operations are inherently time-sensitive, time series analysis is indispensable for understanding how these operations evolve over time. For example, analyzing seasonal sales trends can help in planning stock levels and marketing campaigns. Time series analysis can also be used to monitor and improve delivery performance over time, ensuring that service levels are maintained or improved. After discussing the architecture and applications, the next section discusses the relationship between IoT analytics in S-ERP for Industry 4.0 and Quality 4.0

7.5 Relating IoT Analytics in S-ERP for Industry 4.0 and Quality 4.0

Quality 4.0 refers to the integration of traditional quality management practices with the technological advancements and principles of Industry 4.0 [11]. It represents a paradigm shift in how organizations approach quality management by leveraging digital technologies such as big data analytics, artificial intelligence (AI), machine learning, the Internet of Things (IoT), and cloud computing. The primary goal of Quality 4.0 is to enhance the accuracy, efficiency, and adaptability of quality processes across the entire value chain. Quality 4.0 relies heavily on data analytics to monitor and improve quality. By collecting and analyzing vast amounts of data from various stages of production and across different operational environments, organizations can gain deeper insights into their processes. This enables predictive analytics, where potential quality issues are identified and mitigated before they occur.

The automation of quality processes through AI and machine learning allows for real-time monitoring and rapid response to quality issues. Automated systems can perform continuous quality checks with minimal human intervention, reducing the likelihood of human error and ensuring consistent quality standards. IoT devices play a crucial role in Quality 4.0 by providing real-time data from connected machines and systems. This connectivity enables seamless communication between different components of the production process, allowing for more precise control and immediate correction of quality deviations. Quality 4.0 fosters a collaborative environment where information flows freely between departments, suppliers, and customers. Digital platforms and tools facilitate this collaboration by ensuring that all stakeholders have access to the same real-time data, which enhances transparency and fosters a culture of continuous improvement. One of the hallmarks of Quality 4.0 is its ability to quickly adapt to changes in the market or production environment. The integration of advanced technologies allows organizations to be more responsive to customer needs and regulatory changes, ensuring that quality standards are maintained even in dynamic conditions. With the enhanced ability to capture and analyze customer feedback through digital channels, Quality 4.0 places a stronger emphasis on meeting and exceeding customer expectations. This customer-centric approach ensures that products and services are aligned with market demands and quality benchmarks.

The adoption of Quality 4.0 has significant implications for both industry and academia. For industry, it presents an opportunity to redefine quality management practices in a way that aligns with the rapid technological advancements of Industry 4.0. For academia, it opens new avenues for research in areas such as the application of AI in quality management, the impact of digital transformation on quality practices, and the development of new quality models that incorporate the principles of Industry 4.0.

The implementation of Quality 4.0 within the framework of S-ERP for Industry 4.0 represents a transformative approach to quality management that aligns with the broader goals of sustainability, efficiency, and technological integration. This integration leverages advanced technologies to enhance quality processes while also supporting the sustainability objectives central to S-ERP. Implement IoT sensors across the production line and supply chain to collect real-time data on product quality, machine performance, and environmental conditions. This data feeds directly into the S-ERP system, allowing for continuous monitoring and immediate feedback loops. Use AI and machine learning algorithms within the S-ERP system to analyze data collected from IoT devices. These algorithms can predict potential quality issues, suggest corrective actions, and optimize processes based on historical and real-time data.

Integrate big data analytics capabilities within the S-ERP to process and analyze large datasets from various stages of production. This allows for advanced statistical analysis, identifying patterns and trends that can inform quality improvements. Use predictive analytics to foresee quality deviations before they occur. The S-ERP system can automatically adjust production parameters or alert operators to intervene, thereby minimizing waste and rework and enhancing overall product quality. Implement quality processes that prioritize resource efficiency. By integrating quality control with sustainability metrics within the S-ERP, the system can optimize resource usage, reduce waste, and ensure that production processes are both high quality and environmentally friendly. Use the S-ERP to conduct lifecycle assessments of products, ensuring that quality management extends beyond the factory floor to include the entire product lifecycle. This includes sourcing sustainable materials, reducing energy consumption, and managing end-of-life product disposal.

Leverage the S-ERP's capability to enhance collaboration across the supply chain. Ensure that all suppliers and partners have access to the same quality data, enabling synchronized quality standards and practices across the entire value chain. Incorporate customer feedback directly into the quality management processes within the S-ERP. Real-time feedback can be analyzed to improve product quality and ensure that customer expectations are met or exceeded. Implement flexible quality management protocols within the S-ERP that can adapt to changes in production requirements, market demands, or regulatory standards. This ensures that the organization can maintain high-quality standards even in a rapidly changing environment. Use the S-ERP to foster a culture of continuous improvement. The system should track and analyze quality performance over time, identifying areas for ongoing enhancement and ensuring that quality management is always evolving. Integrate automated compliance checks within the S-ERP. The system can continuously monitor production processes to ensure adherence to industry standards and regulatory requirements, automatically generating reports and alerting management to any non-compliance issues. Include sustainability metrics in quality reporting. The S-ERP can generate comprehensive reports that combine quality performance with sustainability goals, providing a holistic view of the organization's operations and supporting strategic decision-making.

Implementation Strategy:

1. **Assess Current Capabilities:** Begin by assessing the current state of quality management practices and the existing S-ERP system. Identify gaps and opportunities where Quality 4.0 principles can be integrated.
2. **Develop a Roadmap:** Create a detailed implementation roadmap that outlines the steps required to integrate Quality 4.0 into the S-ERP system. This roadmap should include timelines, resource allocation, and key performance indicators (KPIs) for success.
3. **PiIoT Implementation:** Start with a piIoT project in a specific area of the organization to test the integration of Quality 4.0 with S-ERP. Use this piIoT to refine the approach, address any challenges, and gather data to support broader implementation.
4. **Full-Scale Deployment:** Once the piIoT is successful, scale the implementation across the entire organization. Ensure that all stakeholders are trained and that the S-ERP system is fully integrated with Quality 4.0 technologies and practices.
5. **Continuous Monitoring and Improvement:** Regularly monitor the performance of the integrated system, using the data to continuously improve both quality and sustainability outcomes. Adapt the implementation as needed to respond to new challenges and opportunities.

The following section discusses advanced topics such as IoT Analytics as a service and the movement from IoT analytics to IIoT analytics.

7.6 IoT Analytics as a Service and the Movement from IoT Analytics to IIoT Analytics

IoT Analytics as a Service (IoT-AaaS) refers to a cloud-based service model that provides organizations with the tools and platforms to collect, store, analyze, and visualize data generated by Internet of Things (IoT) devices [13]. Rather than investing in and maintaining the necessary infrastructure in-house, businesses can access IoT analytics capabilities through service providers, allowing them to leverage advanced analytics without the complexities of managing the underlying technology. IoT-AaaS platforms are designed to seamlessly integrate with a variety of IoT devices and sensors, collecting data from diverse sources and formats. The service typically includes tools for data ingestion, integration, and preprocessing to ensure that the data is ready for analysis.

These services provide scalable cloud storage solutions capable of handling large volumes of data generated by IoT devices. The storage systems are designed to accommodate the high-velocity, high-volume nature of IoT data, ensuring that data can be accessed and retrieved quickly when needed. IoT-AaaS platforms come equipped with advanced analytics tools, including machine learning algorithms, statistical analysis, predictive modeling, and real-time data processing. These tools allow organizations to derive actionable insights from their IoT data. The service includes dashboards and visualization tools that enable users to explore IoT data interactively. Customized reports can be generated to track key performance indicators (KPIs) and other metrics relevant to the business. IoT-AaaS providers ensure that data is secure and that the service complies with relevant data protection regulations. This includes encryption, access control, and monitoring to protect sensitive data. IoT-AaaS can be deployed in various models, including public cloud, private cloud, or hybrid configurations, depending on the organization's needs and regulatory requirements.

S-ERP systems are increasingly being integrated with IoT technologies to enhance operational efficiency, sustainability, and decision-making. Implementing IoT-AaaS within an S-ERP framework for Industry 4.0 requires a strategic approach to ensure that the system can fully leverage the capabilities of IoT analytics. We present nine implementation steps. The nine steps are as follows:

Step 1 Define Objectives and Use Cases:
Begin by identifying the specific objectives and use cases for IoT analytics within the S-ERP system. Common use cases might include predictive maintenance, energy management, real-time monitoring of production processes, and supply chain optimization. Clearly defining these objectives will guide the selection of appropriate IoT-AaaS platforms and tools.

Step 2 Select an IoT-AaaS Provider:
Choose a service provider that offers IoT analytics capabilities aligned with your organization's needs. Consider factors such as the provider's experience with Industry 4.0 applications, scalability, data security, integration capabilities, and cost-effectiveness. Providers like AWS IoT Analytics, Microsoft Azure IoT, and Google Cloud IoT offer comprehensive solutions that can be tailored to S-ERP requirements.
Step 3 Integration of IoT devices and Sensors:
Integrate IoT devices and sensors across the enterprise's operations, such as manufacturing floors, supply chain logistics, and energy management systems. Ensure that these devices are configured to transmit data to the IoT-AaaS platform securely and reliably.

Step 4 Data Ingestion and Preprocessing:
Set up data ingestion pipelines to collect data from IoT devices in real time. The IoT-AaaS platform will handle data preprocessing, including cleaning, filtering, and transforming data into formats suitable for analysis. This step is crucial to ensure that the data is of high quality and ready for analytics.

Step 5 Implement Advanced Analytics and Machine Learning Models:
Leverage the advanced analytics tools provided by the IoT-AaaS platform to analyze IoT data. Implement machine learning models for predictive maintenance, anomaly detection, demand forecasting, and other relevant applications. These models can be continuously refined and improved as more data is collected.

Step 6 Integrate IoT Analytics with S-ERP:
Integrate the IoT-AaaS platform with the existing S-ERP system. This integration enables the seamless flow of IoT data into the S-ERP, allowing for real-time updates and automated decision-making processes. The S-ERP can then use this data to optimize resource allocation, reduce waste, and improve overall operational efficiency.

Step 7 Develop Custom Dashboards and Reports:
Use the visualization tools within the IoT-AaaS platform to create custom dashboards that display key metrics and KPIs relevant to the S-ERP system. These dashboards should be accessible to decision-makers and operators, providing them with real-time insights and actionable information.

Step 8 Ensure Security and Compliance:
Implement robust security measures to protect IoT data, including encryption, access controls, and regular security audits. Ensure that the IoT-AaaS implementation complies with industry-specific regulations and standards, particularly in sectors with stringent data protection requirements.

Step 9 Monitor, Evaluate, and Optimize:
Continuously monitor the performance of the IoT-AaaS implementation. Use the data collected to evaluate the effectiveness of the analytics models and make necessary adjustments. Optimization should be an ongoing process, with the system evolving as more data becomes available and business needs change.

Implementing IoT Analytics as a Service within an S-ERP framework for Industry 4.0 enables organizations to harness the power of real-time data and advanced analytics to drive operational efficiency, sustainability, and innovation. By leveraging IoT-AaaS, businesses can focus on deriving insights from their IoT data without the burden of managing complex analytics infrastructure, ensuring that they remain competitive in the rapidly evolving Industry 4.0 landscape. The following section discusses the impact of 5G and 6G technologies on IoT analytics.

7.7 Impact of 5G and 6G Technologies on IoT Analytics of S-ERP in Industry 4.0

5G and 6G technologies represent the fifth and sixth generations of mobile network connectivity, each offering progressively enhanced features to support the growing demands of modern communication systems [10]. Fifth-generation (5G) networks provide significantly faster data transmission speeds, lower latency, and higher capacity than 4G. These advancements enable the seamless integration of Internet of Things (IoT) devices, supporting applications that require real-time data processing and ultrareliable communication. Industries leveraging IoT benefit from 5G's ability to handle massive device connectivity and ensure reliable, low-latency communication, which is essential for applications like autonomous vehicles, smart factories, and telemedicine.

Sixth-generation (6G) networks, though still in the conceptual and early research stages, are expected to offer even greater enhancements, including terahertz frequency communication, advanced AI-driven network management, and significantly improved data rates. 6G is anticipated to support even more sophisticated IoT applications, enabling innovations such as holographic communications, advanced robotics, and ubiquitous computing. These networks will be integral in realizing the full potential of IoT in industries, fostering deeper integration and automation, and enabling entirely new paradigms in industrial operations. Both 5G and 6G technologies are pivotal in advancing the capabilities of IoT in industrial settings, with 5G currently driving IoT adoption and 6G poised to unlock future innovations.

IoT analytics in the context of 5G and 6G technologies, when implemented in S-ERP for Industry 4.0, plays a transformative role in optimizing industrial operations. The implementation of 5G in S-ERP systems enables real-time analytics by providing ultralow latency and high-speed data transmission. This allows for the continuous collection and processing of data from IoT devices across an industrial environment. With 5G, S-ERP systems can effectively manage large-scale data from connected sensors, machines, and devices, enabling predictive maintenance, real-time monitoring, and optimization of supply chains. The high bandwidth of 5G supports the integration of AI-driven analytics within S-ERP, leading to more accurate and timely decision-making processes.

With the advent of 6G, IoT analytics in S-ERP systems will become even more advanced, leveraging terahertz communication and AI-native networks. 6G's higher data rates and broader spectrum will facilitate the handling of even larger data volumes, enabling more sophisticated analytics. This will allow S-ERP systems to integrate advanced IoT applications like digital twins, immersive experiences, and ultra-precise robotics. The real-time, AI-enhanced analytics possible with 6G will lead to unparalleled levels of automation, efficiency, and sustainability in Industry 4.0 environments, ensuring that operations are optimized and adaptive to changing conditions in real time.

When 5G and 6G technologies are implemented in S-ERP for Industry 4.0, the architectures for IoT analytics must be designed to exploit the capabilities of these advanced communication networks fully. The brief overview of the architectures is as follows:

1. Edge-Fog-Cloud Hybrid Architecture:
In this architecture, IoT devices and sensors operate at the edge of the network, directly connected to local edge computing nodes. The low latency of 5G and 6G allows for real-time processing at the edge, where preliminary data filtering, aggregation, and basic analytics are performed. This reduces the data load sent to higher layers and enables quick responses to time-sensitive tasks. The fog layer acts as an intermediary situated between the edge and the cloud. It leverages 5G/6G for efficient data transfer and performs more complex analytics closer to where the data is generated. The fog layer is critical for tasks requiring more computational power than the edge can provide but still demands low latency, such as dynamic resource allocation or predictive maintenance analytics. The cloud provides centralized, high-capacity storage and advanced analytics capabilities. Data from the fog layer is aggregated and sent to the cloud for deeper analysis, long-term storage, and integration with other enterprise data within the S-ERP system. The high-speed and high-capacity nature of 5G/6G ensures that even large datasets can be transmitted to the cloud quickly for more comprehensive analytics and AI-driven insights.

2. AI-Driven Decentralized Architecture:
Leveraging the distributed nature of 5G and future 6G networks, IoT analytics can be decentralized across various nodes, each equipped with AI capabilities. This architecture allows different nodes in the S-ERP system to independently analyze and act upon data, reducing the need for centralized control and enabling real-time, location-specific optimizations. In a 6G-enhanced S-ERP environment, multiple edge nodes equipped with AI collaborate in real time to share insights and predictions. This collaborative intelligence allows the entire system to operate more efficiently by distributing processing loads and minimizing the need for extensive data movement. To ensure data integrity and secure transactions across the decentralized system, blockchain technology can be integrated into the architecture. The immutability and transparency of blockchain support trust in distributed analytics, which is crucial for sensitive industrial operations.

3. Digital Twin-Enhanced Architecture:
A digital twin is a virtual model of physical assets, processes, or systems used for real-time monitoring and simulation. In a 5G/6G-powered S-ERP architecture, digital twins can be created for various industrial system components, offering real-time analytics and predictive insights. The high-speed data transmission capabilities of 5G/6G allow for continuous synchronization between the physical and digital twins. This enables real-time simulation and predictive analytics, where any change in the physical environment is immediately reflected in the digital twin and vice versa, optimizing operations and reducing downtime.

4. AI and ML-Enhanced Cloud-Native Architecture:
In this architecture, IoT analytics is primarily processed in a cloud-native environment, where the cloud is the central hub for AI and machine learning (ML) algorithms. The architecture is designed to scale easily with the growing data demands of Industry 4.0, utilizing the high-speed connectivity of 5G and 6G to handle large-scale data efficiently. The architecture supports AutoML (automated machine learning) capabilities, where models are continuously trained and updated based on the latest data from IoT devices. This ensures that the S-ERP system remains adaptive and can optimize operations dynamically.

5. Service-Oriented Architecture (SOA) with Microservices:
IoT analytics can be managed through a microservices architecture, where individual services handle specific analytics tasks. These microservices communicate through APIs and can be deployed across different layers (edge, fog, cloud) depending on the required latency and processing power. This architecture is highly scalable and modular, allowing different services to be updated or replaced without affecting the entire system. The implementation of 5G and 6G enhances the communication between these microservices, ensuring efficient data flow and rapid processing.

Overall, the architectures for IoT analytics in S-ERP for Industry 4.0, when integrated with 5G and 6G technologies, emphasize real-time processing, distributed intelligence, and advanced AI capabilities. These architectures are designed to leverage the high-speed, low-latency, and large-scale connectivity offered by 5G and 6G, enabling industries to optimize operations, reduce costs, and enhance sustainability. In summary, the chapter answers the following research questions.

Research Questions Addressed
The chapter titled"Analytics for Harnessing IoT and Industry 4.0: Data Flows in S-ERP" addresses the following key research questions:

1. **How can IoT analytics be integrated into Sustainable Enterprise Resource Planning (S-ERP) systems within Industry 4.0?**

- The chapter explores integrating IoT-generated data with S-ERP systems to support sustainable practices and enhance operational efficiency, decision-making, and sustainability across industrial operations.

2. **What are IoT data analytics lifecycle phases in the context of S-ERP for Industry 4.0?**

- The chapter outlines the phases involved in the IoT data analytics lifecycle, including data generation, collection, transformation, analysis, deployment, and operational steps, highlighting how each phase contributes to actionable insights and operational improvements in Industry 4.0.

3. **What are the applications of IoT analytics in smart metering, smart transportation, and smart logistics within Industry 4.0?**

- The chapter examines specific applications of IoT analytics in areas such as smart metering, smart transportation, and smart logistics, detailing how these applications enhance efficiency, sustainability, and customer satisfaction within S-ERP systems in Industry 4.0.

4. **How do IoT analytics contribute to Quality 4.0 within the S-ERP and Industry 4.0 framework?**

- The chapter discusses the role of IoT analytics in supporting Quality 4.0 by integrating real-time data from IoT devices to monitor and improve quality processes, ensuring that products and services meet higher quality and sustainability standards.

5. **How can IoT analytics be offered as a service (IoT-AaaS), and what are the implications for S-ERP systems in Industry 4.0?**

- The chapter explores the concept of IoT Analytics as a Service (IoT-AaaS), describing how organizations can leverage cloud-based IoT analytics platforms to enhance their S-ERP systems without complex infrastructure, focusing on implementation steps and benefits.

6. **How do 5G and 6G technologies impact IoT analytics in S-ERP for Industry 4.0?**

- The chapter analyzes how the advancements in 5G and 6G technologies will transform loT analytics by enabling real-time processing, enhanced connectivity, and more sophisticated data analysis, optimizing industrial operations and supporting advanced loT applications in S-ERP systems.

These research questions guide the chapter's in-depth exploration of how IoT analytics can be harnessed within S-ERP systems to achieve greater efficiency, sustainability, and innovation in the context of Industry 4.0.

References

1. Building Blocks for IoT Analytics Internet-of-Things Analytics. https://www.routledge.com/Building-Blocks-for-IoT-Analytics-Internet-of-Things-Analytics/Soldatos/p/book/9788770044493
2. Anjaria K (2024) Enhancing sustainability integration in Sustainable Enterprise Resource Planning (S-ERP) system: application of transaction cost theory and case study analysis. Int J Inf Manage Data Insights 4(2):100243. ISSN 2667-0968. https://doi.org/10.1016/j.jjimei.2024.100243. https://www.sciencedirect.com/science/article/pii/S2667096824000326
3. Bi Z, Jin Y, Maropoulos P, Zhang W-J, Wang L (2013) Internet of things (IoT) and big data analytics (BDA) for digital manufacturing (DM). Int J Product Res 61(12):4004–4021. ISSN 0020-7543. https://doi.org/10.1080/00207543.2021.1953181. https://doi.org/10.1080/00207543.2021.1953181. Publisher: Taylor and Francis _eprint: https://doi.org/10.1080/00207543.2021.1953181
4. Chataut R, Phoummalayvane A, Akl R (2023) Unleashing the power of IoT: a comprehensive review of IoT applications and future prospects in healthcare, agriculture, smart homes, smart cities, and Industry 4.0. Sensors 23(16):7194. ISSN 1424-8220. https://doi.org/10.3390/s23167194. https://www.mdpi.com/1424-8220/23/16/7194. Number: 16 Publisher: Multidisciplinary Digital Publishing Institute
5. Dahlqvist F, Patel M, Rajko A, Shulman J, Growing opportunities in the Internet of Things
6. Nilashi M, Baabdullah AM, Abumalloh RA, Ooi KB, Tan GW, Giannakis M, Dwivedi YK (2023) How can big data and predictive analytics impact the performance and competitive advantage of the food waste and recycling industry? Ann Operat Res. ISSN 1572–9338. https://doi.org/10.1007/s10479-023-05272-y
7. Qi Q, Xu Z, Rani P (2023) Big data analytics challenges to implementing the intelligent Industrial Internet of Things (IIoT) systems in sustainable manufacturing operations. Technol Forecast Soc Change 190:122401. ISSN 0040-1625. https://doi.org/10.1016/j.techfore.2023.122401. https://www.sciencedirect.com/science/article/pii/S0040162523000860
8. Rath KC, Khang A, Roy D (2024) The role of Internet of Things (IoT) technology in Industry 4.0 economy. In Advanced IoT technologies and applications in the Industry 4.0 digital economy. CRC Press, p 28. ISBN 978-1-00-343426-9
9. Rosati R, Romeo L, Cecchini G, Tonetto F, Viti P, Mancini A, Frontoni E (2023) From knowledge-based to big data analytic model: a novel IoT and machine learning based decision support system for predictive maintenance in Industry 4.0. J Intell Manufact 34(1):107–121. ISSN 1572-8145. https://doi.org/10.1007/s10845-022-01960-x. https://doi.org/10.1007/s10845-022-01960-x
10. Rossi MA (2024) EU technology-specific industrial policy. The case of 5G and 6G. Telecommun Policy 48(2):102639. ISSN 0308-5961. https://doi.org/10.1016/j.telpol.2023.102639. https://www.sciencedirect.com/science/article/pii/S0308596123001507
11. Singh J, Ahuja IS, Singh H, Singh A (2023) Application of Quality 4.0 (Q4.0) and Industrial Internet of Things (IIoT) in agricultural manufacturing industry. AgriEngineering 5(1):537–565. ISSN 2624-7402. https://doi.org/10.3390/agriengineering5010035. https://www.mdpi.com/2624-7402/5/1/35. Number: 1 Publisher: Multidisciplinary Digital Publishing Institute
12. Soori M, Arezoo B, Dastres R (2023) Internet of things for smart factories in industry 4.0, a review. Int Things Cyber-Phys Syst 3:192–204. ISSN 2667-3452. https://doi.org/10.1016/j.iotcps.2023.04.006. https://www.sciencedirect.com/science/article/pii/S2667345223000275
13. Xu X, Huang S, Chen Y, Browny K, Halilovicy I, Lu W (2014) TSAaaS: time series analytics as a service on IoT. In: 2014 IEEE international conference on web services, pp 249–256. https://doi.org/10.1109/ICWS.2014.45. https://ieeexplore.ieee.org/abstract/document/6928905
14. Yesodha KRK, Jagadeesan A, Logeshwaran J (2023) IoT applications in modern supply chains: enhancing efficiency and product quality. In: 2023 IEEE 2nd international conference on industrial electronics: developments & applications (ICIDeA), pp 366–371. https://doi.org/10.1109/ICIDeA59866.2023.10295273. https://ieeexplore.ieee.org/abstract/document/10295273

Chapter 8
Role of Sustainable Enterprise Resource Planning (S-ERP) in Digital Transformation

8.1 Introduction

Digital transformation in an organization refers to the integration of digital technologies into all areas of business, fundamentally changing how the organization operates and delivers value to customers. It involves the adoption of emerging technologies such as cloud computing, artificial intelligence, and big data to streamline processes, enhance efficiency, and foster innovation. Beyond technological implementation, digital transformation requires a cultural shift where the organization embraces continuous learning, agility, and customer-centric approaches. It often leads to the redefinition of business models, products, and services, enabling the organization to remain competitive in an increasingly digital economy.

The above paragraph and definition may indicate to readers that digital transformation is similar to business process reengineering, as discussed in Chap. 2 of the present work. However, it is noteworthy that digital transformation and business process reengineering (BPR) are distinct concepts, though they share some similarities in their goals of improving organizational efficiency and effectiveness. Digital transformation involves the holistic integration of digital technologies across all facets of an organization, leading to a fundamental change in how the organization operates, engages with customers, and competes in the market. It is a continuous process that requires not only technological adoption but also cultural and organizational change. On the other hand, business process reengineering focuses on the radical redesign of specific business processes to achieve dramatic improvements in critical performance metrics such as cost, quality, service, and speed. BPR is often a one-time, top-down approach aimed at optimizing existing processes, whereas digital transformation is an ongoing, broader initiative that encompasses not only process improvement but also innovation and the creation of new business models. Thus, while both aim to enhance organizational performance, digital transformation is broader in scope, encompassing more than just process improvement, whereas BPR is specifically targeted at rethinking and redesigning existing processes.

K. Anjaria, *Sustainable Enterprise Resource Planning (S-ERP) for Industry 4.0*, Contributions to Management Science, https://doi.org/10.1007/978-981-96-7734-4_8

Further, readers may also think that digital transformation is very similar to change management and roadmapping concepts discussed in Chaps. 3 and 4. However, interestingly, digital transformation is distinct from roadmapping and change management, although these concepts are interrelated. Digital transformation refers to the comprehensive integration of digital technologies across an organization, fundamentally altering its operations, business models, and customer interactions. It is a broad, strategic initiative that encompasses the entire organization and drives long-term change. Roadmapping, on the other hand, is a planning tool used to outline the steps, timelines, and resources required to achieve specific strategic goals. In the context of digital transformation, roadmapping helps to create a clear path for implementing digital initiatives, but it is a component of the broader transformation process rather than a transformation itself. Change management refers to the systematic approach to dealing with the transition or transformation of an organization's goals, processes, or technologies. It focuses on managing the human aspects of change, ensuring that employees adapt to new ways of working and that the organization can effectively implement and sustain changes. While roadmapping and change management are essential elements that support digital transformation, they are distinct in their focus. Roadmapping provides a structured plan, and change management facilitates the human and organizational transition, whereas digital transformation encompasses the overall shift toward a digitally driven organization.

Digital transformation significantly impacts the value chain by enhancing efficiency, flexibility, and connectivity across all stages, from procurement to customer service. Digital transformation reshapes the value chain by enabling real-time data sharing, automating processes, and integrating advanced technologies such as artificial intelligence and IoT. These advancements optimize operations, reduce costs, and improve decision-making. For example, in procurement, digital tools streamline supplier management and sourcing. In production, automation and data analytics enhance quality and reduce downtime. In distribution, digital platforms improve logistics and inventory management. Finally, in customer service, personalized digital interactions increase customer satisfaction.

Overall, the literature suggests that digital transformation is not a uniform process but rather a strategic choice that organizations make, leading to diverse outcomes. Our argument here is digital transformation is a complex procedure for most organizations, and S-ERP for Industry 4.0 can provide an exciting approach to performing and supporting digital transformation for the business. The shift from traditional product-based models to platform-based models presents a significant challenge. While platforms offer advantages like network effects and data-driven insights, they also require a different strategic approach, focusing on attracting users and facilitating transactions. Established companies may struggle to adapt their existing resources and capabilities to succeed in the platform economy. The literature mentions that "Charting strategy for digital transformation in such firms is complicated," highlighting the difficulty that established companies face in navigating this tension.

The second type of complexity is the firm vs ecosystem. The rise of ecosystems, where multiple actors collaborate to deliver value, adds another layer of complexity. While ecosystems offer access to new capabilities and markets, they require firms to navigate complex interdependencies and potential conflicts. The final layer of the complexity revolves around the people aspect. Adopting digital tools like AI and automation can improve efficiency and decision-making. However, it also raises concerns about job displacement and the need for new skills. Thus, "digital transformation is a broad and complex phenomenon that does not fit easily into any given theory" further reinforces the argument that it is a complex and demanding process for most organizations.

S-ERP integrates advanced technologies like IoT, AI, and big data analytics, which are central to Industry 4.0, enabling seamless automation, real-time data processing, and intelligent decision-making. By adopting S-ERP, organizations can streamline operations, enhance sustainability, and adapt to dynamic market demands, all of which are critical aspects of digital transformation. S-ERP thus acts as a foundational platform that supports and drives the broader digital transformation initiatives within an organization. The following section discusses how to find the digital blind spot.

8.2 Finding Digital Blind Spots for Digital Transformation Using S-ERP in Industry 4.0

Finding digital blind spots is crucial in digital transformation because they represent areas within an organization's digital ecosystem that are overlooked or insufficiently monitored, potentially leading to significant risks and inefficiencies. These blind spots can impede the effectiveness of digital initiatives by introducing vulnerabilities that may be exploited, reducing operational resilience, and undermining data integrity. Additionally, they can obscure opportunities for optimization, resulting in suboptimal resource allocation and reduced competitive advantage. Identifying and addressing digital blind spots ensures a more comprehensive and secure transformation, aligning digital capabilities with strategic objectives and enhancing overall organizational performance.

The present section discusses **three ways** to identify digital blind spots. **The first way** that we present is through Computer Vision (CV). It is evident that blind spots in a business can manifest in various forms, such as unmonitored interfaces between different modules, undetected changes in user interfaces (UIs), or unrecognized elements within digital content-based applications.

In the context of Industry 4.0, digital blind spots may arise from several factors, including the limitations of standard automation toolkits that fail to detect certain elements on non-standard devices or complex desktop applications. For instance, traditional automation tools are well-suited for automating web and mobile applications but may struggle with detecting elements in content-rich applications, such as mobile games or custom desktop software. This gap in detection capability can lead to unaddressed vulnerabilities, thereby creating digital blind spots that compromise the effectiveness of business processes.

To address these challenges, integrating Computer Vision (CV) technology into S-ERP can enhance digital blind spot detection by replicating human visual capabilities through deep learning [21]. CV can analyze screenshots or images of the application interface, identify elements within these images, and determine their spatial orientation. This allows the system to detect UI changes, recognize objects that standard tools might miss, and automate the testing of non-standard devices and applications. For example, CV can be used to detect UI changes in complex applications, ensuring that the most current components are in use and preventing potential mismatches or errors that might go unnoticed by traditional methods. However, implementing CV within S-ERP is not without challenges. It requires additional research, specialized datasets, and customized algorithms, as well as ongoing monitoring to ensure accuracy. The complexity of CV algorithms and their dependency on image quality can also lead to inaccuracies, particularly in dynamic environments where UI elements frequently change.

The second way to find the digital blind spot is Root Cause Analysis (RCA). RCA helps in pinpointing the underlying factors that contribute to operational inefficiencies by drilling down through various variables and sub-variables. For example, if a production delay is identified, RCA can be used to determine whether the issue stems from external factors like vendor delivery failures or internal factors such as communication breakdowns between departments. Through this analysis, S-ERP can identify patterns and correlations that may indicate the presence of digital blind spots, such as an over-reliance on specific personnel or outdated processes. To effectively manage digital blind spots, S-ERP systems must incorporate technology that facilitates real-time data capture and sharing. By implementing collaboration and task management software, organizations can ensure that data is accessible to all relevant stakeholders, reducing the likelihood of blind spots going undetected. Additionally, defining and refining algorithms based on Key Result Areas (KRAs) and Key Performance Indicators (KPIs) allows for continuous monitoring and optimization of processes [19], thereby minimizing operational risks. The use of S-ERP not only aids in blind spot detection but also enhances Business Process Reengineering (BPR) by providing actionable insights into existing gaps. This ongoing analysis enables organizations to adapt quickly to changing conditions, maintain compliance with industry standards, and achieve certifications like CMMI by demonstrating effective process management.

The third and final way is to find the past practices on which businesses over-relied. In the context of S-ERP, digital blind spots can emerge as the organization scales and integrates more sophisticated technologies. These blind spots may be the result of over-reliance on past practices, a lack of updated monitoring tools, or insufficient feedback mechanisms within the system. As leaders grow more experienced, there is a risk of becoming complacent, assuming that their knowledge and past successes are sufficient to navigate current challenges. This assumption can lead to the broadening of digital blind spots, particularly as new technologies and processes are introduced into the S-ERP framework.

To effectively detect and manage these digital blind spots, a multifaceted approach is necessary:

1. **Awareness:** The first step in addressing digital blind spots is recognizing their existence. Leaders and stakeholders must acknowledge that even the most robust S-ERP systems are not immune to oversight. This awareness should be coupled with a willingness to critically assess the current state of digital integration and identify potential gaps in the system.
2. **Feedback Mechanisms:** Just as leaders benefit from seeking feedback on their personal blind spots, organizations should establish robust feedback loops within their S-ERP systems. These mechanisms allow for continuous monitoring and real-time data sharing across departments, ensuring that no aspect of the digital infrastructure is overlooked. Encouraging open communication and input from various stakeholders can help in identifying areas where the system may be vulnerable or underperforming.
3. **Assessment Tools:** Utilizing advanced analytics and assessment tools within S-ERP can provide valuable insights into areas that may be prone to digital blind spots. For example, automated audits, performance metrics, and machine learning algorithms can help to pinpoint inefficiencies or inconsistencies within the system. These tools offer a "second opinion," revealing potential issues that may not be immediately apparent through traditional monitoring methods.
4. **Accountability Structures:** Establishing accountability within the S-ERP framework is essential for maintaining long-term effectiveness. Leaders should empower specific individuals or teams to act as "digital back seat drivers," responsible for overseeing the system's performance and alerting leadership to any emerging blind spots. This accountability ensures that digital blind spots are not only identified but also promptly addressed, minimizing the risk of operational disruptions.

By implementing these strategies, S-ERP systems in Industry 4.0 can achieve more comprehensive digital blind spot detection, leading to improved operational outcomes and sustained organizational success. Just as continuous advancements in vehicle technology have enhanced safety, the proactive management of digital blind spots within S-ERP will ensure that organizations can navigate the complexities of Industry 4.0 with greater confidence and agility.

Once the blind spots are detected, the digital transformation affects three business domains, as shown in Fig. 8.1. The following sections discuss these three domains and how S-ERP helps achieve digital transformation in these domains in detail. In the next section, we discuss customer-centric digital transformation with the help of S-ERP.

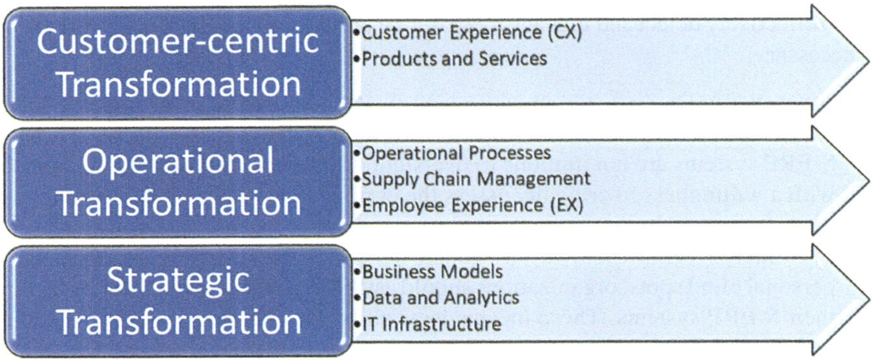

Fig. 8.1 Business Domains of Digital Transformation

8.3 Customer-Centric Digital Transformation with S-ERP in Industry 4.0

In the context of Industry 4.0, customer experience (CX) and customer-centric digital transformation are increasingly being supported and enhanced by S-ERP systems. These systems integrate advanced technologies to create flexible, responsive, customer-focused production environments. In Industry 4.0, CX is redefined through the integration of technologies such as Cyber-Physical Systems (CPS), Industrial Internet of Things (IIoT), and Big Data Analytics (BDA). These technologies enable businesses to understand better and meet the diverse needs of their customers by providing personalized and real-time interactions throughout the customer journey. For instance, CPS allows for the dynamic adjustment of product specifications, extending the period during which customers can modify their orders, thus enhancing the overall customer experience. Additionally, S-ERP systems support these capabilities by providing a unified platform that integrates data across various functions, enabling a seamless flow of information and more personalized customer interactions.

As we advance in the integration of S-ERP systems within Industry 4.0, we are witnessing a significant shift toward an "everyone-to-everyone" (E2E) economy [5]. This emerging economic model is characterized by the democratization of production and consumption, where the boundaries between producers, consumers, and service providers blur, enabling a more inclusive and participatory marketplace.

In the E2E economy, S-ERP systems play a pivotal role by facilitating real-time data exchange and collaboration across the entire value chain. These systems empower all stakeholders—customers, suppliers, manufacturers, and service providers—to engage directly with one another, creating a networked ecosystem where value can be co-created and shared more equitably. For instance, customers are no longer passive recipients of products and services; instead, they actively participate in the design, customization, and production processes enabled by the connectivity and transparency of S-ERP systems.

In the context of Industry 4.0, S-ERP facilitates a comprehensive transformation of customer-centric products and services, which unfolds across both primary and secondary levels [4]. The primary level encompasses a series of strategic activities and initiatives that organizations can employ to effectively engage with customers and other stakeholders. The first phase at the primary level involves engaging customers through mobility solutions and enhanced collaboration via digital mechanisms. This phase emphasizes the utilization of digital platforms and technologies to facilitate real-time interactions with customers, offering mobility solutions that ensure seamless and efficient communication. The second phase focuses on customer support, co-creation of value, predictive marketing, and digitally enhanced selling. This phase underscores the importance of providing robust support across multiple channels, actively involving customers in the value-creation process, leveraging analytics for predictive marketing, and optimizing sales processes through digital tools. In the third phase, the emphasis shifts to the development of novel product designs through collaboration with partners, efficient handling of transactional details, and digital onboarding processes. These activities highlight the critical role of external partner collaboration in crafting products that align with customer needs, alongside the importance of streamlined communication and onboarding practices. The final phase of the primary level is dedicated to ensuring seamless device interoperability for stakeholders, facilitating collaborative efforts across various digital platforms. This phase is crucial in enabling all stakeholders, including customers and partners, to interact and collaborate efficiently, regardless of the devices or platforms they are using.

The secondary level encompasses broader thematic phases that synthesize the foundational activities articulated in the primary level. These themes offer a strategic lens through which the overall process of customer-centric digital transformation can be understood and implemented. The first phase at the secondary level involves enhancing the customer experience from an outside-in perspective. This phase underscores the criticality of comprehending and addressing customer needs by systematically integrating external feedback and insights into the design and delivery processes of products and services. The second phase focuses on reaching and engaging customers. This phase is concerned with the methodologies by which organizations can effectively connect with and engage their customers, ensuring that their experiences remain consistent, cohesive, and positive across all interaction points. The third phase shifts the focus toward reaching and engaging other stakeholders, particularly external partners. Acknowledging the essential role that partners play in delivering customer-centric products and services, this phase emphasizes the necessity of active engagement with partners to co-create value and optimize the holistic customer experience. The final phase at the secondary level involves blending physical and digital experiences. This phase highlights the importance of seamlessly integrating physical and digital interactions, thereby ensuring that customers and stakeholders can transition effortlessly between these environments, resulting in a cohesive and unified experience.

At the highest level, these levels converge into an aggregate dimension focused on external stakeholder-centric activities. This level encapsulates the overarching goal of aligning all organizational activities and processes with the needs and expectations of external stakeholders, particularly customers and partners. In the context of S-ERP for Industry 4.0, achieving customer-centric products and services involves leveraging advanced digital technologies to create a seamless, responsive, and collaborative environment. By integrating customer feedback, enhancing communication, and enabling seamless interactions across digital and physical platforms, organizations can ensure that their products and services are tailored to meet the evolving demands of their customers. Furthermore, engaging partners in this process ensures the entire value chain is optimized for delivering superior customer experiences, thereby driving successful digital transformation in the Industry 4.0 landscape. The following section discusses operational digital transformation.

8.4 Operational Digital Transformation with S-ERP in Industry 4.0

The operational digital transformation can be divided into **three components** from the operational process perspective. **The first component** is sensing. Sensing plays a critical role in enabling organizations to adapt to dynamic environments. Sensing, as a fundamental capability, involves the continuous monitoring of both internal and external conditions, which is essential for identifying emerging trends, potential disruptions, and opportunities for innovation. This capability is further subdivided into key activities, each contributing to the overall effectiveness of the sensing process. The first subcomponent, Monitoring, involves the systematic observation of various data sources, including market trends, technological advancements, and changes in customer behavior. This activity ensures that the organization remains vigilant and responsive to the evolving landscape, thereby positioning itself to anticipate and react to shifts that could impact its operations. Identifying is the next critical activity within the sensing capability. It focuses on pinpointing specific opportunities or threats that have been uncovered through the monitoring process. This identification process requires a deep understanding of the organization's strategic goals, as well as an ability to discern which signals from the environment are most relevant and warrant further exploration. Following identification, the Formulation activity is crucial for developing hypotheses or strategic options based on the identified signals. This involves crafting potential scenarios or action plans that the organization can pursue in response to the detected opportunities or threats. Formulating provides a structured approach to converting raw data into actionable insights that align with the organization's objectives. Collectively, these sensing activities—monitoring, identifying, and formulating—serve as the foundation for a robust digital transformation capability. By effectively leveraging these processes, organizations utilizing S-ERP in Industry 4.0 can enhance their agility, improve decision-making, and maintain a competitive edge in an increasingly digital and interconnected world.

The second component is organizing. In the realm of operational digital transformation, particularly within the framework of S-ERP in Industry 4.0, organizing plays a pivotal role in structuring and aligning resources to effectively implement transformation strategies. The organizing capability involves a systematic approach to coordinating and integrating various components of the organization to ensure that digital transformation efforts are both coherent and aligned with overarching strategic objectives. The first subcomponent of organizing is Aligning, which refers to the process of ensuring that all organizational activities, resources, and strategies are in harmony with the digital transformation goals. Aligning involves the careful coordination of departmental objectives, ensuring that each unit within the organization is moving toward the same strategic direction. This alignment is critical for eliminating silos and fostering a unified approach to digital transformation. Sourcing is the next essential activity within the organizing capability. This involves the acquisition of the necessary resources, including technology, talent, and partnerships, required to support the digital transformation initiatives. Sourcing ensures that the organization has access to the right tools, expertise, and external relationships that are crucial for the successful execution of its digital transformation strategies. Effective sourcing is particularly important in the rapidly evolving technological landscape of Industry 4.0, where timely access to cutting-edge resources can significantly impact the outcome of transformation efforts.

Following sourcing, the Combining activity focuses on the integration of sourced resources into the existing organizational framework. This involves the careful assembly of new technologies, processes, and capabilities with the current operations to create a cohesive and efficient system. Combining is crucial for ensuring that newly acquired resources are effectively utilized and that they enhance rather than disrupt existing workflows. This integrative approach is essential for maximizing the value derived from digital transformation efforts and ensuring that the organization can operate smoothly during the transition to more advanced digital capabilities.

Together, these organizing activities, sourcing, and combining form the backbone of a successful operational digital transformation. Organizations can leverage S-ERP in Industry 4.0 to create a more agile, efficient, and competitive operational framework by effectively coordinating resources and ensuring alignment with strategic goals.

The third component of operational digital transformation from the operational process perspective is restructuring. In the context of operational digital transformation within Industry 4.0, supported by S-ERP systems, the concept of restructuring is integral to adapting and evolving organizational processes to meet the demands of a rapidly changing technological environment. Restructuring involves reconfiguring existing organizational structures, processes, and practices to enhance efficiency, innovation, and responsiveness. This capability is composed of several subcomponents, each playing a critical role in the overall transformation process.

The first subcomponent, Innovating, focuses on fostering a culture of innovation within the organization. This involves encouraging the development and implementation of new ideas, processes, and technologies that can drive the organization forward. Innovating is essential for ensuring that the organization remains competitive and can continuously improve its operations by adopting cutting-edge technologies and methodologies. Coordinating is another vital subcomponent within restructuring. This activity involves the synchronization of various departments, teams, and resources to ensure that the transformation initiatives are implemented smoothly and effectively. Coordinating ensures that all parts of the organization are working toward the same goals and that there is clear communication and collaboration across different functional areas. This is particularly important in complex organizational structures where multiple stakeholders must work together to achieve successful digital transformation.

The final subcomponent, Facilitating, is concerned with creating an environment that supports the restructuring process. Facilitating involves providing the necessary tools, resources, and support systems that enable the organization to adapt to changes efficiently. This includes training employees, adjusting workflows, and ensuring that the infrastructure is in place to support new processes and technologies. Facilitating is crucial for reducing resistance to change and ensuring that the restructuring efforts are sustainable in the long term.

Together, these restructuring activities, coordinating and facilitating enable organizations to effectively realign their operations and processes to leverage the full potential of S-ERP systems within Industry 4.0. By embracing these activities, organizations can ensure that their digital transformation efforts lead to enhanced operational efficiency, greater innovation, and improved adaptability in an increasingly dynamic business environment.

After the operational processes of digital transformation, we focus on the operational digital transformation from the supply chain management perspective. Operational digital transformation in the supply chain can be significantly enhanced through the implementation of S-ERP in Industry 4.0, particularly when guided by lean principles. Integrating S-ERP systems with lean methodologies such as Lean Six Sigma enables organizations to streamline processes, reduce waste, and optimize resource allocation, which are key tenets of lean thinking. S-ERP supports the lean approach by providing real-time data visibility and advanced analytics capabilities, which are critical for identifying inefficiencies and areas for improvement in the supply chain. For instance, the S-ERP system can facilitate the continuous monitoring of supply chain activities, allowing organizations to promptly detect and address issues such as overproduction, excess inventory, or transportation delays common forms of waste are identified in lean methodologies.

Moreover, S-ERP systems enhance the Define-Measure-Analyze-Improve-Control (DMAIC) problem-solving framework by integrating big data analytics and predictive modeling. This integration allows for more precise measurement and analysis of supply chain processes, leading to data-driven decision-making. By using S-ERP to support DMAIC, organizations can more effectively control and sustain improvements, ensuring that the supply chain remains efficient and responsive to changing market demands. Additionally, S-ERP systems enable the automation of routine tasks and the optimization of workflows, which further reduces waste and improves operational efficiency. For example, automation can minimize the manual handling of materials, reduce errors in inventory management, and streamline order processing, all of which contribute to a more efficient supply chain. By aligning with lean concepts, S-ERP in Industry 4.0 not only supports the operational digital transformation of the supply chain but also ensures that this transformation leads to sustained improvements in efficiency, cost-effectiveness, and customer satisfaction. This alignment helps organizations maintain a competitive edge in the rapidly evolving digital economy.

Further, with the help of operational digital transformation using S-ERP, organizations can achieve a smart green supply chain. By leveraging a configurational approach, S-ERP integrates both digital transformation strategies and green supply chain management (GSCM) practices [22], facilitating the creation of a smart green supply chain that enhances environmental performance while maintaining operational efficiency. S-ERP supports the implementation of front-end technologies, such as collaborative robotics and augmented reality, which optimize supply chain operations by reducing waste and improving efficiency. These technologies are instrumental in green operations, such as eco-friendly packaging and sustainable manufacturing practices. For example, sensor technologies connected through IoT can ensure that packaging solutions use minimal materials, aligning with green performance goals. The configurational approach further emphasizes the horizontal integration of internal and external green supply chain activities. S-ERP facilitates this integration by aligning internal green operations, such as green purchasing and manufacturing, with external green relationships, including collaboration with suppliers and customers. This alignment is essential for achieving a cohesive green supply chain where all stakeholders work together to enhance environmental sustainability. Through this integrated approach, S-ERP in Industry 4.0 not only supports the digital transformation of supply chain operations but also ensures that these transformations contribute to the development of a smart green supply chain. This approach enables organizations to achieve higher levels of green performance, thereby aligning their operational goals with broader environmental objectives.

After the supply chain, we focus on the operational digital transformation using S-ERP in Industry 4.0 for employee experience (EX). The integration of S-ERP systems facilitates a more personalized and responsive work environment by leveraging digital tools that enhance cognitive, social, emotional, and behavioral experiences. For instance, S-ERP systems can support cognitive experience by providing employees with access to advanced analytics, automation tools, and real-time data, enabling them to engage in critical thinking, problem-solving, and continuous learning. This cognitive stimulation is essential for fostering innovation and creativity, allowing employees to contribute more effectively to the organization's goals. Moreover, S-ERP systems enhance social experience by enabling better communication and collaboration among employees through integrated digital platforms. These systems support enterprise social networks and other collaborative tools that allow employees to connect, share knowledge, and build professional relationships across different functions and departments. This interconnectedness not only improves team dynamics but also strengthens the overall organizational culture, making employees feel more engaged and valued.

Emotional experience, another critical aspect of employee experience, is also positively influenced by S-ERP systems. By streamlining workflows and reducing the complexity of routine tasks, these systems help reduce stress and enhance job satisfaction. Employees are more likely to experience positive emotions, such as joy and pride, when they feel supported by intuitive and efficient digital tools that make their work more manageable and fulfilling. Finally, S-ERP systems contribute to the behavioral experience by enabling flexibility and autonomy in work processes. These systems allow employees to tailor their work routines, experiment with new approaches, and take proactive actions within a digitally empowered environment. The autonomy provided by S-ERP systems fosters a sense of ownership and responsibility, further enhancing employee engagement and commitment to the organization's digital transformation efforts. Thus, S-ERP systems in Industry 4.0 are instrumental in driving operational digital transformation by significantly enhancing employee experience across cognitive, social, emotional, and behavioral dimensions. This holistic improvement in employee experience not only supports the successful implementation of digital transformation initiatives but also ensures that these initiatives are sustainable and aligned with the long-term objectives of the organization. After the extensive discussion on operational digital transformation, the following section discusses strategic digital transformation with S-ERP in Industry 4.0.

8.5 Strategic Digital Transformation with S-ERP in Industry 4.0

This section discusses the concept of strategic digital transformation with the help of S-ERP, where we take the first aspect, i.e., business model. S-ERP systems in Industry 4.0 serve as a vital catalyst for achieving strategic digital transformation related to business models. The transformation of business models in today's digital economy necessitates a shift from traditional structures to more dynamic, technology-driven frameworks that can adapt to rapid changes in consumer behavior, technological advancements, and competitive landscapes. S-ERP systems facilitate this shift by providing a flexible and integrated platform that supports the reconfiguration of key business model components, enabling organizations to innovate and compete effectively in a digital-first world.

One of the core contributions of S-ERP to strategic digital transformation is its ability to integrate digital technologies such as IoT, big data analytics, AI, and blockchain into the business model, thereby enhancing the organization's ability to create and deliver value in new and innovative ways. These technologies enable the organization to collect and analyze vast amounts of data in real time, providing insights that can be used to optimize operations, personalize customer experiences, and develop new revenue streams. This data-driven approach allows organizations to shift from a product-centric to a customer-centric business model [12], where the focus is on delivering tailored solutions that meet the specific needs of individual customers. Moreover, S-ERP systems support transitioning from traditional linear value chains to more networked and collaborative ecosystems. S-ERP fosters the development of multi-sided platforms and ecosystems essential for thriving in the digital economy by enabling seamless collaboration with external partners, suppliers, and customers. These platforms allow organizations to leverage the capabilities of their partners to co-create value and offer integrated solutions that are more comprehensive and competitive.

Furthermore, S-ERP facilitates the shift toward more flexible and scalable business models, such as subscription-based models [17], on-demand services [24], and platform-based models [20]. These models are better suited to the dynamic nature of the digital economy, where consumer preferences and market conditions can change rapidly. The agility provided by S-ERP allows organizations to quickly pivot their strategies and offerings in response to these changes, ensuring that they remain relevant and competitive.

S-ERP systems within the framework of Industry 4.0 are pivotal in facilitating strategic digital transformation, particularly in the realm of big data and business analytics. The second aspect of strategic digital transformation is data and big data analytics. These systems enable organizations to harness the vast amounts of data generated within and outside the enterprise, transforming this data into actionable insights that drive innovation and strategic decision-making. One of the key contributions of S-ERP systems is their ability to integrate and manage big data from diverse sources, including IoT devices, customer interactions, and market trends. By consolidating this data into a unified platform, S-ERP systems provide organizations with a comprehensive view of their operations, enabling more informed and timely decisions. This capability is crucial for developing predictive models that anticipate market changes, optimize supply chains, and enhance customer experiences.

Moreover, S-ERP systems support advanced business analytics by offering tools that can analyze large datasets to uncover patterns, trends, and correlations. These insights allow organizations to refine their business models, identify new revenue streams, and tailor their offerings to meet evolving customer demands. For example, predictive analytics facilitated by S-ERP can help companies anticipate customer needs, reduce operational risks, and enhance product development processes. The integration of big data analytics within S-ERP systems also fosters a data-driven culture within organizations. By making data accessible and actionable, these systems empower employees at all levels to base their decisions on empirical evidence rather than intuition. This shift toward a data-driven approach not only improves efficiency and accuracy but also enhances the organization's agility, enabling it to quickly adapt to changes in the market or competitive landscape. Furthermore, S-ERP systems enable organizations to leverage the power of big data for strategic planning. By analyzing historical data and forecasting future trends, companies can develop long-term strategies that are resilient to disruptions and aligned with their business objectives. This strategic use of data analytics is essential for maintaining a competitive edge in today's fast-paced digital economy.

S-ERP systems in Industry 4.0 are instrumental in achieving strategic digital transformation related to SMART analytics by providing a robust infrastructure that supports the collection, processing, and analysis of vast amounts of data [8]. SMART analytics, characterized by Specific, Measurable, Assignable, Realistic, and Time-related attributes, are essential for enabling organizations to make informed, data-driven decisions that enhance operational efficiency and drive innovation. By integrating advanced analytics capabilities within S-ERP, organizations can harness the power of big data to generate actionable insights that inform strategic decisions across various domains, such as customer engagement, supply chain management, and product development. S-ERP systems facilitate the seamless flow of data across the organization, ensuring that decision-makers have access to real-time information that is both accurate and relevant.

Moreover, S-ERP systems enable organizations to implement predictive analytics, which is a critical component of SMART analytics. Predictive analytics leverages historical data to forecast future trends, allowing organizations to anticipate changes in the market, optimize resource allocation, and proactively address potential challenges. This forward-looking approach is essential for maintaining a competitive edge in an increasingly dynamic business environment. In addition to predictive capabilities, S-ERP systems support prescriptive analytics, which provides recommendations on the best course of action based on data analysis. This enhances decision-making processes by offering data-driven guidance that aligns with the organization's strategic goals. By integrating prescriptive analytics into their operations, organizations can optimize outcomes, reduce risks, and improve overall performance. Furthermore, S-ERP systems enhance the scalability of SMART analytics by providing a flexible and adaptable platform that can accommodate the growing volume and complexity of data generated in Industry 4.0 environments. This scalability ensures that organizations can continue to leverage data analytics as they expand and evolve without being constrained by technological limitations.

S-ERP systems in Industry 4.0 play a pivotal role in facilitating strategic digital transformation by modernizing and optimizing IT infrastructure. The integration of S-ERP with advanced digital technologies such as cloud computing and big data analytics ensures that IT infrastructure is robust, scalable, and adaptable, which is critical for supporting the dynamic needs of modern businesses. One of the primary ways S-ERP supports IT infrastructure is through enhanced data management capabilities. By consolidating data from various sources into a unified platform, S-ERP systems enable real-time data processing and analytics, essential for informed decision-making and strategic planning. This integration allows organizations to manage large volumes of data efficiently, ensuring that the IT infrastructure can handle the increasing demands of data-driven operations. Furthermore, S-ERP systems support the flexibility and scalability of IT infrastructure, which is crucial for adapting to rapid technological advancements and changes in the business environment. The modular nature of S-ERP allows organizations to customize their IT infrastructure to meet specific needs, enabling them to scale their operations quickly and efficiently in response to market demands.

Additionally, S-ERP systems enhance the security and reliability of IT infrastructure. By providing robust security protocols and ensuring compliance with industry standards, S-ERP systems protect critical business data from cyber threats and ensure the IT infrastructure remains secure and resilient. This is particularly important in Industry 4.0, where integrating IoT devices and cloud-based services can introduce new vulnerabilities. The following section discusses the impact of digital transformation on the environment and sustainability from the business perspective and how S-ERP in Industry 4.0 helps achieve this digital transformation.

8.6 Digital Transformation and Sustainability with S-ERP in Industry 4.0

We begin this section by discussing the contribution or impact of digital transformation on business. Digital transformation contributes to environmental sustainability by leveraging advanced digital technologies to optimize resource use, reduce waste, and improve the efficiency of business processes. This transformation integrates digital tools such as the Internet of Things (IoT), artificial intelligence (AI), Big Data, cloud computing, and cyber-physical systems (CPS) into organizational strategies, resulting in more sustainable business practices.

Digital technologies, such as IoT and Big Data, allow organizations to monitor and manage resources in real time [18], leading to more efficient use of materials and energy. IoT devices can collect data on resource consumption, enabling companies to identify inefficiencies and reduce waste. For example, in smart manufacturing, IoT sensors can optimize energy use by adjusting machine operations based on demand, thus lowering overall energy consumption and reducing carbon emissions. The existing technologies in most organizations do not manage sustainable practices because they can not monitor them. They can not monitor because they can't measure sustainable practices. So, digital transformation, especially with the help of S-ERP, will allow organizations to manage sustainable business practices.

AI and machine learning algorithms can analyze large datasets to predict and prevent waste in production processes. By optimizing production schedules, managing supply chains more effectively, and predicting maintenance needs, these technologies help reduce material waste and enhance product lifecycle management. This contributes to the development of a circular economy, where resources are reused and recycled, minimizing the environmental impact of production.

Much research is currently going on to discuss how technologies like cloud and quantum computing can conserve energy [2]. Cloud computing contributes to environmental sustainability by reducing physical IT infrastructure needs [3]. Moving to the cloud decreases the energy required to maintain data centers, as cloud providers often use more energy-efficient and renewable energy sources. This shift helps reduce the carbon footprint associated with traditional IT infrastructure, aligning with sustainability goals. Cyber-physical systems (CPS) and 3D printing technologies enable the creation of sustainable products through more efficient design and manufacturing processes. CPS integrates physical and digital systems, allowing for real-time data exchange and decision-making, leading to less waste and more efficient use of materials. 3D printing, or additive manufacturing, further reduces waste by building products layer by layer, using only the necessary material. The cloud and CPS are critical aspects of S-ERP. Hence, we suggest that digital transformation using S-ERP would benefit the environment.

The integration of digital technologies is instrumental in advancing global sustainability initiatives, notably the United Nations' Sustainable Development Goals (SDGs). Through the enhancement of energy efficiency, waste reduction, and the promotion of renewable resource use, digital transformation enables organizations to make significant contributions to broader environmental sustainability objectives. Specifically, the implementation of S-ERP in digital transformation initiatives can drive the achievement of several SDGs. For instance, SDG 12 (Responsible Consumption and Production) [10] is addressed through the efficient use of resources, while SDG 9 (Industry, Innovation, and Infrastructure) [7] is supported by the reduction and management of waste. The adoption of green IT infrastructure contributes to SDG 7 (Affordable and Clean Energy) [23], and the development of sustainable products via cyber-physical systems (CPS) and data analytics aligns with SDG 13 (Climate Action) [11]. Furthermore, by promoting global sustainability across organizations through Industry 4.0 technologies, these efforts also contribute to SDG 17 (Partnerships for the Goals) [16], fostering collaboration and collective action toward sustainable development. Thus, S-ERP in Industry 4.0 can help achieve five SDG goals.

Sustainable innovation is possible with the digital transformation supported by S-ERP in Industry 4.0. To achieve sustainable innovation using S-ERP in Industry 4.0, organizations must focus on key strategies that leverage digital transformation to drive both environmental and economic sustainability. Sustainable innovation involves the development of new products, processes, and business models that not only meet market needs but also contribute to environmental goals. Innovation ambidexterity, which involves balancing exploitation and exploration innovation, is critical for sustainability. Exploitation innovation focuses on refining and improving existing products and processes to enhance sustainability, while exploration innovation involves seeking new technologies and solutions that align with sustainability goals. By fostering both types of innovation, organizations can ensure continuous improvement in their sustainability performance while also pursuing groundbreaking innovations that lead to long-term environmental benefits.

Agile and flexible digital architectures enable organizations to quickly adapt to changing environmental conditions and market demands. By adopting open standards and ensuring that digital systems are scalable and upgradable, companies can maintain their competitive edge while also advancing their sustainability agenda.

Agile methodologies allow for the rapid development and implementation of new sustainable solutions, ensuring that organizations can respond effectively to emerging challenges and opportunities. Next, the following section explains the paradoxes associated with digital transformation. In the following section, we also explain how each of the paradoxical situations can be navigated with the help of S-ERP in Industry 4.0.

8.7 Navigating Digital Transformation Paradoxes with S-ERP in Industry 4.0

Digital transformation in business is fraught with several paradoxes that arise from the complex interplay between technology, organizational processes, and human factors. These paradoxes can create challenges for businesses as they navigate the integration of digital technologies into their operations. Below are some of the key paradoxes related to digital transformation, which can be navigated with the help of S-ERP in Industry 4.0.

8.7.1 Innovation Paradox

While digital transformation drives innovation, it also creates a tension between the need for continuous innovation and the stability of existing processes. Companies are often caught between the need to innovate rapidly (exploration) and the necessity to optimize and refine existing technologies and processes (exploitation) [15]. Let I_t represent the innovation output at time t, and R_t represent the resources allocated to innovation at time t. The innovation paradox can be modeled as a trade-off between exploration E_t and exploitation X_t :

$$I_t = f(E_t, X_t)$$

where E_t drives new innovation but increases uncertainty, and X_t optimizes existing processes but limits breakthrough innovation. The challenge is balancing E_t and X_t to maximize I_t.

S-ERP in Industry 4.0 can play a crucial role in navigating this paradox by integrating advanced technologies that enable both exploration and exploitation within a unified framework. S-ERP systems in Industry 4.0 are equipped with advanced data analytics and real-time monitoring capabilities that allow organizations to gather and analyze vast amounts of data from across the enterprise. This data-driven approach supports both exploration and exploitation by providing insights that help managers make informed decisions about where to innovate and how to optimize existing processes.

Let D_t represent the decision-making quality at time t, influenced by both exploration E_t and exploitation X_t

$$D_t = f(E_t, X_t, A_t)$$

where A_t represents the analytics capabilities provided by S-ERP. A_t helps in balancing E_t and X_t by providing actionable insights that enhance decision-making, thus optimizing the innovation output I_t. S-ERP systems enable dynamic resource allocation, allowing organizations to shift resources between exploration and exploitation activities as needed. This flexibility is critical for responding to changing market conditions or technological advancements, ensuring that innovation efforts are both sustainable and aligned with strategic goals. The resources allocated to innovation R_t can be dynamically adjusted using the resource management capabilities of S-ERP:

$$R_t = f(E_t, X_t, F_t)$$

where F_t represents the flexibility provided by S-ERP to reallocate resources based on real-time data and organizational priorities. This flexibility ensures that resources are optimally used to balance exploration and exploitation. S-ERP in Industry 4.0 supports process optimization through automation, real-time monitoring, and predictive maintenance. These capabilities allow organizations to refine existing processes (exploitation) while freeing up resources and time for innovation (exploration). This dual focus ensures that while the organization continues to innovate, it also maintains and improves the efficiency of existing operations. Let P_t represent process efficiency at time t. The S-ERP system optimizes P_t through continuous improvement initiatives, which can be modeled as:

$$P_t = g(X_t, M_t)$$

where M_t represents the maintenance and monitoring functions of S-ERP. By optimizing P_t, S-ERP ensures that exploitation does not hinder exploration, allowing for a balanced innovation strategy. S-ERP systems facilitate collaboration and knowledge sharing across different departments and stakeholders. This integration is vital for fostering innovation (exploration) while ensuring that new knowledge and technologies are effectively implemented and optimized (exploitation). Collaboration C_t at time t can be enhanced by S-ERP's communication and knowledge management tools, which support both exploration and exploitation:

$$C_t = h(E_t, X_t, K_t)$$

where K_t represents the knowledge sharing facilitated by S -ERP. Enhanced C_t ensures that both new ideas (exploration) and existing processes (exploitation) benefit from collective organizational knowledge. The following paradox is the efficiency paradox.

8.7.2 Efficiency Paradox

Digital transformation is expected to increase efficiency by automating processes and improving decision-making. However, the introduction of new technologies often leads to increased complexity and the need for higher levels of oversight, which can reduce overall efficiency [14]. Let $\eta(t)$ represent the efficiency of the organization at time t. The paradox can be expressed as:

$$\eta(t) = \frac{O_t}{C_t + M_t}$$

where O_t is the output, C_t is the cost of implementation, and M_t is the cost of managing complexity. As C_t and M_t increase, $\eta(t)$ may decrease, even if O_t increases. The efficiency paradox in digital transformation highlights the challenge that, while new technologies are expected to enhance efficiency through automation and improved decision-making, they often introduce additional complexity and oversight needs that can counteract these benefits. S-ERP in Industry 4.0 provides tools and strategies to navigate this paradox by streamlining processes, reducing complexity, and optimizing resource use. S-ERP systems in Industry 4.0 can automate not only individual tasks but also entire workflows across the enterprise. By doing so, they reduce the complexity associated with managing multiple disparate systems. The integration of processes within a single S-ERP platform simplifies operations and reduces the need for extensive oversight, thereby addressing the complexity that often undermines efficiency gains. Let $\eta(t)$ represent the efficiency at time t. With S-ERP, the costs C_t and M_t related to implementation and complexity management can be optimized:

$$\eta(t) = \frac{O_t}{C_t + M_t}$$

where $C_t = C_{t,\text{base}} - R_t$ and $M_t = M_{t,\text{base}} - S_t$. Here, R_t represents the reduction in implementation costs due to process simplification, and S_t represents the reduction in complexity management costs due to system integration. By minimizing C_t and M_t, $\eta(t)$ is maximized. S-ERP systems provide real-time data analytics and business intelligence tools that improve decision-making accuracy and speed. This reduces the inefficiencies that arise from poor or delayed decisions, which are common in complex systems. By providing clear insights and actionable information, S-ERP reduces the need for extensive manual oversight, thus mitigating the negative impact on efficiency. The decision-making process $D(t)$ can be modeled as a function of data availability A_t and analysis L_t :

$$D(t) = f(A_t, L_t)$$

S-ERP enhances A_t and L_t, leading to better decisions that increase output O_t without a proportional increase in complexity M_t, thereby improving $\eta(t)$. S-ERP systems enable adaptive resource management by dynamically allocating resources where they are most needed, based on real-time data and predictive analytics. This adaptability helps to minimize wasted resources and reduce the overhead associated with managing complex systems, which can otherwise diminish efficiency. The adaptive management of resources R_t can reduce costs C_t and complexity M_t through better allocation:

$$R_t = g\,(D_t, P_t)$$

where D_t represents decision quality and P_t represents process efficiency. S-ERP optimizes R_t, leading to reduced C_t and M_t, thus increasing $\eta(t)$. S-ERP systems support continuous improvement initiatives by providing tools for monitoring performance and implementing feedback loops. These tools help identify inefficiencies and areas where complexity can be reduced, enabling the organization to refine processes continuously. This ongoing refinement helps to counteract the tendency for complexity to erode efficiency gains over time. Let F_t represent the feedback loop effectiveness at time t. S-ERP enhances F_t which reduces the negative impact of complexity M_t on efficiency:

$$M_t = M_{t,\,\text{base}} - F_t$$

Thus, as F_t increases, M_t decreases, leading to higher overall efficiency $\eta(t)$. S-ERP systems in Industry 4.0 are designed to be scalable and flexible, meaning they can grow and adapt as the organization's needs evolve. This scalability reduces the need for frequent, costly overhauls and minimizes the introduction of new complexities as the organization expands, helping maintain high efficiency. Let $S(t)$ represent the system's scalability at time t. Higher scalability $S(t)$ reduces the impact of complexity M_t by enabling the system to handle growth without proportional increases in oversight or complexity:

$$M_t = M_{t,\,\text{base}} - S(t)$$

As $S(t)$ increases, M_t decreases, contributing to greater efficiency $\eta(t)$. The following paradox is the control paradox.

8.7.3 Control Paradox

Digital transformation enables greater control over business processes through data analytics and real-time monitoring. However, this increased control can lead to a loss of flexibility and autonomy, creating a paradox where the systems designed to improve management can stifle innovation and responsiveness [13]. Let $C(t)$ represent the level of control at time t and $F(t)$ represent flexibility. The paradox can be modeled as:

$$F(t) = \frac{1}{C(t)}$$

indicating an inverse relationship where increasing control $C(t)$ decreases flexibility $F(t)$, potentially leading to rigidity in operations. The control paradox in digital transformation highlights the challenge that while enhancing control over business processes through data analytics and real-time monitoring can improve management, it often comes at the cost of flexibility and autonomy. This paradox can lead to increased control, stifling innovation and reducing the organization's ability to respond quickly to changes. S-ERP in Industry 4.0 can help navigate this paradox by balancing control with flexibility, ensuring that organizations maintain effective oversight and the agility needed to innovate and respond to market dynamics. S-ERP systems in Industry 4.0 are often designed with a modular architecture that allows for scalability and flexibility in system configuration. This modularity enables organizations to maintain control over essential processes while allowing other areas of the business to remain flexible and adaptable to change. Let $C(t)$ represent the level of control at time t and $F(t)$ represent the level of flexibility. By using a modular architecture, the S-ERP system can enable different modules to operate with varying degrees of control:

$$F(t) = \frac{1}{C(t) + \mu_t}$$

where μ_t represents the flexibility introduced by modularity. Even as $C(t)$ increases, the modularity μ_t can prevent a significant decrease in $F(t)$, maintaining a balance between control and flexibility. S-ERP systems provide configurable workflows that allow organizations to customize processes based on their specific needs. This configurability ensures that while core processes are tightly controlled, there is still room for adaptation and flexibility in other areas, enabling the organization to respond quickly to new opportunities or challenges. The flexibility $F(t)$ can be enhanced by the configurability of workflows W_t within the S-ERP system:

$$F(t) = \frac{1}{C(t)} + W_t$$

where W_t represents the degree of workflow configurability. Higher W_t mitigates the loss of flexibility as control $C(t)$ increases, allowing the organization to maintain responsiveness. S-ERP systems in Industry 4.0 come equipped with real-time data analytics and predictive capabilities that help organizations anticipate changes and adapt their strategies accordingly. This predictive insight allows businesses to retain control over operations while being proactive rather than reactive, thus preserving flexibility. Predictive capabilities P_t can offset the rigidity introduced by higher control $C(t)$:

$$F(t) = \frac{1}{C(t)} + P_t$$

where P_t represents the predictive capabilities of the S-ERP system. Higher P_t allows the organization to remain flexible and responsive, even as $C(t)$ increases. 4. Decentralized Decision-Making: S-ERP systems can support decentralized decision-making by providing relevant data and insights at different levels of the organization. This decentralization allows teams to make decisions autonomously within their domains while the overall system retains centralized control over critical processes. This balance helps maintain flexibility without compromising control. Let D_t represent the degree of decentralization at time t. Decentralized decision-making can reduce the impact of centralized control $C(t)$ on flexibility:

$$F(t) = \frac{1}{C(t)} + D_t$$

where D_t enhances flexibility by allowing autonomous decision-making within controlled environments, ensuring that the organization remains agile. S-ERP systems enable continuous monitoring and feedback loops that allow organizations to assess the effectiveness of their control mechanisms and make adjustments as needed. This ongoing assessment helps prevent control mechanisms from becoming overly rigid and allows for timely adjustments that enhance flexibility. Feedback loops L_t introduced by continuous monitoring can dynamically adjust the balance between control and flexibility:

$$F(t) = \frac{1}{C(t)} + L_t$$

where L_t represents the effectiveness of feedback loops. As L_t increases, it allows for continuous refinement of control mechanisms, thereby preserving flexibility. The following paradox is the transparency paradox.

8.7.4 Transparency Paradox

Digital transformation enhances transparency through better data availability and sharing. However, increased transparency can lead to information overload and decision paralysis, where the abundance of data complicates decision-making rather than simplifying it [9]. Let $T(t)$ represent the level of transparency at time t and $D(t)$ represent the decision effectiveness. The paradox can be expressed as:

$$D(t) = f(T(t), I(t))$$

where $I(t)$ represents information overload. As $T(t)$ increases, $I(t)$ may increase, potentially reducing $D(t)$. The transparency paradox in digital transformation arises from the increased availability and sharing of data, which, while intended to improve decision-making, can lead to information overload and decision paralysis. S-ERP in Industry 4.0 can navigate this paradox by providing mechanisms that enhance decision-making efficiency without overwhelming decision-makers with excessive information. S-ERP systems in Industry 4.0 are equipped with intelligent data filtering and prioritization tools that help reduce information overload. These tools can automatically sift through vast amounts of data, highlighting only the most relevant information needed for decision-making. By prioritizing critical data, S-ERP ensures that transparency does not lead to decision paralysis. Let F_t represent the filtering effectiveness at time t. The decision effectiveness $D(t)$ can be modeled as:

$$D(t) = f\,(T(t), I(t) - F_t)$$

where $I(t)$ represents the information overload, and F_t reduces this overload by filtering out unnecessary data. As F_t increases, $I(t)$ decreases, leading to improved $D(t)$. S-ERP systems can present data in a contextualized manner, meaning that information is tailored to the specific needs of the user or the decision at hand. This contextualization reduces cognitive load and helps decision-makers focus on the most pertinent data, thereby improving decision effectiveness despite high levels of transparency. The impact of transparency $T(t)$ on decision-making $D(t)$ can be mediated by contextualization C_t :

$$D(t) = f\left(T(t), \frac{I(t)}{C_t}\right)$$

where C_t represents the degree of contextualization. Higher C_t reduces the impact of $I(t)$, thus enhancing $D(t)$. S-ERP in Industry 4.0 includes automated decision support systems (DSS) that can analyze large datasets and provide recommendations or even make decisions autonomously. These systems help manage the transparency paradox by reducing the burden on human decision-makers, allowing them to focus on strategic decisions rather than being overwhelmed by data. Let A_t represent the effectiveness of automated decision support systems at time t. The decision effectiveness $D(t)$ can then be expressed as:

$$D(t) = f\,(T(t), I(t) \times (1 - A_t))$$

where A_t reduces the direct impact of information overload $I(t)$ on decision-making, allowing $D(t)$ to remain high even as $T(t)$ increases. S-ERP systems provide role-based access controls and data segmentation, ensuring that individuals receive only the data that is relevant to their roles and responsibilities. This segmentation helps in managing the volume of information that any single user has to process, reducing the risk of overload and improving decision-making efficiency. Role-based access R_t and data segmentation S_t can modulate the relationship between transparency $T(t)$ and information overload $I(t)$:

$$I(t) = \frac{T(t)}{R_t \times S_t}$$

Higher R_t and S_t reduce the effective $I(t)$, thus improving $D(t)$. S-ERP systems often include advanced real-time analytics and data visualization tools that make complex data easier to understand and interpret. By converting raw data into intuitive visual formats, these tools help decision-makers quickly grasp the implications of the data, thus mitigating the negative effects of information overload. Let V_t represent the effectiveness of visualization tools at time t. The relationship between transparency $T(t)$ and decision effectiveness $D(t)$ can be expressed as:

$$D(t) = f\left(T(t), \frac{I(t)}{V_t}\right)$$

where higher V_t reduces the cognitive load associated with $I(t)$, thereby enhancing $D(t)$. The following paradox is the skill paradox.

8.7.5 Skill Paradox

Digital transformation requires a workforce with advanced digital skills. However, the rapid pace of technological change can lead to a skills gap, where the existing workforce is unable to keep up with the new demands, resulting in underutilization of digital tools [6]. Let S_t represent the skill level of the workforce at time t and T_t represent the technological complexity. The paradox can be modeled as:

$$U_t = f(S_t - T_t)$$

where U_t is the utilization of technology. If $S_t < T_t$, then U_t decreases, indicating underutilization. The skill paradox in digital transformation occurs when the rapid pace of technological change creates a skills gap within the workforce, leading to the underutilization of advanced digital tools. As technology evolves, the existing workforce may struggle to keep pace with new demands, resulting in inefficiencies and decreased productivity. S-ERP in Industry 4.0 can navigate this paradox by providing mechanisms to upskill the workforce, simplify technology adoption, and enhance the utilization of digital tools. S-ERP systems in Industry 4.0 can integrate continuous learning and training modules directly into the platform, allowing employees to upskill in real-time as they interact with new technologies. This on-the-job training ensures that the workforce's skill level S_t keeps pace with technological complexity T_t, thereby reducing the skills gap and enhancing technology utilization. Let L_t represent the effectiveness of learning and training programs at time t. The skill level S_t can be modeled as:

$$S_t = S_{t-1} + L_t$$

where S_{t-1} is the previous skill level, and L_t is the learning input. By increasing L_t, the organization can ensure that S_t grows in line with T_t, thereby maintaining or improving the utilization of technology U_t :

$$U_t = f\ (S_t - T_t)$$

S-ERP systems can incorporate user-friendly interfaces and automation features that reduce the complexity of interactions with advanced technologies. This simplification means that even if the workforce's skill level S_t is slightly lower than the technological complexity T_t, the automation and intuitive design of the S-ERP system can compensate, preventing a decline in technology utilization U_t. Let A_t represent the level of automation and interface simplicity at time t. The impact of T_t on U_t can be mitigated by A_t :

$$U_t = f\ ((S_t - T_t) + A_t)$$

Higher A_t reduces the negative impact of high T_t, ensuring that U_t remains stable or even increases despite a potential skills gap. S-ERP systems in Industry 4.0 can employ adaptive learning algorithms that tailor training and support to the specific needs of individual employees. These algorithms can assess the current skill level S_t of each employee and provide personalized learning paths to bridge the gap with T_t. This approach ensures that training is efficient and targeted, maximizing the impact on technology utilization. Let P_t represent the personalized training effectiveness at time t. The skill level S_t can be dynamically adjusted by P_t to better match T_t :

$$S_t = S_{t-1} + P_t$$

Higher P_t leads to a faster and more effective increase in S_t, minimizing the skills gap and improving U_t. S-ERP systems can offer real-time support and knowledge-sharing platforms where employees can seek help and share insights with colleagues. This collaborative approach helps to quickly resolve issues related to new technologies and fosters a culture of continuous improvement, thereby narrowing the skills gap and improving the overall utilization of digital tools. Let K_t represent the knowledge sharing and support effectiveness at time t. The collective skill level S_t of the workforce can be enhanced by K_t :

$$S_t = S_{t-1} + K_t$$

By increasing K_t, organizations can rapidly increase S_t, ensuring that the workforce can effectively use the technology, thereby enhancing U_t. S-ERP systems designed with modularity and scalability allow organizations to introduce new technologies gradually, giving employees time to adapt and upskill before more complex modules are implemented. This phased approach ensures that the workforce is not overwhelmed by sudden increases in technological complexity T_t, allowing for smoother transitions and better technology utilization. Let M_t represent the modularity and scalability of the system at time t. The rate of change in T_t can be controlled by M_t :

$$T_t = T_{t-1} + (1 - M_t) \times \Delta T$$

where ΔT represents the potential increase in complexity. By controlling the pace of change ΔT through M_t, organizations can ensure that S_t keeps pace with T_t, maintaining or improving U_t. The following paradox is the security paradox.

8.7.6 Security Paradox

Digital transformation often includes the adoption of cloud services, IoT, and other networked technologies, which increase the surface area for potential security breaches. Efforts to secure these systems can create friction in operations, thus hindering the efficiencies and agility that digital transformation aims to achieve [1]. Let $S(t)$ represent the security level at time t and $A(t)$ represent agility. The paradox can be expressed as:

$$A(t) = \frac{1}{S(t)}$$

indicating that as security measures $S(t)$ increase, agility $A(t)$ decreases due to the added constraints. The security paradox in digital transformation arises from the need to balance the implementation of robust security measures with maintaining operational agility. As organizations adopt cloud services, IoT, and other networked technologies, they increase their exposure to potential security breaches. While enhancing security is crucial, doing so can introduce friction in operations, potentially reducing the agility that digital transformation aims to achieve. S-ERP in Industry 4.0 can navigate this paradox by implementing strategies that balance security and agility, ensuring that one does not come at the expense of the other. S-ERP systems in Industry 4.0 can implement adaptive security measures that dynamically adjust based on the current risk level and operational needs. This approach allows organizations to maintain high security $S(t)$ when necessary while relaxing constraints when the risk is lower, thereby preserving operational agility $A(t)$. Let $R(t)$ represent the real-time risk assessment at time t. The security level $S(t)$ can be dynamically adjusted based on $R(t)$, influencing agility $A(t)$:

$$S(t) = f(R(t), B_t)$$

where B_t represents the baseline security level. As $R(t)$ increases, $S(t)$ increases, potentially decreasing $A(t)$. However, adaptive measures can be designed to minimize the impact on $A(t)$ by optimizing the response to $R(t)$:

$$A(t) = \frac{1}{S(t)}$$

with $S(t)$ being adaptive, allowing $A(t)$ to remain as high as possible. S-ERP systems can automate many security protocols, such as threat detection, incident response, and compliance monitoring. Automation reduces the operational burden of maintaining security, thereby minimizing its impact on agility. This ensures that security measures do not become a bottleneck for operations. Let M_t represents the degree of automation in security protocols at time t. The impact of security on agility can be modeled as:

$$A(t) = \frac{1}{S(t) \times (1 - M_t)}$$

where M_t reduces the negative impact of $S(t)$ on $A(t)$. Higher M_t leads to a smaller reduction in $A(t)$ even as $S(t)$ increases, allowing for strong security without compromising agility. S-ERP systems can implement risk-based access controls, where the level of access granted to users is dynamically adjusted based on the current security risk. This approach ensures that stringent security measures are applied only when necessary, allowing the system to remain agile under normal conditions. Let C_t represent the conditional access control based on risk at time t. The security level $S(t)$ can be modulated by C_t to maintain agility:

$$S(t) = S_{\text{base}} \times C_t$$

where S_{base} is the baseline security level. C_t increases $S(t)$ when risk is high and reduces it when risk is low, allowing $A(t)$ to remain stable:

$$A(t) = \frac{1}{S(t)}$$

By adjusting C_t, the system can ensure that security measures are only as restrictive as necessary, preserving operational agility. S-ERP systems in Industry 4.0 can utilize network segmentation and micro-segmentation to isolate critical assets and limit the spread of potential breaches. This strategy enhances security without significantly impacting the agility of the broader network, as only the most sensitive areas are subject to the strictest controls. Let G_t represents the segmentation level at time t. The security impact $S(t)$ on agility $A(t)$ can be reduced by effectively segmenting the network:

$$S(t) = S_{\text{base}} \times (1 - G_t)$$

where G_t reduces the need for widespread security controls, thereby minimizing the impact on $A(t)$

$$A(t) = \frac{1}{S(t)}$$

Higher G_t (more effective segmentation) reduces $S(t)$, preserving agility while maintaining security in critical areas. S-ERP systems provide continuous monitoring and automated response capabilities that allow for quick identification and mitigation of security threats. This proactive approach reduces the need for static, restrictive security measures, thereby preserving agility while maintaining a high level of security. Let R_t represents the real-time monitoring effectiveness at time t. The security level $S(t)$ and its impact on agility $A(t)$ can be optimized by R_t :

$$S(t) = S_{\text{base}} \times (1 - R_t)$$

where R_t allows for dynamic adjustments to $S(t)$, maintaining high agility $A(t)$:

$$A(t) = \frac{1}{S(t)}$$

Effective monitoring R_t allows for responsive security measures that do not overly restrict operations, ensuring that agility is preserved. In summary, the chapter answers the following research questions.

Research Questions Addressed

The chapter titled "Role of S-ERP in Digital Transformation" addresses the following key research questions:

1. **How does Sustainable Enterprise Resource Planning (S-ERP) facilitate digital transformation within Industry 4.0?**

    - The chapter explores the role of S-ERP in integrating advanced technologies like loT, AI, and big data analytics to enable seamless automation, real-time data processing, and intelligent decision-making, which are critical aspects of digital transformation.

2. **What are digital blind spots, and how can they be identified and managed using S-ERP in Industry 4.0?**

    - The chapter discusses the concept of digital blind spots, their implications for digital transformation, and the methods for detecting and addressing them using S-ERP, including Computer Vision (CV), Root Cause Analysis (RCA), and feedback mechanisms.

3. **How does S-ERP support customer-centric digital transformation in Industry 4.0?**
   - The chapter examines how S-ERP systems enhance customer experience by integrating technologies like Cyber-Physical Systems (CPS), Industrial loT (I-loT), and Big Data Analytics, enabling personalized, real-time customer interactions throughout the customer journey.
4. **What is the role of S-ERP in operational digital transformation from an operational process perspective in Industry 4.0?**
   - The chapter outlines the components of operational digital transformation, such as sensing, organizing, and restructuring, and discusses how S-ERP supports these processes to enhance agility, efficiency, and innovation.
5. **How does S-ERP contribute to strategic digital transformation, particularly in business model innovation, data analytics, and IT infrastructure within Industry 4.0?**
   - The chapter explores how S-ERP enables the transformation of business models, supports big data analytics for strategic decision-making and modernizes IT infrastructure to support the dynamic needs of businesses in the digital economy.
6. **How does digital transformation impact sustainability, and how does S-ERP in Industry 4.0 help achieve environmental and economic sustainability goals?**
   - The chapter discusses the role of S-ERP in promoting sustainability through efficient resource use, waste reduction, and support for sustainable business practices, aligning with global sustainability initiatives like the United Nations' Sustainable Development Goals (SDGs).
7. **How can S-ERP help navigate the paradoxes associated with digital transformation in Industry 4.0, such as the innovation, efficiency, and control paradox?**
   - The chapter identifies several paradoxes during digital transformation and provides insights into how S-ERP systems can help balance these competing demands, ensuring successful digital transformation.

These research questions guide the chapter's comprehensive exploration of the role of S-ERP in enabling, managing, and sustaining digital transformation within Industry 4.0.

References

1. Anderson R, Moore T (2009) Information security: where computer science, economics and psychology meet. Philos Trans R Soc A: Math, Phys Engin Sci 367(1898):2717–2727. https://doi.org/10.1098/rsta.2009.0027. Publisher: Royal Society
2. Anjaria K (2024) Quantum model regression for generating fuzzy numbers in adiabatic quantum computing. Inf Sci 678:121018. ISSN 0020-0255. https://doi.org/10.1016/j.ins.2024.121018. https://www.sciencedirect.com/science/article/pii/S0020025524009320
3. Anjaria K, Patel N (2021) Attainment of green computing in cloudlet-based mobile cloud computing model using squirrel search algorithm. In: Mahapatra RP, Panigrahi BK, Kaushik BK, Roy S (eds) Proceedings of 6th international conference on recent trends in computing. Springer, Singapore, pp 457–470. ISBN 978-981-334-501-0
4. Balakrishnan R, Das S (2020) How do firms reorganize to implement digital transformation? Strat Change 29(5):531–541. ISSN 1099–1697. https://onlinelibrary.wiley.com/doi/abs/10.1002/jsc.2362
5. Berman S, Marshall A (2014) The next digital transformation: from an individual-centered to an everyone-to-everyone economy. Strat & Leader 42(5):9–17. ISSN 1087-8572. https://doi.org/10.1108/SL-07-2014-0048. Publisher: Emerald Group Publishing Limited
6. Brynjolfsson E, McAfee A, Sponsored by the Harvard Institute for learning in retirement
7. Costa E (2024) Industry 5.0 and SDG 9: a symbiotic dance towards sustainable transformation. Sustain Earth Rev 7(1):4. ISSN 2520-8748. https://doi.org/10.1186/s42055-024-00073-y
8. El Khatib MM, Alzoubi HM, Ahmed G, Kazim HH, Al Falasi SA, Mohammed F, Al Mulla M (2022) Digital transformation and SMART-the analytics factor. In 2022 international conference on business analytics for technology and security (ICBATS), pp 1–11. https://doi.org/10.1109/ICBATS54253.2022.9759084. https://ieeexplore.ieee.org/abstract/document/9759084
9. Eppler MJ, Mengis J (2004) The concept of information overload - a review of literature from organization science, accounting, marketing, MIS, and related disciplines. In: Meckel M, Schmid BF (eds) Kommunikationsmanagement im Wandel: Beiträge aus 10 Jahren =mcminstitute. Gabler, Wiesbaden, pp 271–305. 978-3-8349-9772-2
10. Foláyan MO, Virtanen JI, Gaffar B, Abodunrin O, Sun IG, Duangthip D, Kemoli A, Masumo RM, Vukovic A, Al-Batayneh OB, Mfolo T, Schroth RJ, El Tantawi M (2024a) Scoping review on the association between early childhood caries and responsible resource consumption and production: exploring Sustainable Development Goal 12. BMC Oral Health 24(1):98. ISSN 1472-6831
11. Foláyan MO, Schroth RJ, Abodunrin O, Al-Batayneh OB, Arheiam A, Mfolo T, Virtanen JI, Duangthip D, Feldens CA, El Tantawi M (2024b) Early childhood caries, climate change and the sustainable development goal 13: a scoping review. BMC Oral Health 24(1):524. ISSN 1472-6831
12. Gupta S, Ramachandran D (2021) Emerging market retail: transitioning from a product-centric to a customer-centric approach. J Retail 97(4):597–620. ISSN 0022-4359. https://doi.org/10.1016/j.jretai.2021.01.008. https://www.sciencedirect.com/science/article/pii/S0022435921000087
13. King JL, Lyytinen K (eds) (2006) Information systems: the state of the field. Wiley series in information systems. Wiley, Chichester, England. 978-0-470-01777-7. OCLC: ocm62179136
14. Lewis M, Slack N (2003) Operations management. Psychology Press. ISBN 978-0-415-24924-9. Google-Books-ID: p6vuyWD8I30C
15. March JG (1991) Exploration and exploitation in organizational learning. Organizat Sci 2(1):71–87. ISSN 1047-7039. https://doi.org/10.1287/orsc.2.1.71. https://pubsonline.informs.org/doi/abs/10.1287/orsc.2.1.71. Publisher: INFORMS
16. Monaco S, Monaco S (2024) SDG 17. Strengthen the means of implementation and revitalize the global partnership for sustainable development. In Identity, territories, and sustainability: challenges and opportunities for achieving the UN sustainable development goals. Emerald Publishing Limited, pp 167–175. ISBN 978-1-83797-549-5 978-1-83797-550-1. https://doi.org/10.1108/978-1-83797-549-520241018

17. Punj G (2015) The relationship between consumer characteristics and willingness to pay for general online content: Implications for content providers considering subscription-based business models. Market Lett 26(2):175–186. ISSN 1573-059X. https://doi.org/10.1007/s11002-013-9273-y
18. Qi W, Sun M, Hosseini SRA (2023) Facilitating big-data management in modern business and organizations using cloud computing: a comprehensive study. J Manage Organizat 29(4), 697–723. ISSN 1833–3672:1839–3527. https://doi.org/10.1017/jmo.2022.17. https://www.cambridge.org/core/journals/journal-of-management-and-organization/article/abs/facilitating-bigdata-management-in-modern-business-and-organizations-using-cloud-computing-a-comprehensive-study/FCD0DBE58A6FEDEBFEF49E70CEDE7AA4
19. Rajapakshe W (2024) Performance management process: performance planning and monitoring. In: Employee performance management for improved workplace motivation. IGI Global, pp 31–56. ISBN 9798369343876. https://doi.org/10.4018/979-8-3693-4387-6.ch002. https://www.igi-global.com/chapter/performance-management-process/www.igi-global.com/chapter/performance-management-process/355453
20. Rohn D, Bican PM, Brem A, Kraus S, Clauss T (2021) Digital platform-based business models – An exploration of critical success factors. J Engin Technol Manage 60:101625. ISSN 0923-4748. https://doi.org/10.1016/j.jengtecman.2021.101625. https://www.sciencedirect.com/science/article/pii/S092347482100014X
21. Waelen RA (2024) The ethics of computer vision: an overview in terms of power. AI Ethics 4(2):353–362. ISSN 2730-5961. https://doi.org/10.1007/s43681-023-00272-x
22. Wang F, Su Q, Zhang Z (2023) The influence of collaborative innovation network characteristics on firm innovation performance from the perspective of innovation ecosystem. Kybernetes 53(4):1281–1305. ISSN 0368-492X. https://doi.org/10.1108/K-04-2022-0553. Publisher: Emerald Publishing Limited
23. Wasnik AP (2023) India's journey toward cleaner and affordable energy for achieving SDG 7: progress and prospects. In: Dutta V, Ghosh P (eds) Sustainability: science, policy, and practice in India: challenges and opportunities. Springer International Publishing, Cham, pp 105–121. 978-3-031-50132-6
24. Röschmann AZ, Erny M, Wagner J (2022) On the (future) role of on-demand insurance: market landscape, business model and customer perception. Geneva Papers Risk Insurance - Issues Pract 47(3):603–642. ISSN 1468-0440. https://doi.org/10.1057/s41288-022-00265-7

Chapter 9
System Integration with Industry 4.0 Technologies in Sustainable Enterprise Resource Planning (S-ERP)

9.1 Introduction

System integration in Industry 4.0 refers to the comprehensive process of connecting various technological systems, both horizontally and vertically, within an industrial setup to create a cohesive and interconnected digital environment. This integration enables the seamless flow of information across different subsystems, departments, and even across the entire supply chain, allowing for enhanced decision-making, real-time monitoring, and improved operational efficiency.

In the context of Industry 4.0, horizontal integration involves linking different functional units, such as production, logistics, and sales, across the value chain. This ensures that data and information are consistently shared among all stakeholders, enabling more informed decision-making and greater agility in responding to market demands. Vertical integration, on the other hand, connects all levels of an organization, from the shop floor to the top management, ensuring that data from operational processes are available to decision-makers at all levels. This holistic approach allows for the automation and optimization of processes, leading to the development of smart factories characterized by high levels of flexibility, efficiency, and adaptability. We have extensively discussed horizontal and vertical system integration in detail in Chap. 2 of the present book.

Despite its significant advantages, system integration in Industry 4.0 presents several challenges. One of the primary difficulties is the interoperability of different systems and technologies. Industrial environments often involve a wide range of legacy systems, proprietary technologies, and varying communication protocols, making it difficult to achieve seamless integration. Moreover, the complexity of integrating loT devices, sensors, and other smart technologies further complicates this process, especially when aiming for real-time data exchange and processing.

Another major challenge is data security. As systems become more interconnected, they also become more vulnerable to cyberattacks. Ensuring the security of integrated systems, especially when dealing with sensitive industrial data, requires

K. Anjaria, *Sustainable Enterprise Resource Planning (S-ERP) for Industry 4.0*, Contributions to Management Science, https://doi.org/10.1007/978-981-96-7734-4_9

robust cybersecurity measures that can safeguard against potential breaches while maintaining system integrity and availability. Lastly, the process of system integration can be resource-intensive, requiring substantial investments in both time and capital. Organizations must invest in new technologies, training, and infrastructure to support the integration process, which can be a significant barrier, particularly for small and medium-sized enterprises (SMEs). Additionally, the transition from traditional systems to an integrated Industry 4.0 environment requires careful planning and change management to ensure a smooth and effective implementation.

In the evolving landscape of Industry 4.0, the increasing automation of shop floors through the integration of advanced technologies, such as robotics and AI, has significantly enhanced productivity and operational efficiency. However, total automation remains impractical in certain industrial contexts due to the complex, nonrepetitive nature of some tasks that require human judgment, creativity, and adaptability. In such scenarios, Operator Assistance Systems (OAS) emerge as a pivotal solution to bridge the gap between human labor and automation, particularly within S-ERP frameworks.

Operator Assistance Systems (OAS) are designed to augment the cognitive and physical capabilities of human operators, enabling them to perform industrial tasks more effectively and efficiently. These systems interact with operators to provide real-time guidance, support decision-making, and enhance situational awareness. For instance, OAS can deliver visual cues through augmented reality (AR) interfaces, assist in complex assembly tasks, or provide ergonomic support through exoskeletons that reduce physical strain. By enhancing the operator's capabilities, OAS not only increases productivity but also empowers the workforce by transforming manual labor into a more skilled, value-added activity.

In environments where total automation is either ineffective or economically unfeasible, OAS plays a crucial role in maintaining high levels of productivity. For example, tasks that involve intricate assembly processes, customized production runs, or unexpected disruptions in the manufacturing process often require the flexibility and problem-solving abilities of human operators. OAS can support these operators by providing them with the necessary tools to perform their tasks with greater precision and less physical effort. This symbiotic relationship between human operators and automated systems enhances overall operational efficiency while preserving the unique value that human labor brings to the industrial setting.

Moreover, the integration of OAS within S-ERP systems aligns with the broader goals of Industry 4.0, which emphasize not only automation but also the sustainable and responsible use of technology. By supporting operators in performing their tasks more effectively, OAS contribute to optimizing resource use, reducing waste, and improving workplace safety. These benefits align with the principles of sustainable development and highlight the role of human-centered automation in achieving long-term industrial sustainability.

In Chap. 2, we have explored the theoretical foundations of technology integration within Industry 4.0. However, it is important to recognize that organizations cannot transition to full automation in a single step. The journey typically involves a phased approach, where organizations initially adopt Operator Assistance Systems (OAS) as an intermediary stage before moving toward complete automation. To ensure that this book offers practical insights relevant to real-world applications of Industry 4.0, this chapter will focus on the strategic implementation of OAS as a critical step in the progression toward full automation. By examining the practical steps involved in integrating OAS, we aim to provide a comprehensive guide that aligns theoretical concepts with actionable strategies for industry professionals.

Following the introduction, the chapter moves into a discussion on Operator Assistance Systems (OAS), which are highlighted as a critical step for implementing S-ERP in Industry 4.0. This section also includes an examination of the strategic policies and tactics required for the effective deployment of OAS, emphasizing their role in enhancing workforce productivity and system efficiency. The chapter then transitions to the integration of low-carbon technologies into S-ERP systems, exploring their importance in promoting sustainability within industrial processes. A practical example is provided, demonstrating how to assess the integration of low-carbon technologies using rough set theory. This is followed by a detailed analysis of cleaner production technologies and their integration into S-ERP systems, with a specific focus on improving environmental sustainability. An example of implementing carbon flow analysis to evaluate cleaner production technologies is presented to illustrate the application of this method within an S-ERP framework.

As the chapter progresses, it examines the integration of quantum technologies within S-ERP systems, discussing their potential to revolutionize industrial processes by enhancing computational capabilities and data security. The chapter then shifts focus to blockchain technology, exploring how its integration into S-ERP systems can improve security, transparency, and efficiency, which are crucial for maintaining the integrity of industrial operations. The chapter concludes with a discussion on the integration of Digital Twin and Additive Manufacturing technologies within S-ERP systems. This final section highlights the synergistic impact of these technologies on optimizing manufacturing processes and driving innovation in Industry 4.0. Through this structured approach, the chapter allows the reader to progressively build their understanding of how each technology contributes to optimizing industrial processes through S-ERP in the context of Industry 4.0. The following section discusses OAS.

9.2 Operator Assistance Systems (OAS): A Critical Step for S-ERP in Industry 4.0

We begin this section by contextualizing the challenges faced by automotive manufacturers such as General Motors (GM) and Tesla in implementing complete automation strategies, specifically "lights-out" factories. These efforts aimed to create fully automated production environments operated exclusively by robots, which were expected to increase efficiency and reduce costs. However, both companies encountered significant operational challenges, resulting in inefficiencies and, ultimately, a failure to achieve the desired productivity improvements.

We argue that industrial work often involves tasks that are not easily automated because they are more complex and nuanced than simple repetitive actions. As a result, despite advances in automation, human operators remain essential on the shop floor, particularly in tasks requiring cognitive and problem-solving abilities. We suggest that the future of manufacturing will likely continue to involve a combination of human and machine workforces, with Operator Assistance Systems (OAS) playing a critical role in enhancing human capabilities rather than replacing them [13].

OAS is described as systems that interact with human operators to modify and enhance their cognitive or physical abilities while performing industrial tasks [11]. The literature emphasizes the need for a human-centered approach to technology integration in industrial settings, which requires understanding the specific contexts in which these systems will be deployed and the implications for the workforce. To address this gap, we need an approach that includes qualitative expert interviews and ethnographic observations across various industrial contexts. The following subsection discusses OAS's policy.

9.2.1 OAS Policy and Tactics

OAS shall be designed and implemented considering the specific industrial environment and tasks. Systems must be contextually appropriate, enhancing both the cognitive and physical capabilities of operators in a manner that reflects the unique needs of the operational setting. The primary focus of OAS development will be on improving operators' cognitive abilities, such as inductive reasoning, problem-solving, and decision-making. These enhancements are critical for managing the increasing complexity of modern industrial processes. The deployment of OAS must prioritize systems that support cognitive functions to ensure operators can effectively navigate and manage advanced production environments.

While cognitive support is prioritized, physical augmentation systems, such as exoskeletons, will be approached with caution. The policy should recognize the reservations expressed by operators regarding body-worn devices and instead promote the use of non-invasive physical assistance or autonomous robotic support where feasible. Physical augmentation solutions must be designed to ensure operator comfort and acceptance, avoiding unnecessary physical intrusion.

To address the differing perspectives on OAS between strategic management and operational workers, the policy should mandate inclusive dialogue and collaboration across all levels of the organization during the development and implementation phases. Ensuring that OAS meets the needs and concerns of all stakeholders is critical for the successful adoption and integration of these systems.

The successful integration of OAS requires a human-centric approach. All OAS designs and implementations must consider the contextual user requirements, ensuring that systems are not only technologically advanced but also aligned with the practical and operational needs of the workforce. The organization commits to ongoing assessment and adjustment of OAS to maintain this alignment as technology and work environments evolve. The perspectives of tactics in the context of OAS in industrial environments are marked by a heterogeneous and sometimes contradictory outlook on the future of manufacturing. While some employees and stakeholders may view the development of the shop floor positively, particularly in improving occupational health and safety and enhancing effectiveness and efficiency through operator support, others may have divergent views. These divergences often reflect stakeholders' varied expectations and experiences regarding the role and implementation of OAS.

In addition to supporting current practices, some stakeholders may focus on the potential of OAS to introduce new methodologies and practices on the shop floor. These could include advanced diagnostic tools, real-time monitoring systems, and other technological innovations that augment the operator's ability to manage complex tasks and workflows. This perspective suggests that OAS can catalyze continuous improvement and innovation in industrial processes. However, the tactical perspectives are not universally aligned. Some participants may express concerns about the implementation of OAS, particularly regarding the potential mismatch between the expectations of different hierarchical levels within the organization. While higher-level management may see OAS as a tool for significant transformation, operators and mid-level managers may be more skeptical or concerned about the practical implications of such systems. This misalignment indicates the need for careful management of expectations and inclusive dialogue during the implementation of OAS to ensure that all stakeholders are on board and that the systems are designed to meet the actual needs of those working on the shop floor.

The critical need for aligning OAS with the specific needs and contexts of the workforce is to enhance productivity and safety effectively. While technological advancements offer significant potential for augmenting human capabilities, their success depends on ensuring these systems are contextually appropriate and aligned with the industrial environment. The potential misalignment of stakeholder expectations across different hierarchical levels is a challenge, as well as the need for robust and intuitive system design and the importance of providing both cognitive and physical support in a way that is acceptable and beneficial to operators. The importance of a human-centric approach is emphasized, focusing on understanding and prioritizing the practical realities and preferences of the workforce. The successful integration of technology in industrial settings will require careful consideration of user requirements, creating flexible and adaptable systems, and fostering collaboration across all levels of the organization. This approach will help ensure that technology deployment projects are aligned with the goals and needs of all stakeholders, leading to a more efficient, safe, and sustainable future for manufacturing. The following section will discuss the integration of low-carbon technology and its assessment.

9.3 Integration of Low-Carbon Technology and Its Assessment

Low-carbon technology integration refers to the systematic incorporation of technologies aimed at reducing carbon emissions within industries, particularly those with high-energy consumption, such as the construction sector [15]. This integration involves managing and evaluating the implementation of these technologies through a comprehensive approach that considers the entire lifecycle of projects, from initial resource input to the final output.

Key aspects of low-carbon technology integration include the development of an evaluation system that assesses the effectiveness of these technologies based on several criteria: resource input, process control, and output performance. The goal is to ensure that the technology not only reduces carbon emissions but also aligns with the economic, social, and environmental objectives of the industry.

The management process for integrating low-carbon technologies is complex and requires collaboration across various stakeholders, ensuring that all parties involved in a project are aligned in their goals and methodologies. The effectiveness of low-carbon technology integration is typically measured through a combination of quantitative and qualitative indicators, which evaluate both the technological innovation and its broader impact on sustainability.

S-ERP in Industry 4.0 can significantly contribute to the integration of low-carbon technologies by providing a comprehensive, systematic approach to managing the resources, processes, and outputs associated with industrial operations. S-ERP enables the effective integration of low-carbon technologies by facilitating real-time data collection, process optimization, and decision-making, all of which are critical for reducing carbon emissions and improving energy efficiency across the entire lifecycle of industrial projects.

S-ERP systems are designed to integrate various technological and management aspects of an organization, ensuring that all elements of the production process are aligned with sustainability goals. In the context of low-carbon technology integration, S-ERP can manage the inputs, such as materials and energy, and optimize their usage to minimize waste and emissions. The system also supports process control, allowing for continuous monitoring and adjustment of operations to ensure that carbon reduction targets are met.

Moreover, S-ERP provides a platform for evaluating the performance of low-carbon technologies by analyzing the outputs of industrial processes. This includes measuring carbon emissions, energy savings, and other sustainability metrics, which are essential for assessing the effectiveness of the technologies implemented. By integrating these evaluations into the overall management system, S-ERP helps organizations achieve a balanced approach to sustainability, where economic, social, and environmental objectives are harmonized. Next, we discuss the assessment index for Low-Carbon Technology (LCT) integration using rough set theory as discussed by Lai et al. in [8]. To develop an assessment index for LCT integration innovation using rough set theory within the framework of S-ERP in Industry 4.0, the methodology involves the following steps:

Step 1: Index Selection and Data Collection via S-ERP
Utilize the S-ERP system to gather relevant data on indicators that reflect the comprehensive performance of low-carbon technology integration. These indicators should cover aspects such as resource input, technology integration, management processes, and output performance. The S-ERP system can streamline the selection process by leveraging data analytics and historical performance records to ensure that the chosen indicators are relevant, comprehensive, and based on real-time data.

Step 2: Application of Rough Set Theory Through S-ERP
Use the S-ERP system to create a knowledge representation system. This will be a digital tabular representation where columns represent selected indicators (attributes) and rows represent different construction projects (objects) tracked by the S-ERP. The S-ERP system can automate the discretization of continuous data into intervals or categories. For example, the system can automatically categorize project performance scores into levels such as "low," "medium," and "high," based on predefined thresholds. Implement rough set theory calculations within the S-ERP system to determine the lower approximation $R(X)$ and upper approximation $R^*(X)$ of the data. The S-ERP can automate these calculations, ensuring real-time classification of data with varying degrees of certainty. Use S-ERP to calculate the dependency degree $\gamma(P, Q)$, which measures how much the decision attribute (e.g., overall project performance) depends on the condition attributes (selected indicators). The S-ERP system can continuously update this dependency degree as new data is collected, providing a dynamic understanding of the relationships between different factors.

Step 3: Weight Calculation Using S-ERP
The S-ERP system can automate the determination of attribute importance by calculating how much the classification of objects changes when an attribute is removed. This importance is defined as:

$$\text{Importance} = \gamma(C, D) - \gamma(C - \{a\}, D)$$

where a is the attribute being evaluated. The system can then automatically rank these attributes based on their calculated importance. The S-ERP system can normalize the importance scores to derive the final weights for each indicator. This process is automated, ensuring that all indicators are weighted appropriately based on their relative importance to the overall evaluation.

Step 4: Performance Evaluation via S-ERP
The S-ERP system can evaluate the overall performance of LCT integration innovation for each project by summing the weighted scores of the selected indicators:

$$P_j = \sum_{i=1}^{n} W_i \times \text{ Score of indicator } i \text{ for project } j$$

Here, P_j represents the performance score for project j. The S-ERP system automates this calculation, providing real-time performance scores for each project, which can be visualized in dashboards or reports.

Step 5: Interpretation and Decision-Making in S-ERP
The S-ERP system can rank the projects based on their performance scores, offering insights into which projects are most effective in achieving low-carbon technology integration. The system can also generate recommendations for improvements, guiding resource allocation and strategic decisions to optimize LCT integration.

9.3.1 Example: Implementation of Low-Carbon Technology Integration Innovation Assessment Using S-ERP in Industry 4.0 with Rough Set Theory

A construction company uses an S-ERP system in an Industry 4.0 environment to evaluate the performance of several ongoing projects based on their low-carbon technology (LCT) integration. The company has selected several key indicators to assess the integration's success and is applying rough set theory to determine the importance of each indicator, which will be used to calculate the overall performance score for each project.

Step 1: Index Selection and Data Collection via S-ERP
The company has chosen the following indicators (attributes) to assess LCT integration:

1. Rate of LCT Investment (Z11)
2. Portion of LCT Engineers (Z12)
3. Assessment of Low-Carbon Standards (Z13)
4. Operation Cost Efficiency (Z14)
5. Rate of LCT Adoption (Z15)

The S-ERP system collects real-time data for these indicators from various projects. Here are the data for three projects:

- ***Project A:***
- Z11: 2.0% (Moderate)
- Z12:60% (Moderate)
- Z13: 2 (Good)
- Z14: 1.5 (Moderate)
- Z15: 2 (Good)
- ***Project B:***
- Z11:3.1% (High)
- Z12: 75% (High)
- Z13: 3 (Excellent)
- Z14: 1.2 (High)
- Z15: 3 (Excellent)

- ***Project C:***
- Z11: 1.5% (Low)
- Z12: 40% (Low)
- Z13: 1 (Basic)
- Z14: 1.8 (Low)
- Z15: 1 (Basic)

The numbers above are indicative of having different ordinal scale values and units.

Step 2: Application of Rough Set Theory Through S-ERP
The S-ERP system organizes the collected data into a tabular form, where each row represents a project, and each column represents an indicator. This table serves as the foundation for applying rough set theory.

The continuous data for each indicator is discretized into categories:

- Z11 (Investment Rate): Low (0-1.5%), Moderate (1.6-3.0%), High (3.1%+)
- Z12 (Portion of LCT Engineers): Low (0-40%), Moderate (41-70%), High (71%+)
- Z13 (Standard Assessment): Basic (1), Good (2), Excellent (3)
- Z14 (Cost Efficiency): Low (1.5-1.8), Moderate (1.2-1.4), High (1.0-1.1)
- Z15 (Rate of Adoption): Basic (1), Good (2), Excellent (3)

Using rough set theory, the S-ERP system calculates the lower approximation $R(X)$ and upper approximation $R^*(X)$ of the data for each project. These approximations help in identifying which projects can be confidently categorized as high performing in LCT integration. The dependency degree $\gamma(P, Q)$ is calculated by the S-ERP system to measure the extent to which the decision attribute (overall project performance) depends on each of the condition attributes (indicators). For instance:

$$\gamma(\text{Z11}, \text{ Performance}) = \frac{|\text{ POS}_{\text{Z11}}(\text{ Performance }) \text{ }|}{|U|}$$

where POS $_{Z11}$ (Performance) is the positive region representing projects that can be classified with certainty based on LCT investment rate.

Step 3: Weight Calculation Using S-ERP
The S-ERP system assesses how much the classification of projects changes when each attribute is removed. The importance of each attribute is calculated as: Importance of $Z11 = \gamma($ All Indicators, Performance $) - \gamma($ All Indicators except Z11, Performance)

Attributes that cause significant changes in classification when removed are assigned higher importance. The importance scores are then normalized to calculate the final weights for each indicator:

$$W_{Z11} = \frac{\text{Importance of } Z11}{\sum \text{ Importance of all indicators}}$$

Assuming the following calculated and normalized importance scores:

- Z11 (Rate of LCT Investment): 0.18
- Z12 (Portion of LCT Engineers): 0.22
- Z13 (Assessment of Standards): 0.30
- Z14 (Cost Efficiency): 0.15
- Z15 (Rate of LCT Adoption): 0.15

Step 4: Performance Evaluation via S-ERP
The overall performance score for each project is calculated by the S-ERP system using the weighted sum of the indicators:

Project A

$$P_A = (0.18 \times 2) + (0.22 \times 2) + (0.30 \times 2) + (0.15 \times 1.5) + (0.15 \times 2)$$
$$= 0.36 + 0.44 + 0.60 + 0.225 + 0.30 = 1.925$$

Project B

$$\mathrm{P}_B = (0.18 \times 3) + (0.22 \times 3) + (0.30 \times 3) + (0.15 \times 1.2) + (0.15 \times 3)$$
$$= 0.54 + 0.66 + 0.90 + 0.18 + 0.45 = 2.73$$

Project C

$$\mathrm{P}_C = (0.18 \times 1) + (0.22 \times 1) + (0.30 \times 1) + (0.15 \times 1.8) + (0.15 \times 1)$$
$$= 0.18 + 0.22 + 0.30 + 0.27 + 0.15 = 1.12$$

Step 5: Interpretation and Decision-Making in S-ERP
Based on the performance scores, Project B ranks highest (2.73), followed by Project A (1.925) and Project C (1.12). The S-ERP system provides the following insights and recommendations:

- ***Project A:*** Focus on improving LCT investment and rate of adoption to close the gap with higher-performing projects.
- ***Project B:*** Continue with the current strategy, as it is effectively integrating low-carbon technologies.

The following section discusses cleaner production technology integration.

9.4 Integration of Cleaner Production Technology and Its Analysis

Cleaner production technology integration refers to the systematic incorporation of environmentally friendly technologies into industrial processes to reduce resource consumption, minimize waste, and lower emissions at the source [16]. This approach is distinct from end-of-pipe technologies, which focus on treating waste after it has been produced. Cleaner Production (CP) technologies aim to achieve sustainable production by optimizing the use of resources such as energy and raw materials, thereby reducing the environmental impact of industrial activities. In industries that produce large amounts of biomass waste, such as the herbal medicine industry, CP technology integration plays a crucial role in promoting low-carbon development. The integration process typically involves adopting technologies that convert biomass waste into valuable products, such as bio-oil, through processes like pyrolysis. This not only helps reduce dependency on fossil fuels but also minimizes the industry's carbon footprint.

The evaluation of CP technology integration is often conducted using tools like carbon flow analysis (CFA), which assesses the flow of carbon throughout the production process. CFA helps quantify the carbon content in inputs, outputs, and wastes, thereby providing a clear picture of the carbon utilization efficiency and the effectiveness of the implemented CP technologies in reducing emissions.

For instance, in the herbal medicine industry, CP technology integration might include the use of pyrolysis to convert herb residue into bio-oil, which can then be used to replace fossil fuels in the production process. This approach not only reduces the amount of solid waste but also lowers CO2 emissions. Other technologies, such as ethanol recovery and steam condensation water recycling, can further enhance the efficiency of resource utilization and contribute to a significant reduction in the environmental impact. The goal is to evaluate the effectiveness of CP technology integration in reducing carbon emissions and improving carbon utilization efficiency in an industrial setting using CFA. This methodology can be implemented using S-ERP systems within Industry 4.0 frameworks. The methodology for the carbon flow analysis, as discussed by Meng et al., [10] is as follows:

Step 1: Establish System Boundaries for Carbon Flow Analysis
Define the system boundary for the CFA, which includes the main production processes, inputs (raw materials, energy), outputs (products, waste), and emissions (CO2, other pollutants). The system boundary typically includes:

1. Inputs: Fossil fuels, raw materials, water, steam, ethanol, etc.
2. Processes: Extraction, concentration, pyrolysis, and other production processes.
3. Outputs: Finished products, solid waste (e.g., herb residue), liquid waste (e.g., wastewater), and gaseous emissions (e.g., CO_2).

The S-ERP system within Industry 4.0 automatically captures data related to these inputs and outputs in real time, facilitating accurate and timely analysis.

Step 2: Data Collection and Carbon Content Calculation
Use the S-ERP system to collect detailed input/output data related to production processes. This includes the quantities of raw materials, energy consumption, products, and waste generated. The S-ERP system can also track the carbon content of these inputs and outputs. Calculate the carbon content of each input/output using the following equation:

$$C_i = M_i \times W_i$$

where:

- C_i is the carbon content (tons).
- M_i is the quantity of the material (tons).
- W_i is the carbon content percentage (obtained from elemental analysis).

For fossil fuels, use the IPCC guidelines to convert the energy consumed into carbon content.

Example Calculation for Fossil Fuels:

$$C_{EC} = M_{\text{Coal}} \times W_{\text{Coal}}$$

where M_{Coal} is the quantity of coal consumed, and W_{Coal} is the carbon content of the coal.

Step 3: Carbon Flow Analysis Using S-ERP
The S-ERP system helps in balancing the carbon flows by tracking all carbon inputs, stocks, and outputs:

$$CEC + CMC = CP + CSR + CLE + CGE$$

where:

- CEC: Carbon from energy consumption.
- CMC : Carbon in raw materials.
- CP: Carbon stored in the product.
- CSR: Carbon in solid waste.
- CLE: Carbon in liquid waste.
- CGE: Carbon in gaseous emissions.

Carbon Emissions Calculation:

$$CW = CSR + CLE + CGE = (CEC + CMC) - CP$$

where CW denotes the sum of carbon in the waste.

Step 4: Low-Carbon Evaluation Indicators
Use the following key performance indicators (KPIs) to evaluate the effectiveness of CP technology integration:

1. Percentage of Carbon Stored in Product (RC-P):

$$RC - P = \frac{CP}{CEC + CMC} \times 100\%$$

2. Percentage of Carbon Emitted as CO 2 (RC-CO2):

$$RC - CO2 = \frac{CEC + CCOD - CO2}{CEC + CMC} \times 100\%$$

3. Percentage of Carbon in COD (RC-COD):

$$RC - COD = \frac{CCOD}{CEC + CMC} \times 100\%$$

4. Percentage of Carbon in Solid Waste (RC-SW):

$$RC - SW = 1 - RC - P - RC - CO2 - RC - COD$$

The S-ERP system automatically calculates these indicators based on real-time data, providing insights into the carbon efficiency of the production processes.

Step 5: Interpretation and Optimization
Use the S-ERP system to compare these KPIs across different production scenarios (e.g., before and after CP technology integration). After that, identify areas for improvement, such as reducing carbon emissions or increasing carbon utilization efficiency. Implement corrective measures based on the insights gained, such as optimizing resource use, recycling waste, or adjusting production processes.

By integrating carbon flow analysis with an S-ERP system in an Industry 4.0 environment, organizations can achieve a comprehensive and real-time evaluation of their cleaner production technology integration efforts. This approach not only enhances carbon efficiency but also supports sustainable industrial practices by providing actionable insights for continuous improvement.

9.4.1 Example Implementation of Carbon Flow Analysis for Cleaner Production Technology Integration Using S-ERP in Industry 4.0

A pharmaceutical company is using an S-ERP system within an Industry 4.0 framework to evaluate the integration of Cleaner Production (CP) technologies in its manufacturing process. The company produces herbal medicines, generating significant amounts of biomass waste. The goal is to assess the effectiveness of integrating pyrolysis technology to convert herb residue into bio-oil and reduce carbon emissions.

Step 1: Establish System Boundaries for Carbon Flow Analysis
The company defines the system boundaries for the CFA to include:

- Inputs: Fossil fuels, raw herbal materials, water, steam, ethanol.
- Processes: Extraction, concentration, pyrolysis of herb residue.
- Outputs: Finished herbal products, bio-oil, solid waste (e.g., ash), liquid waste (e.g., wastewater), gaseous emissions (e.g., CO2).

The S-ERP system tracks and logs data on all these inputs and outputs in real time.

Step 2: Data Collection and Carbon Content Calculation
The S-ERP system collects the following data:

- Fossil Fuels (Coal): 50 tons, carbon content = 70%.
- Raw Herbal Material: 200 tons, carbon content = 40%.
- Products (Herbal Medicine): 120 tons, carbon content = 20%.
- Bio-oil Produced: 30 tons, carbon content = 80%.
- Solid Waste (Ash): 50 tons, carbon content = 5%.
- Liquid Waste: 20 tons, carbon content = 2%.
- Gaseous Emissions (CO2): 50 tons.

The S-ERP system calculates the carbon content for each input/output using the equation:

$$C_i = M_i \times W_i$$

where:

- C_i is the carbon content (tons).
- M_i is the quantity (tons).
- W_i is the carbon content percentage.

Calculations:

- Carbon from Fossil Fuels (Coal):

$$C_{\text{Coal}} = 50 \text{ tons } \times 0.70 = 35 \text{ tons}$$

- Carbon in Raw Herbal Material:

$$C_{\text{Herb}} = 200 \text{ tons } \times 0.40 = 80 \text{ tons}$$

- Carbon in Products (Herbal Medicine):

$$C_{\text{Product}} = 120 \text{ tons } \times 0.20 = 24 \text{ tons}$$

- Carbon in Bio-oil:

$$C_{\text{Bio-oil}} = 30 \text{ tons } \times 0.80 = 24 \text{ tons}$$

- Carbon in Solid Waste (Ash):

$$C_{\text{Ash}} = 50 \text{ tons } \times 0.05 = 2.5 \text{ tons}$$

- Carbon in Liquid Waste:

$$C_{\text{LiquidWaste}} = 20 \text{ tons } \times 0.02 = 0.4 \text{ tons}$$

- Carbon in Gaseous Emissions (CO2):

$$C_{CO2} = 50 \text{ tons}$$

Step 3: Carbon Flow Analysis Using S-ERP
Carbon Balance Equation: The S-ERP system uses the carbon balance equation to ensure that the carbon inputs equal the sum of carbon stored in the products and waste:

$$CEC + CMC = CP + CSR + CLE + CGE$$

Substituting the values:

$$35 + 80 = 24 + 24 + 2.5 + 0.4 + 50$$

The system confirms:

$$115 = 100.9$$

A discrepancy indicates a need for further data refinement or process adjustment, which is typical in real-time evaluations.

$$CW = CSR + CLE + CGE = (35 + 80) - (24 + 24) = 115 - 48 = 67 \text{ tons}$$

where CW is the sum of carbon in the waste streams.

Step 4: Low-Carbon Evaluation Indicators
The S-ERP system calculates key performance indicators (KPIs) to evaluate the effectiveness of CP technology integration:
1. Percentage of Carbon Stored in Product (RC-P):

$$RC - P = \frac{CP}{CEC + CMC} \times 100\% = \frac{24 + 24}{115} \times 100\% = \frac{48}{115} \times 100\% \approx 41.7\%$$

2. Percentage of Carbon Emitted as CO 2 (RC-CO2):

$$RC - CO2 = \frac{CGE}{CEC + CMC} \times 100\% = \frac{50}{115} \times 100\% \approx 43.5\%$$

3. Percentage of Carbon in Solid Waste (RC-SW):

$$RC - SW = \frac{CSR}{CEC + CMC} \times 100\% = \frac{2.5}{115} \times 100\% \approx 2.17\%$$

4. Percentage of Carbon in Liquid Waste (RC-LW):

$$RC - LW = \frac{CLE}{CEC + CMC} \times 100\% = \frac{0.4}{115} \times 100\% \approx 0.35\%$$

Step 5: Interpretation and Optimization
The S-ERP system identifies that a significant portion of carbon (43.5%) is emitted as CO2, which suggests a need to improve the pyrolysis process to reduce gaseous emissions. The relatively high percentage of carbon stored in products (41.7%) is positive but could be further optimized by increasing the efficiency of carbon conversion into bio-oil. The following section discusses quantum technology integration in S-ERP.

9.5 Quantum Technology Integration in S-ERP for Industry 4.0

Quantum technology integration in the context of Industry 4.0 refers to the incorporation of quantum computing and other quantum-based technologies into industrial processes to enhance computational capabilities, improve security, and optimize operations [12]. Industry 4.0 represents the fourth industrial revolution, characterized by the fusion of physical, digital, and biological systems, with advanced technologies like the Internet of Things (IoT), artificial intelligence (AI), and big data playing central roles. Quantum technology integration is poised to further revolutionize this landscape by providing solutions to complex problems that classical computing systems cannot efficiently solve.

Quantum computing, one of the core components of quantum technology, leverages the principles of quantum mechanics to perform calculations at unprecedented speeds [3]. This capability is particularly valuable in scenarios requiring large-scale data processing, complex optimization, and simulation tasks that are beyond the reach of classical computers. In Industry 4.0, quantum computing can be applied to optimize supply chains, design new materials, enhance machine learning models, and solve intricate logistical problems.

Additionally, quantum communication technologies, such as quantum key distribution (QKD), offer a new level of security for data transmission, which is crucial for protecting sensitive information in interconnected industrial environments. QKD uses the principles of quantum mechanics to create encryption keys that are theoretically impossible to break, providing robust protection against cyber threats.

The integration of quantum technologies also includes quantum sensing, which enhances the precision of measurements and data collection in industrial processes. Quantum sensors can detect minute changes in physical conditions, such as temperature, pressure, or magnetic fields, with extreme accuracy, improving the quality control and efficiency of manufacturing processes.

However, integrating quantum technology into Industry 4.0 comes with challenges. These include the current maturity level of quantum technologies, the need for specialized infrastructure, and the interoperability between quantum and classical systems. Addressing these challenges requires ongoing research, development, and collaboration between industry and academia to fully realize the potential of quantum technologies in industrial applications. Now, let's discuss the conceptual framework of the integration of quantum technology in Industry 4.0 with S-ERP.

Quantum technology integration within Industry 4.0 involves embedding quantum computing, quantum communication, and quantum sensing technologies into industrial processes to enhance computational power, data security, and operational efficiency. The integration aims to address the growing complexity and demand for real-time processing in smart manufacturing environments. The conceptual framework for this integration focuses on three core areas: Technological Foundations, Integration and Implementation, and Societal and Policy Considerations. Technology Foundation addresses the fundamental technological requirements for integrating quantum technologies into Industry 4.0 environments.

- **Quantum Computing:** Quantum computing plays a pivotal role by offering unprecedented computational capabilities that can solve complex optimization problems, simulate new materials, and enhance machine learning algorithms used in manufacturing processes [3].
- **Quantum Communication:** This involves using quantum cryptography, particularly quantum key distribution (QKD), to secure industrial communication networks. The goal is to protect sensitive data from cyber threats that could compromise smart factories and connected devices [4].
- **Quantum Sensing:** Quantum sensors provide highly accurate measurements, which are crucial for quality control, predictive maintenance, and environmental monitoring in industrial settings. These sensors outperform classical sensors in terms of sensitivity and precision, contributing to higher operational efficiency [6].

Integration and implementation focus on the practical aspects of embedding quantum technologies into existing Industry 4.0 infrastructures.

- **Scalability and Interoperability:** One of the key challenges is scaling quantum technologies to meet the demands of large-scale industrial operations. Additionally, ensuring interoperability between quantum and classical systems is crucial for seamless integration.
- **Quantum Hardware Integration:** The integration of quantum hardware, such as quantum processors and communication devices, into industrial systems requires overcoming challenges related to size, environmental sensitivity, and the need for specialized operating conditions.

- **System Architecture:** Industry 4.0 systems must be re-architected to accommodate quantum technologies. This involves redesigning networks to incorporate quantum communication channels and integrating quantum computing units with traditional IT infrastructure.
- **Coexistence with Classical Systems:** During the transition to fully quantum-enhanced Industry 4.0 environments, it is essential to ensure that quantum systems can coexist and operate alongside classical computing systems without conflicts.

Finally, societal and policy considerations address the broader implications of quantum technology integration, focusing on ethical, regulatory, and societal impacts.

- **Regulatory and Ethical Frameworks:** Developing regulatory frameworks that govern the use of quantum technologies in industry is essential to ensure responsible innovation. This includes creating standards for quantum communications, data privacy, and security.
- **Talent and Skills Development:** The successful integration of quantum technologies requires a workforce skilled in both quantum mechanics and industrial applications. Developing educational programs and training initiatives is crucial to bridging this skills gap.
- **Public Perception and Trust:** Building public trust in quantum-enhanced industrial systems involves transparent communication about the benefits and risks associated with these technologies. It also requires addressing concerns related to privacy, security, and the potential societal impact of widespread quantum technology adoption.

The conceptual framework for quantum technology integration in Industry 4.0 highlights the multifaceted challenges and opportunities associated with embedding quantum technologies into industrial environments. By addressing technological foundations, practical implementation, and societal considerations, this framework provides a comprehensive approach to realizing the full potential of quantum technologies in transforming industrial processes, enhancing security, and driving innovation. This integration is poised to significantly advance the capabilities of Industry 4.0, paving the way for more intelligent, secure, and efficient industrial systems. The following section discusses the integration of the blockchain.

9.6 Blockchain Integration for S-ERP in Industry 4.0

Blockchain integration within the context of S-ERP in Industry 4.0 refers to incorporating blockchain technology to enhance enterprise operations' reliability, transparency, and security. S-ERP systems, designed to manage and optimize the complex processes of modern industrial organizations, benefit significantly from blockchain's immutable and decentralized ledger system.

Blockchain technology ensures that all data recorded in the S-ERP system is secure, transparent, and tamper-proof [2]. Each transaction or data entry is stored in a block that is cryptographically secured and linked to the previous block, forming a chain of records that cannot be altered without consensus from the network. This is particularly important in maintaining the integrity of financial records, inventory data, and other critical enterprise information. Blockchain integration allows for real-time tracking and verification of goods and materials throughout the supply chain [14]. Each stage of the supply chain, from procurement to delivery, can be recorded on the blockchain, providing an immutable record that enhances transparency and accountability. This is especially useful for verifying the authenticity of materials, ensuring compliance with regulations, and detecting any disruptions in the supply chain.

S-ERP systems can leverage blockchain-based smart contracts to automate business processes [5]. These smart contracts are self-executing agreements with the terms directly written into code. In an industrial context, smart contracts can automate tasks such as order fulfillment, payment processing, and regulatory compliance, reducing the need for manual intervention and minimizing errors. Traditional S-ERP systems rely on centralized databases that can be vulnerable to attacks or failures. Blockchain's decentralized architecture distributes data across multiple nodes, ensuring that the system remains operational even if some nodes fail. This decentralization also enhances data privacy and reduces the risk of data breaches, as there is no single point of failure.

Blockchain technology simplifies compliance with regulatory requirements by providing a transparent and easily auditable trail of transactions [1]. S-ERP systems integrated with blockchain can automatically generate compliance reports and audit trails, ensuring that all operations meet legal and regulatory standards. This is particularly beneficial in industries with stringent compliance requirements, such as pharmaceuticals, aerospace, and finance. Blockchain enables secure and transparent collaboration between different entities in the value chain, including suppliers, manufacturers, and customers. By providing a shared, trusted ledger, blockchain reduces the potential for disputes and enhances trust among all participants, leading to more efficient and cooperative business operations. Integrating blockchain with S-ERP systems in Industry 4.0 presents challenges such as the need for significant computational resources, the integration of legacy systems, and ensuring scalability. Overcoming these challenges requires careful planning, technological innovation, and a strong alignment between business objectives and technological capabilities. The following section explains the integration of digital twin and additive manufacturing technologies.

9.7 Integration of Digital Twin and Additive Manufacturing Technologies

Digital Twin (DT) technology is a core component of Industry 4.0, enabling the creation of digital replicas of physical systems, processes, and products [9]. In the context of S-ERP for Industry 4.0, DTs serve as real-time digital counterparts of physical assets, capturing and analyzing data from various sensors embedded within the physical systems. This integration allows for continuous monitoring, predictive maintenance, and process optimization, leading to enhanced decision-making and operational efficiency.

In S-ERP systems, DTs facilitate the simulation of production scenarios, forecasting potential issues and optimizing manufacturing processes before they occur in the physical world [7]. This predictive capability is critical in maintaining sustainability goals, as it reduces waste, improves resource efficiency, and minimizes downtime. Furthermore, DTs enable a more responsive and agile manufacturing environment by providing real-time feedback, allowing for immediate adjustments to be made in the production process.

Additive Manufacturing (AM), commonly known as 3D printing, is another pivotal technology in Industry 4.0. AM allows for the layer-by-layer construction of complex and customized products, offering significant advantages in terms of design flexibility, material efficiency, and production speed. Integrating AM into S-ERP systems within Industry 4.0 environments enhances the capabilities of traditional manufacturing by enabling the production of highly intricate components that would be challenging or impossible to create using conventional methods.

Within an S-ERP framework, AM is leveraged to streamline the production process, reduce lead times, and minimize material waste. The S-ERP system can manage the entire AM process, from design and prototyping to final production, ensuring that all stages are optimized for sustainability. This includes tracking the use of materials and energy consumption and ensuring that the production aligns with the company's environmental goals.

The integration of Digital Twin and Additive Manufacturing technologies within S-ERP systems for Industry 4.0 creates a powerful synergy that enhances the overall efficiency and sustainability of manufacturing operations. By combining the real-time monitoring and predictive capabilities of DTs with the design flexibility and material efficiency of AM, organizations can achieve a more responsive, efficient, and sustainable manufacturing process. This integration enables the continuous improvement of production processes, as the data collected by DTs can be used to refine and optimize AM operations. For instance, real-time data from DTs can inform adjustments in the AM process to reduce defects, improve material usage, and enhance the overall quality of the final product. Moreover, this combination supports the development of customized products, as the S-ERP system can quickly adapt production parameters based on DT feedback, ensuring that the AM process remains aligned with customer specifications and sustainability goals. The method of integration is as follows:

Step 1. System Architecture Design

Design an integration layer connecting Digital Twin (DT) and Additive Manufacturing (AM) technologies with the S-ERP system. This layer serves as the communication backbone, allowing real-time data exchange between physical manufacturing processes, their digital replicas, and the S-ERP system. Establish a bidirectional data flow architecture where data from the physical AM process is captured by sensors and fed into the Digital Twin. This data is then processed, analyzed, and sent to the S-ERP system. Conversely, the S-ERP system can send optimization commands or updates back to the Digital Twin, which in turn updates the physical system.

Step 2. Digital Twin Implementation
Develop a digital replica of the physical AM systems, including machines, production lines, and processes. This Digital Twin should be capable of real-time monitoring, simulation, and analysis of the AM processes. The twin must include models for process parameters, material properties, and machine behaviors. Equip the physical AM systems with a network of sensors to continuously monitor key parameters such as temperature, pressure, material flow, and machine status. These sensors should be integrated into the Digital Twin, providing the necessary data for accurate real-time replication of the physical processes. Enable the Digital Twin to perform real-time simulations of the AM processes based on the data received from the physical sensors. These simulations should predict potential issues such as defects, delays, or resource inefficiencies. The feedback from these simulations can then be used to adjust the physical processes via the S-ERP system.

Step 3. Additive Manufacturing Optimization
Utilize the data and simulations from the Digital Twin to optimize the AM processes. This can include adjusting machine parameters, improving material usage, and refining production schedules. The S-ERP system should support these optimizations by managing resource allocations, production planning, and order management. Implement predictive maintenance strategies using the Digital Twin's real-time monitoring capabilities. By analyzing sensor data and simulation results, the S-ERP system can predict when maintenance is required, minimizing downtime and improving machine longevity.

Step 4. S-ERP Integration and Control
Ensure that all data generated by the Digital Twin and the AM processes are securely stored and managed within the S-ERP system. This data should be easily accessible for analysis, reporting, and decision-making. Leverage the S-ERP system to automate production processes based on the real-time insights provided by the Digital Twin. This includes automatic scheduling, resource management, and quality control measures. Implement a continuous improvement loop where the S-ERP system uses historical data and machine learning algorithms to refine production processes over time. This loop should be informed by ongoing feedback from the Digital Twin and AM systems.

Step 5. Example Implementation
Consider an AM facility producing aerospace components. The facility integrates DT and AM technologies with an S-ERP system to enhance production efficiency and quality. Sensors installed on AM machines collect data on temperature, material feed rate, and layer quality. This data is sent to the Digital Twin, where real-time simulations predict potential issues such as layer defects or material wastage. The S-ERP system processes this data and adjusts production schedules and machine parameters to optimize material usage and reduce defects. It also triggers predictive

maintenance alerts based on machine wear data from the Digital Twin. By integrating DT and AM technologies, the facility reduces material waste by 15%, decreases production time by 10%, and improves product quality by 20%, all while minimizing unplanned downtime through predictive maintenance.

This method demonstrates how Digital Twin and Additive Manufacturing technologies can be effectively integrated using S-ERP within an Industry 4.0 framework. By leveraging real-time data, simulations, and predictive analytics, organizations can significantly improve efficiency, sustainability, and product quality. In summary, the chapter answers the following research questions.

Research Questions Addressed

The chapter titled "System Integration with Industry 4.0 Technologies in S-ERP" addresses the following key research questions:

1. **How can Digital Twin (DT) and Additive Manufacturing (AM) technologies be integrated into Sustainable Enterprise Resource Planning (S-ERP) systems within Industry 4.0?**

    - The chapter explores the systematic approach required for integrating DT and AM technologies into S-ERP systems, focusing on enhancing real-time monitoring, predictive maintenance, and process optimization in modern industrial processes.

2. **What are the challenges and benefits of integrating advanced Industry 4.0 technologies like DT and AM within S-ERP systems?**

    - The chapter examines the potential challenges of interoperability, data security, and the need for specialized infrastructure while also highlighting the significant benefits of improved efficiency, sustainability, and product quality through the synergistic application of these technologies.

3. **How can Operator Assistance Systems (OAS) bridge the gap between human labor and automation in S-ERP systems for Industry 4.0?**

    - The chapter discusses the role of OAS in enhancing human operators' cognitive and physical capabilities, enabling them to perform tasks more effectively within an industrial setting where full automation may not be practical or cost-effective.

4. **How can low-carbon and cleaner production technologies be integrated into S-ERP systems to promote sustainability within Industry 4.0?**

    - The chapter provides a methodology for integrating low-carbon and cleaner production technologies into S-ERP systems, including assessing and optimizing these technologies using tools like carbon flow analysis and rough set theory.

5. **What is the potential of quantum technologies in revolutionizing industrial processes through their integration with S-ERP systems in Industry 4.0?**
   - The chapter explores how quantum computing, quantum communication, and quantum sensing can be integrated into S-ERP systems to enhance computational capabilities, data security, and operational efficiency in complex industrial environments.
6. **How can blockchain technology enhance the transparency, security, and efficiency of S-ERP systems within Industry 4.0?**
   - The chapter discusses integrating blockchain technology into S-ERP systems, emphasizing its role in securing data, automating business processes through smart contracts, and ensuring compliance with regulatory standards.

These research questions guide the chapter's comprehensive exploration of integrating advanced Industry 4.0 technologies with S-ERP systems to optimize industrial processes, enhance sustainability, and drive innovation.

References

1. Huson YA, Sierra-García L, Garcia-Benau MA (2024) A bibliometric review of information technology, artificial intelligence, and blockchain on auditing. Total Quality Manage Busin Excell 35(1-2):91–113. ISSN 1478-3363. https://doi.org/10.1080/14783363.2023.2256260. Publisher: Routledge _eprint: https://doi.org/10.1080/14783363.2023.2256260
2. Akanfe O, Lawong D, Raghav Rao H (2024) Blockchain technology and privacy regulation: reviewing frictions and synthesizing opportunities. Int J Inf Manage 76:102753. ISSN 0268-4012. https://doi.org/10.1016/j.ijinfomgt.2024.102753. https://www.sciencedirect.com/science/article/pii/S026840122400001X
3. Anjaria K (2024) Quantum model regression for generating fuzzy numbers in adiabatic quantum computing. Inf Sci 678:121018. ISSN 0020-0255. https://doi.org/10.1016/j.ins.2024.121018. https://www.sciencedirect.com/science/article/pii/S0020025524009320
4. Conlon LO, Shajilal B, Walsh A, Zhao J, Janousek J, Lam PK, Assad SM (2024) Verifying the security of a continuous variable quantum communication protocol via quantum metrology. npj Quantum Inf 10(1):1–11. ISSN 2056-6387. https://doi.org/10.1038/s41534-024-00834-9. https://www.nature.com/articles/s41534-024-00834-9. Publisher: Nature Publishing Group
5. Datta S, Namasudra S (2024) Blockchain-based smart contract model for securing healthcare transactions by using consumer electronics and mobile-edge computing. IEEE Trans Consumer Electron 70(1):4026–4036. ISSN 1558-4127. https://doi.org/10.1109/TCE.2024.3357115. https://ieeexplore.ieee.org/abstract/document/10411994. Conference Name: IEEE Transactions on Consumer Electronics
6. DeMille D, Hutzler NR, Rey AM, Zelevinsky T (2024) Quantum sensing and metrology for fundamental physics with molecules. Nat Phys 20(5):741–749. ISSN 1745-2481. https://doi.org/10.1038/s41567-024-02499-9. https://www.nature.com/articles/s41567-024-02499-9. Publisher: Nature Publishing Group

7. Guerra-Zubiaga D, dos Santos MC, Voicu RC, Richards G, Gosnell S, Barbosa GF (2024) A digital twin approach to support a multi-task industrial robot operation using design of experiments. J Brazilian Soc Mech Sci Engin 46(8):516. ISSN 1806-3691. https://doi.org/10.1007/s40430-024-05114-3
8. Lai X, Liu J, Georgiev G (2016) Low carbon technology integration innovation assessment index review based on rough set theory – an evidence from construction industry in China. J Cleaner Product 126:88–96. ISSN 0959-6526. https://doi.org/10.1016/j.jclepro.2016.03.035. https://www.sciencedirect.com/science/article/pii/S0959652616301020
9. Liu S, Zheng P, Bao J (2024) Digital Twin-based manufacturing system: a survey based on a novel reference model. J Intell Manufact 35(6):2517–2546. ISSN 1572-8145. https://doi.org/10.1007/s10845-023-02172-7
10. Meng X, Wen Z, Qian Y, Yu H (2017) Evaluation of cleaner production technology integration for the Chinese herbal medicine industry using carbon flow analysis. J Cleaner Product 163:49–57. ISSN 0959-6526. https://doi.org/10.1016/j.jclepro.2015.10.067. https://www.sciencedirect.com/science/article/pii/S0959652615015383
11. Moencks M, Roth E, Bohné T, Kristensson PO (2022) Human-computer interaction in industry: a systematic review on the applicability and value-added of operator assistance systems. Foundat Trends® Human-Comput Interact 16(2–3), 65–213. ISSN 1551–3955:1551–3963. https://doi.org/10.1561/1100000088. https://www.nowpublishers.com/article/Details/HCI-088. Publisher: Now Publishers Inc
12. Rafi S, Akbar MA, Mahmood S (2024) A conceptual framework for quantum integration challenges in 6G technology. In: Proceedings of the 1st ACM international workshop on quantum software engineering: the next evolution, QSE-NE 2024. Association for Computing Machinery, pp 19–26. ISBN 9798400706738. https://doi.org/10.1145/3663531.3664754
13. Roth E, Moencks M, Bohné T (2023) The augmented workforce: a systematic review of operator assistance systems. In: Handbook of human-machine systems. Wiley, Ltd, pp 267–279. ISBN 978-1-119-86366-3. https://doi.org/10.1002/9781119863663.ch23. Section: 23 _eprint: https://onlinelibrary.wiley.com/doi/pdf/10.1002/9781119863663.ch23
14. Soori M, Jough FKG, Dastres R, Arezoo B (2024) Blockchains for industrial Internet of Things in sustainable supply chain management of industry 4.0, a review. Sustain Manufact Serv Econ 3:100026. ISSN 2667-3444. https://doi.org/10.1016/j.smse.2024.100026. https://www.sciencedirect.com/science/article/pii/S2667344424000094
15. Wang R, Wen X, Wang X, Fu Y, Zhang Y (2022) Low carbon optimal operation of integrated energy system based on carbon capture technology, LCA carbon emissions and ladder-type carbon trading. Appl Energy 311:118664. ISSN 0306-2619. https://doi.org/10.1016/j.apenergy.2022.118664. https://www.sciencedirect.com/science/article/pii/S0306261922001313
16. Yin S, Shao Y, Wu A, Wang H, Liu X, Wang Y (2020) A systematic review of paste technology in metal mines for cleaner production in China. J Cleaner Product 247:119590. ISSN 0959-6526. https://doi.org/10.1016/j.jclepro.2019.119590. https://www.sciencedirect.com/science/article/pii/S0959652619344609

Chapter 10
Enhancing Organizational Resilience and Agility in S-ERP for Industry 4.0

10.1 Introduction

Organizational Resilience refers to an organization's capacity to anticipate, prepare for, respond to, and adapt to incremental changes and sudden disruptions to survive and prosper. It embodies the ability to absorb shocks, recover from crises, and maintain continuous functioning, which includes but is not limited to maintaining production, serving customers, and managing supply chains in the face of adversity. This resilience is not merely about surviving disruptions but also about actively learning from these experiences to strengthen the organization against future challenges. Factors contributing to organizational resilience include strategic human resource management, a supportive organizational culture, and robust risk management practices.

Organizational Agility, on the other hand, is the dynamic capability that allows an organization to rapidly adjust and respond to changes in its internal and external environments [9]. Agility involves the flexibility to shift strategies, reconfigure resources, and adapt operations swiftly in response to market demands, technological advancements, or other significant changes. In the context of Industry 4.0, organizational agility is often enhanced through the adoption of digital technologies such as the Internet of Things (IoT), cyber-physical systems, and big data analytics. These technologies enable quicker decision-making, predictive analysis, and more efficient operations, thereby enhancing organizational agility.

Together, resilience and agility are critical for organizations to thrive in the increasingly volatile and complex environment of Industry 4.0. While resilience ensures an organization's survival and continuous operation during disruptions, agility allows it to seize opportunities and maintain competitiveness in dynamic markets proactively. These concepts are interrelated, as a resilient organization is better equipped to maintain or regain its agility in the face of challenges, and an agile organization can more effectively build and enhance its resilience over time.

K. Anjaria, *Sustainable Enterprise Resource Planning (S-ERP) for Industry 4.0*,
Contributions to Management Science, https://doi.org/10.1007/978-981-96-7734-4_10

As we delve into the world of Industry 4.0, we are confronted with a new set of challenges that organizations must navigate. These challenges, viewed through the lens of volatility, uncertainty, complexity, and ambiguity (VUCA), underscore the need for organizational resilience and agility.

1. **Volatility:** The rapid pace of technological advancements, such as the proliferation of IoT, artificial intelligence (AI), and automation, creates a volatile environment where organizations must continuously adapt to new tools and processes. This volatility can disrupt existing business models and require frequent reconfiguration of organizational resources [9].
2. **Uncertainty:** Industry 4.0's reliance on emerging technologies and data-driven decision-making increases uncertainty, as organizations may struggle to predict market trends, technological developments, and competitive dynamics. This uncertainty can complicate strategic planning and increase the risk of misalignment between technological investments and business goals [8].
3. **Complexity:** The integration of multiple digital technologies, such as cyber-physical systems, cloud computing, and big data analytics, adds layers of complexity to organizational operations. Managing these interconnected systems requires specialized skills and sophisticated coordination, which can overwhelm organizations and lead to inefficiencies if not properly managed.
4. **Ambiguity:** The transformative impact of Industry 4.0 technologies creates ambiguity in organizational roles, processes, and outcomes. As traditional business practices are redefined by digital transformation, organizations may face ambiguity in understanding the implications of these changes, leading to potential resistance and misalignment within the workforce.

Despite these challenges, Industry 4.0 also presents significant opportunities for organizations:

1. **Enhanced Agility:** The adoption of Industry 4.0 technologies enables organizations to become more agile, allowing them to rapidly respond to changes in the market and technological environment. This agility is crucial for maintaining competitiveness in a VUCA world.
2. **Improved Decision-Making:** The availability of real-time data through advanced analytics and IoT devices enhances decision-making capabilities. Organizations can leverage this data to gain deeper insights into market trends, operational efficiencies, and customer preferences, thereby reducing uncertainty and making more informed strategic decisions [7].
3. **Innovation and Competitive Advantage:** Industry 4.0 fosters innovation by providing new tools and platforms for developing products, services, and business models. Organizations harnessing these technologies effectively can gain a significant competitive advantage, differentiating themselves in increasingly complex and ambiguous markets.
4. **Resilience Building:** The integration of advanced technologies enhances organizational resilience by enabling more effective risk management and recovery strategies. Organizations can better anticipate and respond to disruptions, ensuring continuity and long-term sustainability in a volatile environment [2].

Thus, the above points state that while Industry 4.0 poses significant challenges related to VUCA, it also offers opportunities for organizations to enhance their agility, decision-making, innovation, and resilience, thereby positioning themselves to thrive in a rapidly evolving business landscape [6].

S-ERP systems are pivotal in helping organizations navigate the challenges associated with Industry 4.0, particularly in addressing volatility, uncertainty, complexity, and ambiguity (VUCA). By integrating sustainability into traditional ERP systems, S-ERP provides a comprehensive framework that enhances operational efficiency and ensures long-term organizational resilience and agility. S-ERP systems enable organizations to manage the volatility inherent in Industry 4.0 by providing real-time data analytics and adaptive resource management capabilities [1]. The integration of IoT and big data analytics within S-ERP allows organizations to continuously monitor and optimize their processes, thus reducing the impact of external and internal fluctuations on operations. This adaptability is critical in volatile environments, where rapid changes in technology and market conditions require swift and informed decision-making.

S-ERP systems contribute to reducing uncertainty by offering predictive analytics and scenario planning tools that help organizations forecast future trends and potential disruptions. By leveraging these tools, organizations can make more informed strategic decisions, aligning their resources and processes with anticipated changes in the market or regulatory environment. This proactive approach to uncertainty helps organizations maintain stability and focus amidst the unpredictability of Industry 4.0.

The complexity of Industry 4.0, characterized by the integration of diverse digital technologies, is effectively managed through S-ERP systems. These systems provide a unified platform that consolidates data from various sources, enabling seamless communication and coordination across different organizational functions. By streamlining complex processes and automating routine tasks, S-ERP systems reduce the cognitive load on employees and improve overall organizational efficiency.

S-ERP systems help organizations navigate ambiguity by offering enhanced visibility and transparency into operations and supply chains. The integration of sustainability metrics within S-ERP systems also clarifies the long-term impact of organizational decisions, aligning them with broader environmental and social goals. This clarity is essential in ambiguous situations where the outcomes of decisions are uncertain, but alignment with sustainability principles provides a guiding framework.

By incorporating sustainability into the core of ERP systems, S-ERP enhances both resilience and agility. Resilience is bolstered through robust risk management features, which allow organizations to anticipate and mitigate potential disruptions agility, on the other hand, is achieved through the system's ability to quickly adapt to changing conditions, whether by reallocating resources, adjusting production schedules, or responding to new market demands—this dual capability of resilience and agility positions organizations to thrive in the dynamic environment of Industry 4.0. S-ERP systems serve as a critical enabler for organizations to effectively manage the challenges posed by Industry 4.0. By integrating sustainability into the ERP framework, these systems enhance the organization's ability to respond to VUCA conditions, thereby achieving greater resilience and agility in a rapidly evolving business landscape.

After the Introduction, Sect. 10.2 focuses on the Impact of S-ERP and Industry 4.0 on Organizational Resilience and Agility. This section delves into how the integration of S-ERP systems with Industry 4.0 technologies significantly influences an organization's ability to remain resilient in the face of disruptions and agile in response to market changes. The discussion highlights the transformative role of digital technologies, such as IoT, AI, and big data analytics, in enhancing both the operational stability and flexibility of organizations. This section underscores how S-ERP systems enable organizations to anticipate, prepare for, and respond effectively to various challenges, thereby reinforcing their long-term resilience and agility. In Sect. 10.3, the chapter presents Strategies for Enhancing Resilience and Agility with S-ERP. This comprehensive section is subdivided into three key areas. Section 10.3.1 introduces the Adoption of Resilient-Agility Matrix, a strategic tool designed to assess and improve an organization's resilience and agility. This matrix provides a framework for evaluating current capabilities and identifying areas for enhancement. Section 10.3.2 discusses the Multi-Criteria Decision Analysis (MCDA) Framework, offering insights into how organizations can apply this analytical approach to make informed decisions that balance various criteria related to resilience and agility. This part of the chapter emphasizes the importance of prioritizing initiatives that enhance these capabilities while aligning with overall strategic goals. Section 10.3.3 explores Human Resources Aspects Related to Agility and Resilience, focusing on the crucial role of HR in fostering a resilient and agile workforce. This discussion includes strategies for employee training, development, and cultural shifts that support the organization's broader objectives of resilience and agility.

Section 10.4, titled Integration of Agility and Resilience, brings the discussion full circle by examining how the concepts of agility and resilience can be seamlessly integrated into broader organizational strategies. This section explores the alignment of these capabilities with digital transformation initiatives, sustainability goals, and risk management frameworks. The section argues that by embedding agility and resilience into the core strategic vision, organizations can ensure that they are not only prepared to face current challenges but are also positioned for sustainable growth and competitive advantage in the evolving landscape of Industry 4.0.

This structure allows the chapter to provide a holistic view of how S-ERP systems can be leveraged to enhance organizational resilience and agility, offering both theoretical insights and practical strategies that organizations can implement to achieve long-term success. The following section discusses the impact of Industry 4.0 on organizational resilience and agility.

10.2 Impact of S-ERP and Industry 4.0 on Organizational Resilience and Agility

S-ERP and Industry 4.0 technologies, including the Internet of Things (IoT), big data analytics, artificial intelligence (AI), and cloud computing, fundamentally transform the business landscape by enhancing organizational resilience and agility. These technologies enable organizations to respond swiftly to disruptions and adapt to changing market conditions, thus improving their ability to thrive in a volatile, uncertain, complex, and ambiguous (VUCA) environment.

The IoT connects physical devices across a network, enabling them to communicate and exchange data in real time. This connectivity allows businesses to monitor and manage operations with unprecedented precision, leading to enhanced situational awareness and faster response times to disruptions. The equation for IoT data flow can be represented as:

$$\text{Data Flow (DF)} = \sum_{i=1}^{n} S_i \times T_i \times R_i$$

where:

- S_i represents the sensors deployed,
- T_i is the time interval for data collection,
- R_i is the reliability of the data transmission.

This real-time data collection and processing enhances resilience by enabling predictive maintenance, reducing downtime, and allowing for immediate corrective actions. It also boosts agility by providing the flexibility to adjust production schedules and supply chain logistics dynamically.

Big data analytics involves processing vast amounts of structured and unstructured data to uncover patterns, correlations, and insights that drive decision-making. This capability is critical for both resilience and agility as it enables businesses to anticipate market trends, customer behavior, and potential disruptions. The formal representation of big data analytics can be given by the following equation:

Insight (I) $= f($ Volume (V), Velocity (V), Variety (V), Veracity (V) $)$
where:

- Volume (V) represents the volume of data,
- Velocity (V) is the speed of data processing,
- Variety (V) indicates the diversity of data types,
- Veracity (V) reflects the accuracy and trustworthiness of the data.

Big data analytics supports agility by enabling rapid and informed decision-making. It also enhances resilience by providing predictive insights that help organizations prepare for potential risks and opportunities.

AI enhances organizational capabilities by automating decision-making processes, optimizing operations, and predicting outcomes with high accuracy. AI models can process and analyze data far faster and more accurately than human capabilities, leading to more resilient and agile business operations. A simple AI model can be represented as:

$$\text{Output (0)} = W \cdot X + B$$

where:

- W represents the weight matrix (learned parameters),
- X represents the input data,
- B is the bias term.

AI-driven systems can identify patterns in data that are not immediately obvious, enabling organizations to anticipate changes and respond proactively. This capability is crucial for maintaining agility in dynamic markets and enhancing resilience against unforeseen disruptions. Cloud computing provides scalable, on-demand computing resources that enhance both resilience and agility. By leveraging cloud platforms, organizations can scale their IT infrastructure according to demand, ensuring continuous operations even during peak loads or disruptions. The elasticity of cloud computing can be expressed as:

$$\text{Elasticity } (E) = \frac{\text{Max Capacity } (C) - \text{ Min Capacity (C)}}{\text{Min Capacity } (C)}$$

This elasticity allows organizations to quickly adapt to changing conditions, whether that involves ramping up resources for a new project or scaling down during a downturn. Cloud computing also enhances resilience by enabling disaster recovery solutions and ensuring data redundancy and security.

The technologies underpinning S-ERP and Industry 4.0, such as loT, big data analytics, Al, and cloud computing, have the dual potential to both disrupt and empower organizations. This duality necessitates a heightened focus on resilience and agility, as organizations must be prepared to navigate both the opportunities and challenges these technologies bring. The integration of IoT, Al, and big data analytics into S-ERP systems introduces significant complexity to organizational operations. These technologies require substantial changes to existing processes, IT infrastructure, and even organizational culture. The complexity can be mathematically represented as:

Complexity $(C) = f($ Number of Technologies (N),Interdependencies (I), Scalability (S), User Adaptation $(U))$
where:

- N is the number of new technologies being integrated,
- I is the degree of interdependence between systems,
- S represents scalability requirements,
- U denotes the ease with which users adapt to the new system.

This complexity can disrupt traditional business models, requiring organizations to overhaul their operations, retrain staff, and adapt to new technological paradigms. The interconnected nature of Industry 4.0 technologies, particularly IoT and cloud computing, increases vulnerability to cyber threats. These vulnerabilities can disrupt operations and compromise data security, leading to significant operational and financial risks. This heightened exposure necessitates stronger cybersecurity measures and resilience strategies. The rapid pace of technological change associated with Industry 4.0 creates uncertainty and ambiguity within organizations. The speed at which new technologies emerge and evolve can lead to difficulties in strategic planning and decision-making as organizations struggle to predict future trends and align their resources accordingly.

Despite the challenges, these technologies empower organizations by providing enhanced decision-making capabilities. Big data analytics and AI enable real-time insights and predictive analytics, allowing organizations to make data-driven decisions that can improve efficiency, customer satisfaction, and market responsiveness. The decision-making process can be modeled as:

$$\text{Decision Quality (DQ)} \\ = f(\text{ Data Volume (V), Data Velocity (V), Analytical Capability (A), Insight Accurracy (I) })$$

where:

- V represents the volume of data,
- V is the speed at which data is processed,
- A denotes the organization's analytical capability,
- I reflect on the accuracy of the insights derived from data.

These capabilities significantly enhance organizational agility, allowing for rapid adjustments to changing conditions. S-ERP systems integrated with IoT and AI can optimize resource management, reduce waste, and improve supply chain efficiency. Cloud computing allows organizations to scale their operations dynamically, ensuring that resources are allocated efficiently and sustainably. This operational efficiency strengthens organizational resilience by reducing costs and enhancing the ability to absorb shocks. Integrating sustainability metrics into S-ERP systems enables organizations to align their operations with broader environmental and social goals. This not only enhances their public image but also drives innovation by encouraging the development of sustainable products and practices. This alignment with sustainability principles fosters resilience by preparing organizations to meet regulatory requirements and consumer expectations.

S-ERP and Industry 4.0 technologies offer opportunities and challenges that can significantly impact organizational resilience and agility. Below are examples highlighting how these technologies can enhance or hinder these critical capabilities.

Enhancing Organizational Resilience and Agility
1. Real-Time Data and Predictive Analytics (Big Data and AI):

- **Example:** In the manufacturing sector, predictive maintenance enabled by loT sensors and AI can prevent equipment failures by analyzing data patterns and predicting when a machine is likely to fail. This capability enhances resilience by reducing downtime and minimizing disruptions to production. Moreover, it increases agility by allowing organizations to schedule maintenance proactively, ensuring that operations continue smoothly without unexpected interruptions.
- **Impact:** This approach not only saves costs associated with unplanned maintenance but also optimizes the production schedule, allowing for quicker responses to market demands and operational adjustments.

2. Supply Chain Optimization (IoT and Cloud Computing):

- **Example:** The integration of IoT and cloud computing in S-ERP systems allows for real-time tracking of goods throughout the supply chain. Based on real-time data, companies like Amazon use these technologies to monitor inventory levels, predict stock shortages, and reroute shipments dynamically [10]. This ability to adapt quickly to changes in supply and demand enhances organizational agility.
- **Impact:** By providing greater visibility into the supply chain, organizations can respond more rapidly to disruptions, such as delays or changes in customer demand, thus maintaining operational continuity and customer satisfaction.

3. Sustainability and Resource Management (S-ERP Systems):

- **Example:** Organizations like Unilever have attempted to integrate sustainability metrics into their ERP systems to track and reduce their environmental impact [4]. S-ERP systems can optimize resource usage, reduce waste, and ensure compliance with environmental regulations, thus enhancing resilience and agility by aligning operations with long-term sustainability goals.
- **Impact:** This proactive approach not only prepares organizations to meet evolving regulatory standards but also fosters innovation in sustainable practices, thereby ensuring long-term viability and flexibility in responding to new challenges.

Hindering Organizational Resilience and Agility
1. Integration Complexity (IoT and AI)

- **Example:** The implementation of loT and AI in legacy ERP systems can introduce significant complexity. For instance, organizations may face challenges in integrating these technologies due to incompatibility with existing systems, lack of expertise, or the need for extensive customization. A notable case is the implementation issues large enterprises like General Electric faced when attempting to integrate advanced analytics into their existing infrastructure, which resulted in delays and cost overruns [12].

- **Impact:** Such integration challenges can hinder agility by delaying the adoption of new technologies and increasing operational costs. The complexity can also compromise resilience if the organization is unable to adapt quickly to the new system or if disruptions occur during the integration process.

2. Cybersecurity Vulnerabilities (IoT and Cloud Computing):

- **Example:** The interconnectedness of devices in an IoT-enabled environment increases the attack surface for cyber threats. For example, the target data breach in 2013, which was facilitated through an insecure third-party vendor, highlights how vulnerabilities in one part of the network can compromise the entire system [14].
- **Impact:** These vulnerabilities can undermine organizational resilience by exposing critical data and systems to potential attacks. The need for constant vigilance and robust cybersecurity measures can also limit agility, as organizations may become more risk-averse in adopting new technologies.

S-ERP and Industry 4.0 technologies can significantly enhance organizational resilience and agility by enabling real-time data analysis, optimizing supply chains, and promoting sustainability. However, they can also pose challenges, such as integration complexity, cybersecurity risks, and data overload, which can hinder these capabilities. Organizations must carefully manage the adoption and integration of these technologies to fully realize their potential while mitigating the associated risks. For all these reasons, we have discussed roadmapping, change management, and security extensively in the book's previous chapter. The following section discusses strategies for enhancing resilience and agility with S-ERP.

10.3 Strategies for Enhancing Resilience and Agility with S-ERP

Organizations can adopt a structured approach that integrates formal methods, logical frameworks, and practical matrices to effectively leverage S-ERP systems to enhance organizational resilience and agility. While agility is often associated with quick, short-term responses to changes, it can also be strategically embedded into an organization's long-term vision and operations. When treated as a long-term strategy, agility involves the continuous development of organizational capabilities that enable sustained responsiveness, adaptability, and resilience in the face of evolving market dynamics, technological advancements, and competitive pressures. Below are key strategies and tools that can guide organizations in this endeavor.

Table 10.1 Resilience-agility matrix

Dimension	Low Resilience, Low Agility	High Resilience, Low Agility	Low Resilience, High Agility	High Resilience, High Agility
Processes	Reactive, siloed	Robust, but inflexible	Flexible, but unprepared	Adaptive, integrated
Technology	Legacy systems, outdated	Stable, but slow to innovate	Innovative, but uncoordinated	Cutting-edge, synchronized
Culture	Risk-averse, hierarchical	Risk-aware, but conservative	Adaptive, but fragmented	Agile, resilient culture
Decision-Making	Slow, centralized,	Thorough, but delayed	Rapid, but inconsistent	Rapid, data-driven

10.3.1 Adoption of Resilient-Agility Matrix

Organizations can utilize a Resilience-Agility Matrix to assess and enhance their capabilities in both areas. This matrix allows organizations to evaluate their current state and identify areas for improvement. The matrix could be structured as follows:

The Resilience-Agility Matrix shown in Table 10.1 is a strategic tool designed to help organizations evaluate their current capabilities in terms of resilience and agility and identify areas for improvement. This matrix provides a structured approach to understanding how an organization can navigate the complexities of modern business environments, particularly in the context of Industry 4.0, where technological advancements and market dynamics demand a high degree of both resilience and agility.

Matrix Structure and Dimensions

The matrix is divided into four quadrants, each representing a different combination of resilience and agility levels:

1. **Low Resilience, Low Agility:** Organizations in this quadrant struggle to cope with disruptions and are slow to respond to changes. Their processes are typically reactive and siloed, meaning that each department or function operates independently with little coordination, which hinders the organization's ability to adapt. Technology is often outdated, consisting of legacy systems that are difficult to update or integrate with newer solutions. The organizational culture in this quadrant is risk-averse and hierarchical, leading to slow, centralized decision-making that further impedes the ability to react promptly to changes.
2. **High Resilience, Low Agility:** Organizations here are robust and capable of withstanding disruptions, but they tend to be inflexible and slow to innovate. Their processes are well-established and reliable, but they lack the adaptability needed to respond quickly to new opportunities or threats. The technology used is stable and secure but may lag in adopting the latest innovations. The culture is risk-aware and conservative, prioritizing stability over change. Decision-making

processes are thorough and deliberate, which supports resilience but can delay the organization's ability to pivot in response to market shifts.

3. **Low Resilience, High Agility:** Organizations in this quadrant are quick to respond to changes and can innovate rapidly, but they lack the robustness needed to survive significant disruptions. Their processes are flexible but may be unprepared for unexpected challenges, leading to potential vulnerabilities. The technology employed is innovative and cutting-edge, but it might be uncoordinated across the organization, leading to inefficiencies or gaps in coverage. The culture is adaptive, with a focus on speed and innovation, but this can sometimes result in fragmentation or inconsistency in how different parts of the organization operate. Decision-making is rapid but may be inconsistent, potentially leading to short-term gains at the expense of long-term stability.
4. **High Resilience, High Agility:** The ideal state for organizations, this quadrant represents a balance where processes are both adaptive and integrated, allowing the organization to both innovate and withstand disruptions effectively. Technology is cutting-edge and synchronized across the organization, enabling seamless integration and coordination. The culture is both agile and resilient, encouraging innovation while maintaining a strong foundation that can absorb shocks. Decision-making in this quadrant is rapid and data-driven, ensuring that the organization can respond quickly and effectively to changes while maintaining strategic alignment.

Guidance for Moving Toward High Resilience, High Agility Organizations should strive to move toward the "High Resilience, High Agility" quadrant. This requires:

- **Investing in Adaptive Processes:** Organizations need to develop processes that are both robust and flexible. This might involve adopting agile methodologies, continuous improvement practices (such as Lean or Six Sigma), and ensuring that processes can quickly adapt to new conditions without sacrificing stability.
- **Integrating Cutting-Edge Technologies:** Embracing the latest technologies, such as IoT, AI, and cloud computing, is crucial. However, it's not just about adopting these technologies but also ensuring they are fully integrated and aligned with the organization's strategic goals. This might involve updating legacy systems, ensuring interoperability, and fostering a culture of continuous technological learning.
- **Fostering a Culture of Resilience and Agility:** Cultural change is often the most challenging but also the most critical. Organizations need to cultivate a culture that values both stability and innovation. This might involve leadership development programs, employee training, and creating an environment that encourages experimentation while also learning from failures.
- **Adopting Data-Driven, Rapid Decision-Making Practices:** Finally, organizations must enhance their decision-making capabilities by leveraging real-time data and analytics. This involves implementing data-driven decision-making frameworks, ensuring that decision-making processes are decentralized yet coordinated, and empowering teams to make rapid, informed decisions that align with the organization's overall strategy.

The Resilience-Agility Matrix provides a valuable framework for organizations to assess their current capabilities and strategically plan for improvement. By focusing on the four key dimensions—processes, technology, culture, and decision-making—organizations can systematically enhance their resilience and agility, positioning themselves to thrive in the dynamic and complex environment of Industry 4.0. The agility can be measured in the organization with parameters like speed of decision-making, time to market, responsiveness to market changes, flexibility in resource allocation, adaptability of processes, employee feedback and engagement, customer satisfaction, innovation rate and project completion time. The resilience can be tracked with the help of parameters like system uptime, recovery time objective (RTO), mean time to recovery (MTTR), incident response time, compliance with regulatory standards, supply chain continuity, employee retention and engagement, financial reserves and liquidity, customer satisfaction and loyalty, risk management maturity, and business continuity plan (BCP) testing frequency and effectiveness.

Along with the Resilience-Agility Matrix, organizations can also implement **a Plan-Do-Check-Act (PDCA) cycle** [13] within their S-ERP system implementation to ensure continuous improvement in resilience and agility:

- **Plan:** Identify key areas where resilience and agility need to be enhanced. Set measurable objectives and establish baseline metrics using data analytics tools within the S-ERP system.
- **Do:** Implement changes such as upgrading technology, streamlining processes, or training employees. Use AI-driven modules in the S-ERP system to simulate outcomes and optimize the implementation.
- Check: Monitor and analyze the outcomes against the set objectives using big data analytics. For instance, track the reduction in downtime, the speed of decision-making, or the adaptability of the supply chain.
- **Act:** Based on the analysis, standardize successful practices across the organization and make necessary adjustments where objectives are not met. Continuously refine the process to adapt to new challenges or opportunities.

The following subsection discusses the Multi-Criteria Decision Analysis (MCDA) framework.

10.3.2 The Multi-Criteria Decision Analysis (MCDA) Framework

An organization in the manufacturing sector is aiming to integrate sustainability into its S-ERP system, with a specific focus on enhancing organizational resilience and agility in the context of Industry 4.0. The goal is to identify and prioritize sustainability initiatives that not only contribute to environmental and social goals but also strengthen the organization's ability to respond quickly to disruptions and adapt to changing market conditions. The organization has identified three potential sustainability initiatives:

Table 10.2 Evaluation of initiatives based on various criteria

Initiative	Environmental impact (C1)	Cost (C2)	Implementation time (C3)	CSR alignment (C4)	Impact on resilience (C5)	Impact on agility (C6)
A	9	4	5	8	7	6
B	7	8	6	9	9	7
C	8	7	7	7	8	8

1. **Initiative A: Green Energy Integration**—Transitioning to renewable energy sources for manufacturing plants.
2. **Initiative B: Waste Reduction Program**—Implementing a comprehensive waste management and reduction strategy.
3. **Initiative C: Sustainable Supply Chain**—Enhancing supply chain sustainability through responsible sourcing and supplier partnerships.

The organization needs to decide which initiative to prioritize, using a Multi-Criteria Decision Analysis (MCDA) framework that evaluates each option based on its impact on sustainability, resilience, and agility.

Criteria for Evaluation Shown in Table 10.2:

1. **Environmental Impact (C1):** The potential reduction in carbon footprint and other environmental benefits. (Scale: 1-10, where 10 is the highest positive impact)
2. **Cost (C2):** The financial investment required for implementation. (Scale: 1-10, where 1 represents the highest cost and 10 represents the lowest cost)
3. **Implementation Time (C3):** The time required to fully implement the initiative. (Scale: 1-10, where 1 represents the longest time and 10 represents the shortest time)
4. **CSR Alignment (C4):** The degree to which the initiative aligns with the organization's corporate social responsibility (CSR) objectives. (Scale: 1-10, where 10 is the highest alignment)
5. **Impact on Resilience (C5):** The extent to which the initiative enhances the organization's ability to withstand disruptions and maintain operations. (Scale: 1-10, where 10 is the highest positive impact)
6. **Impact on Agility (C6):** The extent to which the initiative improves the organization's ability to adapt quickly to changes and seize new opportunities. (Scale: 1-10, where 10 is the highest positive impact)

The indicative data is as follows.

Criteria Weights: The organization assigns weights to each criterion based on their strategic importance, with a specific emphasis on resilience and agility:

- Environmental Impact (C1): 0.2
- Cost (C2): 0.1
- Implementation Time (C3): 0.1

- CSR Alignment (C4): 0.1
- Impact on Resilience (C5): 0.3
- Impact on Agility (C6): 0.2

The Weighted Sum Model (WSM) is applied to evaluate each initiative based on the weighted criteria. For each initiative i, the score is calculated as:

$$S_i = \sum_{j=1}^{6} w_j \times x_{ij}$$

where:

- w_j is the weight of criterion j,
- x_{ij} is the score of initiative i on criterion j.

1. Initiative A :

$$S_A = (0.2 \times 9) + (0.1 \times 4) + (0.1 \times 5) + (0.1 \times 8) + (0.3 \times 7) + (0.2 \times 6)$$
$$= 1.8 + 0.4 + 0.5 + 0.8 + 2.1 + 1.2 = 6.8$$

2. Initiative B :

$$S_B = (0.2 \times 7) + (0.1 \times 8) + (0.1 \times 6) + (0.1 \times 9) + (0.3 \times 9) + (0.2 \times 7)$$
$$= 1.4 + 0.8 + 0.6 + 0.9 + 2.7 + 1.4 = 7.8$$

3. Initiative C:

$$SC = (0.2 \times 8) + (0.1 \times 7) + (0.1 \times 7) + (0.1 \times 7) + (0.3 \times 8) + (0.2 \times 8)$$
$$= 1.6 + 0.7 + 0.7 + 0.7 + 2.4 + 1.6 = 7.7$$

Decision: Based on the scores calculated using the Weighted Sum Model:

- **Initiative B (Waste Reduction Program)** has the highest score of 7.8, making it the most favorable option.
- **Initiative C (Sustainable Supply Chain)** follows closely with a score of 7.7.
- **Initiative A (Green Energy Integration)** scores 6.8, making it the least favorable option among the three.

Resilience and Agility Considerations:

- **Initiative B** scores highly on resilience (9) and has a good impact on agility (7), making it an excellent choice for enhancing the organization's ability to withstand and quickly recover from disruptions.

- **Initiative C** also performs well, particularly in agility (8) and resilience (8), suggesting that it would enhance the organization's ability to adapt and thrive in a dynamic environment.

The MCDM approach, utilizing the Weighted Sum Model, provides a comprehensive framework for integrating sustainability into S-ERP systems while simultaneously enhancing organizational resilience and agility. Initiative B (Waste Reduction Program) emerges as the top choice, offering the best balance between sustainability, resilience, and agility. However, **Initiative C** (Sustainable Supply Chain) is also a strong contender, particularly for organizations prioritizing agility. By carefully considering these factors, organizations can make informed decisions that align with their strategic goals in Industry 4.0, ensuring long-term sustainability, resilience, and adaptability in a rapidly changing environment.

Further, with the help of the MCDA framework, organizations may develop a Resilience-Agility Dashboard within their S-ERP systems to continuously monitor key performance indicators (KPIs) related to both resilience and agility:

- **KPIs for Resilience:** System uptime, recovery time after disruptions, compliance with regulatory standards, etc.
- **KPIs for Agility:** Speed of response to market changes, time to market for new products, flexibility in resource allocation, etc.

The dashboard aggregates data from various S-ERP modules and presents it in real time, allowing leadership to make informed decisions quickly. For instance, if the dashboard indicates a decrease in supply chain resilience due to a disruption, decision-makers can rapidly allocate resources or adjust strategies to restore balance. The following subsection discusses human resources aspects related to agility and resilience of the organization for S-ERP in Industry 4.0.

Leveraging the predictive analytics capabilities within S-ERP systems represents a strategic approach to enhancing an organization's ability to anticipate and respond to disruptions, particularly in the context of Industry 4.0. Predictive analytics, a core component of advanced S-ERP systems, utilizes historical data, machine learning algorithms, and statistical techniques to forecast potential disruptions before they occur. This foresight is crucial for maintaining operational continuity, optimizing resource allocation, and enhancing both organizational resilience and agility.

S-ERP systems equipped with predictive analytics can analyze vast amounts of data from various sources, including IoT devices, supply chain networks, and market trends. By identifying patterns and anomalies within this data, predictive analytics can forecast disruptions such as equipment failures, supply chain delays, or fluctuations in demand. For instance, an S-ERP system might analyze temperature and vibration data from manufacturing equipment to predict potential machinery breakdowns. This proactive identification allows organizations to schedule maintenance before a failure occurs, thereby avoiding costly downtime and production delays.

Beyond anticipation, predictive analytics empowers organizations to respond more effectively to disruptions when they do occur. By simulating various response scenarios, S-ERP systems can recommend optimal strategies for mitigating the impact of disruptions. For example, in the event of a predicted supply chain disruption, the system can suggest alternative suppliers or logistical routes that minimize delays and maintain production schedules. This capability not only enhances the organization's agility by enabling rapid adjustments but also strengthens resilience by reducing the operational and financial impact of disruptions.

The integration of predictive analytics into S-ERP systems significantly enhances decision-making processes by providing data-driven insights that are both timely and actionable. Decision-makers can rely on predictive models to assess the likelihood and potential impact of various disruption scenarios, allowing them to allocate resources more effectively and prioritize actions that align with the organization's strategic objectives. This data-driven approach ensures that responses to disruptions are not only rapid but also aligned with long-term sustainability goals.

In the broader context of Industry 4.0, leveraging S-ERP's predictive analytics capabilities is not merely a tactical advantage but a strategic necessity. The increasing complexity and interconnectivity of modern supply chains, coupled with the accelerated pace of technological change, make traditional reactive approaches to disruption management insufficient. Organizations that can anticipate and swiftly respond to disruptions are better positioned to maintain competitive advantage, ensure customer satisfaction, and achieve sustainable growth. Furthermore, the integration of predictive analytics within S-ERP systems fosters a culture of proactive risk management, where potential threats are systematically identified and addressed before they escalate into significant challenges.

10.3.3 Human Resources Aspects Related to Agility and Resilience

Empowering employees with the necessary information and tools to make informed decisions about resilience and agility is crucial in fostering a responsive and adaptable organizational culture. From an employee perspective, this empowerment is not only a matter of access to data but also involves cultivating a work environment where knowledge, skills, and resources are readily available and continuously enhanced. Such empowerment is integral to enabling employees to contribute effectively to the organization's resilience and agility, particularly in the dynamic context of Industry 4.0.

One of the foundational elements of employee empowerment in decision-making is ensuring access to relevant, accurate, and timely information. Employees must be provided with data that is directly applicable to their roles and responsibilities, enabling them to make decisions that align with the organization's strategic objectives. For instance, access to real-time data through S-ERP systems allows employees to monitor key performance indicators (KPIs) related to both resilience and agility, such as system uptime, supply chain performance, or customer demand fluctuations.

This information enables employees to identify potential disruptions early and take proactive measures to mitigate risks.

To make informed decisions, employees must be equipped with the necessary skills to interpret and utilize the information available to them [11]. This requires ongoing training and skill development programs that focus not only on technical competencies but also on critical thinking and decision-making skills. For example, training in data analytics tools and techniques can empower employees to analyze trends, predict outcomes, and make decisions that enhance both resilience and agility. Additionally, scenario-based training can help employees practice decision-making in simulated environments, improving their ability to respond effectively to real-world challenges.

In addition to information, employees need access to advanced tools that facilitate effective decision-making. This includes access to predictive analytics platforms, decision support systems, and other digital tools embedded within the S-ERP system. For instance, tools that provide predictive insights into potential disruptions allow employees to anticipate issues and develop strategies to address them before they escalate. The availability of such tools empowers employees to act swiftly and decisively, enhancing the organization's overall agility.

Empowerment is also about fostering a culture of autonomy and accountability, where employees feel confident in making decisions and are supported in taking initiative. From an employee perspective, this involves having the authority to make decisions within their scope of work and being trusted by leadership to act in the best interests of the organization. For example, in the event of a supply chain disruption, an empowered employee would have the authority to make adjustments to procurement or logistics plans without waiting for approval from higher management. This autonomy enables faster response times, which is critical for maintaining agility in a rapidly changing environment.

Finally, empowering employees involves establishing mechanisms for feedback and continuous improvement. Employees should be encouraged to share their insights and experiences, contributing to the refinement of processes and tools that support resilience and agility. Regular feedback loops ensure that the tools and information provided are meeting employees' needs and that any gaps in knowledge or resources are addressed promptly. This collaborative approach not only improves decision-making processes but also fosters a sense of ownership and engagement among employees.

Fostering a culture of continuous improvement and innovation focused on resilience and agility is essential in creating an environment where employees are motivated and equipped to contribute proactively to the organization's success [5]. From an employee perspective, this culture represents a commitment to ongoing learning, adaptability, and creative problem-solving, all of which are critical in the dynamic context of Industry 4.0. By embedding these values into the organizational fabric, employees are empowered to enhance their capabilities and contribute to the organization's ability to navigate uncertainty and change.

A culture of continuous improvement and innovation begins with a commitment to lifelong learning. Employees need to be encouraged and supported in their pursuit of new knowledge and skills that enhance their ability to respond to challenges and opportunities. This commitment can be manifested through regular training programs, access to educational resources, and opportunities for professional development. For example, workshops on the latest industry trends or courses in advanced data analytics can equip employees with the knowledge needed to drive innovation in processes and systems that bolster resilience and agility. From an employee perspective, this commitment signals that the organization values their growth and is invested in their long-term success.

A key component of fostering a culture of innovation is encouraging experimentation and calculated risk-taking [3]. Employees should feel empowered to explore new ideas and approaches without fear of failure. This requires creating a safe environment where experimentation is viewed as a vital part of the learning process and where failures are seen as opportunities for growth rather than setbacks. For instance, an employee might propose a novel method for optimizing supply chain processes; even if the initial results are not as expected, the lessons learned from the experiment can lead to further innovations that enhance the organization's agility. Encouraging this mindset helps employees to think creatively and take initiative, driving continuous improvement.

Collaboration is another critical element in fostering a culture of continuous improvement and innovation. When employees are encouraged to work together across departments and disciplines, they bring diverse perspectives and expertise to bear on challenges related to resilience and agility. From an employee perspective, collaborative problem-solving not only enhances the quality of solutions but also builds a sense of shared purpose and collective ownership of the organization's success. Structured opportunities for collaboration, such as cross-functional teams or innovation workshops, can facilitate this process, enabling employees to jointly develop and refine strategies that strengthen organizational resilience.

To sustain a culture of continuous improvement and innovation, it is important to recognize and reward employees for their contributions. This recognition can take various forms, from formal awards and incentives to informal acknowledgments of effort and creativity. For employees, knowing that the organization values their innovative ideas and commitment to improvement motivates them to continue pushing boundaries and seeking better ways to enhance resilience and agility. Moreover, a system of recognition that highlights successful innovations can serve as a model for other employees, demonstrating the tangible impact of creative problem-solving and continuous improvement on the organization's overall performance.

Continuous feedback is essential for maintaining momentum in a culture of improvement and innovation. Employees need regular, constructive feedback on their efforts, as well as opportunities to provide input on processes, tools, and strategies. This feedback loop ensures that improvement initiatives are aligned with the organization's goals and that employees have the support they need to refine their approaches. From an employee perspective, continuous feedback not only aids in personal and professional development but also enhances their ability to contribute meaningfully to the organization's resilience and agility. Feedback mechanisms might include regular performance reviews, peer evaluations, and open forums for discussing new ideas.

Finally, fostering a culture of continuous improvement and innovation requires integrating these values into the daily work of employees. This integration can be achieved by embedding innovation-related goals into performance metrics, encouraging regular reflection on processes, and providing time and resources for creative thinking. For example, employees might be given dedicated time each week to work on innovation projects or to propose improvements to existing processes. From an employee perspective, when innovation and continuous improvement are part of the everyday work environment, these activities become a natural and integral part of their role rather than an additional task or burden.

Fostering a culture of continuous improvement and innovation focused on resilience and agility requires a holistic approach that engages employees at all levels of the organization. From an employee perspective, this culture is characterized by a commitment to learning, encouragement of experimentation, promotion of collaboration, recognition of contributions, continuous feedback, and integrating innovation into daily work. By embedding these principles into the organizational culture, employees are empowered to actively enhance resilience and agility, ensuring that the organization remains adaptive and competitive in the ever-evolving landscape of Industry 4.0. The following section discusses the integration of agility and resilience.

10.4 Integration of Agility and Resilience

Enhancing resilience and agility through Sustainable Enterprise Resource Planning (S-ERP) systems offers a strategic alignment with broader organizational goals, particularly in the realms of digital transformation, sustainability, and risk management. The integration of these concepts within an organization's operational framework not only enhances its ability to adapt and thrive in a dynamic environment but also ensures that the organization remains competitive and sustainable in the long term.

Digital transformation represents a fundamental shift in how organizations operate, leveraging advanced technologies to optimize processes, enhance customer experiences, and drive innovation. S-ERP systems, with their robust data analytics, automation capabilities, and integration with Industry 4.0 technologies such as IoT and AI, play a crucial role in facilitating this transformation. By embedding resilience and agility within the digital transformation strategy, organizations can ensure that their technological advancements are not only innovative but also adaptable to changing conditions.

For instance, the agility embedded within an S-ERP system enables organizations to rapidly reconfigure business processes in response to new digital initiatives or emerging market trends. This flexibility is critical in a digital transformation context, where organizations must continuously adapt to evolving technologies and customer expectations. Moreover, the resilience provided by S-ERP systems ensures that digital infrastructures are robust and capable of withstanding disruptions, thereby protecting the organization's digital assets and maintaining operational continuity.

Sustainability has become a cornerstone of modern organizational strategy, driven by increasing regulatory requirements, consumer expectations, and the global imperative to address environmental challenges. S-ERP systems are uniquely positioned to support sustainability initiatives by integrating sustainability metrics into core business processes, thereby enhancing both resilience and agility.

For example, an organization that incorporates sustainability goals into its S-ERP system can track and optimize resource usage, reduce waste, and ensure compliance with environmental regulations. This not only contributes to the organization's sustainability objectives but also enhances resilience by reducing dependencies on scarce resources and mitigating risks associated with environmental impacts. Furthermore, agility is improved as the organization can quickly adjust its operations to align with new sustainability standards or to capitalize on emerging green market opportunities.

Risk management is a critical organizational function that aims to identify, assess, and mitigate risks that could potentially disrupt operations or harm the organization's reputation. S-ERP systems contribute to a comprehensive risk management strategy by providing real-time visibility into operational data, enabling predictive analytics, and facilitating rapid response mechanisms. Enhancing resilience through S-ERP systems involves embedding risk management processes directly into the organizational workflow, allowing for continuous monitoring and assessment of potential risks. This integration ensures that the organization can quickly identify and respond to risks, thereby minimizing their impact on operations. Additionally, the agility provided by S-ERP systems allows organizations to rapidly reallocate resources and adjust strategies in response to emerging risks, ensuring that they remain resilient in the face of uncertainty.

The integration of resilience and agility with broader organizational strategies such as digital transformation, sustainability, and risk management positions organizations to achieve long-term success. By aligning these elements within the S-ERP framework, organizations can create a synergistic effect where each component reinforces the others. For example, digital transformation efforts that are agile and resilient are more likely to succeed and deliver sustainable competitive advantages. Similarly, sustainability initiatives that are supported by agile and resilient processes are more adaptable to changing environmental regulations and market demands. The integration of agility and resilience within the context of Industry 4.0 presents several paradoxes that organizations must navigate to achieve a balanced and effective strategy. These paradoxes highlight the inherent tensions between agility and resilience and the complexities of managing them in an increasingly dynamic and technologically driven environment. The paradoxes are as follows:

1. The Speed-Stability Paradox:
Agility demands rapid responses and quick adaptations to changing conditions, often requiring organizations to prioritize speed and flexibility. However, resilience is built on stability, robustness, and the ability to absorb shocks, sometimes requiring slower,

more deliberate processes. The paradox lies in the need to be both a fast and stable organization—organizations must find a way to maintain operational stability while still being able to pivot quickly in response to new challenges.

S-ERP systems can enable organizations to achieve speed without sacrificing stability by integrating real-time data analytics and predictive modeling. These capabilities allow for rapid decision-making based on accurate, up-to-date information while maintaining stable operations through continuous monitoring and automated controls. S-ERP systems can provide dashboards that show both current operational status and potential risks, helping organizations make quick adjustments while preserving overall stability.

2. The Flexibility-Rigidity Paradox:
Agility thrives on flexibility, allowing organizations to reconfigure resources, processes, and strategies on the fly. On the other hand, resilience often depends on the establishment of rigid, reliable systems that can withstand disruptions without failing. This paradox highlights the tension between maintaining the flexibility needed for agility and the rigidity that supports resilience, challenging organizations to create structures that can be both adaptable and robust.

S-ERP systems support the flexibility needed for agility through modular architectures and customizable workflows. At the same time, they enforce the necessary rigidity for resilience by standardizing core processes and ensuring compliance with regulatory requirements. Organizations can use S-ERP to create flexible process templates that can be quickly adapted to changing conditions while core functions remain standardized to ensure reliability and consistency.

3. The Innovation-Consistency Paradox
Agility is closely associated with innovation—the ability to generate and implement new ideas quickly. However, resilience requires consistency in processes and outcomes to ensure that the organization can weather disruptions. The paradox here is that constant innovation can introduce variability and unpredictability, which might undermine the consistency that resilience relies on. Organizations must, therefore, balance the pursuit of innovation with the need to maintain consistent, reliable operations.

S-ERP systems can foster innovation without compromising consistency by providing a controlled environment for experimentation. Organizations can use S-ERP's sandbox environments to test new ideas and processes before rolling them out across the enterprise. This allows for innovation to be rigorously evaluated and refined, ensuring that when implemented, it integrates seamlessly with existing operations and maintains the consistency required for resilience.

4. The Efficiency-Redundancy Paradox
Agility often seeks to optimize efficiency, reduce waste, and streamline operations to enable quick responses to change. Resilience, however, may require redundancy, such as having backup systems, additional inventory, or excess capacity to handle unexpected disruptions. This paradox challenges organizations to determine how to optimize efficiency without sacrificing the redundancies necessary for resilience.

S-ERP systems can optimize efficiency while maintaining necessary redundancies through intelligent resource planning and inventory management. For instance, S-ERP systems can use AI and machine learning to predict demand and optimize stock levels, ensuring that redundancies are strategically placed to handle disruptions without incurring unnecessary costs. This balance allows organizations to operate efficiently while retaining the buffers needed to withstand unexpected events.

5. The Autonomy-Control Paradox
Agility benefits from granting autonomy to teams and individuals, allowing them to make decisions quickly and act independently. In contrast, resilience may require centralized control and coordination to ensure a unified response during crises. The paradox lies in the need to empower teams with autonomy to foster agility while maintaining enough control to ensure coordinated and resilient actions across the organization.

S-ERP systems can manage the tension between autonomy and control by providing role-based access and decentralized decision-making capabilities within a unified framework. Employees and teams can be empowered to make decisions relevant to their specific functions, supported by real-time data provided by the S-ERP system. At the same time, the system ensures that all actions align with overall organizational goals and are subject to centralized oversight, ensuring coordinated and resilient responses.

6. The Short-Term-Long-Term Paradox
Agility often focuses on short-term actions—quick wins, rapid adaptations, and immediate responses. Resilience, however, is inherently a long-term concept concerned with sustainability and the ability to endure over time. The paradox is that focusing too much on short-term agility may compromise long-term resilience and vice versa. Organizations must balance the need for immediate responsiveness with the strategic foresight required for long-term resilience.

S-ERP systems facilitate the balancing of short-term agility with long-term resilience by enabling organizations to manage both tactical and strategic initiatives simultaneously. With S-ERP, organizations can track short-term performance metrics while also monitoring long-term trends and sustainability indicators. This dual focus allows organizations to respond quickly to immediate challenges while staying aligned with their long-term strategic objectives, ensuring sustained growth and resilience.

7. The Complexity-Simplicity Paradox
Agility can involve embracing complexity, such as integrating advanced technologies, managing multiple strategies simultaneously, or operating in diverse markets. Resilience, on the other hand, often benefits from simplicity—clear processes, straightforward decision-making, and easily understood contingency plans. The paradox here is that the complexity needed for agility may complicate the simplicity that resilience thrives on, requiring organizations to find ways to manage complexity without sacrificing the straightforwardness that supports resilience.

S-ERP systems can help manage the complexity required for agility by simplifying and automating routine tasks while providing advanced tools for managing complex operations. For example, S-ERP can automate supply chain management, reducing complexity in day-to-day operations while also offering sophisticated analytics and scenario planning tools to handle more complex, strategic decisions. This combination allows organizations to maintain operational simplicity where needed without sacrificing the complexity required for agile responses.

8. The Growth-Protection Paradox
Agility is closely linked to growth—expanding into new markets, developing new products, and scaling operations. Resilience, however, often emphasizes protection, safeguarding existing assets, preserving operational integrity, and ensuring continuity. The paradox is that the aggressive pursuit of growth can expose the organization to risks that undermine resilience, while an overemphasis on protection may stifle growth and innovation. Organizations must strike a balance between advancing and protecting their interests.

S-ERP systems enable organizations to pursue growth while maintaining protection through integrated risk management and scenario planning tools. These tools allow organizations to assess the risks associated with growth initiatives and implement protective measures proactively. S-ERP can also track and analyze the impact of growth strategies in real time, allowing organizations to adjust their approach if necessary to protect their core assets and ensure long-term resilience. In summary, the chapter answers the following research questions.

Research Questions Addressed

The chapter titled "Enhancing Organizational Resilience and Agility in S-ERP for Industry 4.0" addresses the following key research questions:

1. **How can Sustainable Enterprise Resource Planning (S-ERP) systems enhance organizational resilience in Industry 4.0?**
   - The chapter explores how S-ERP systems can help organizations anticipate, prepare for, respond to, and adapt to incremental changes and sudden disruptions, ensuring continuous operation and long-term sustainability.
2. **How do S-ERP systems enhance organizational agility within Industry 4.0?**
   - The chapter examines how S-ERP systems enable organizations to rapidly adjust and respond to changes in internal and external environments, facilitating quick decision-making and operational flexibility in dynamic markets.
3. **How do S-ERP systems help organizations navigate the challenges of volatility, uncertainty, complexity, and ambiguity (VUCA) in Industry 4.0?**
   - The chapter discusses how S-ERP systems address VUCA challenges by providing real-time data analytics, adaptive resource management, and predictive analytics, enhancing resilience and agility.

4. **What strategies can organizations adopt to enhance resilience and agility through S-ERP systems?**
   - The chapter presents various strategies to improve resilience and agility within organizations, including adopting the Resilient-Agility Matrix, Multi-Criteria Decision Analysis (MCDA) framework, and human resource management techniques.
5. **How can S-ERP systems integrate resilience and agility into broader organizational strategies such as digital transformation, sustainability, and risk management?**
   - The chapter explores how integrating resilience and agility into S-ERP systems aligns with and supports digital transformation initiatives, sustainability goals, and comprehensive risk management frameworks, ensuring long-term success in Industry 4.0.
6. **What are the paradoxes associated with integrating agility and resilience in S-ERP systems, and how can organizations navigate these paradoxes?**
   - The chapter identifies and discusses paradoxes such as the Speed-Stability Paradox, Flexibility-Rigidity Paradox, and Innovation-Consistency Paradox, providing insights into how organizations can balance these competing demands to achieve a harmonious integration of agility and resilience.

These research questions guide the chapter's exploration of how S-ERP systems can be leveraged to enhance organizational resilience and agility, ensuring that organizations can thrive in the complex and rapidly changing environment of Industry 4.0.

References

1. Anjaria K (2024) Enhancing sustainability integration in Sustainable Enterprise Resource Planning (S-ERP) system: application of transaction cost theory and case study analysis. Int J Inf Manage Data Insights 4(2):100243. ISSN 2667-0968. https://doi.org/10.1016/j.jjimei.2024.100243. https://www.sciencedirect.com/science/article/pii/S2667096824000326
2. Georgescu I, Bocean CG, Vărzaru AA, Rotea CC, Mangra MG, Mangra GI (2024) Enhancing organizational resilience: the transformative influence of strategic human resource management practices and organizational culture. Sustainability 16(10):4315. ISSN 2071-1050. https://doi.org/10.3390/su16104315. https://www.mdpi.com/2071-1050/16/10/4315. Number: 10 Publisher: Multidisciplinary Digital Publishing Institute
3. Giaccone SC, Magnusson M (2022) Unveiling the role of risk-taking in innovation: antecedents and effects. R&D Management 52(1):93–107. ISSN 1467-9310. https://doi.org/10.1111/radm.12477. https://onlinelibrary.wiley.com/doi/abs/10.1111/radm.12477. _eprint: https://onlinelibrary.wiley.com/doi/pdf/10.1111/radm.12477

4. Hack S, Berg C (2014) The potential of IT for corporate sustainability. Sustainability 6(7):4163–4180. ISSN 2071-1050. https://doi.org/10.3390/su6074163. https://www.mdpi.com/2071-1050/6/7/4163. Number: 7 Publisher: Multidisciplinary Digital Publishing Institute
5. Lepeley M-T, Beutell NJ, Abarca N, Majluf N (2021) Soft skills for human centered management and global sustainability. Routledge. ISBN 978-1-00-036894-9. Google-Books-ID: KcIZEAAAQBAJ
6. Lubowiecki-Vikuk A, Budzanowska-Drzewiecka M, Borzyszkowski J, Taheri B (2023) Critical reflection on VUCA in tourism and hospitality marketing activities. Int J Contemp Hospit Manage 35(8):2983–3005. ISSN 0959-6119. https://doi.org/10.1108/IJCHM-04-2022-0479. Publisher: Emerald Publishing Limited
7. Marinagi C, Reklitis P, Trivellas P, Sakas D (2023) The impact of Industry 4.0 technologies on key performance indicators for a resilient supply chain 4.0. Sustainability 15(6):5185. ISSN 2071-1050. https://doi.org/10.3390/su15065185. https://www.mdpi.com/2071-1050/15/6/5185. Number: 6 Publisher: Multidisciplinary Digital Publishing Institute
8. Miceli A, Hagen B, Riccardi MP, Sotti F, Settembre-Blundo D (2021) Thriving, not just surviving in changing times: how sustainability, agility and digitalization intertwine with organizational resilience. Sustainability 13(4):2052. ISSN 2071-1050. https://doi.org/10.3390/su13042052. Number: 4 Publisher: Multidisciplinary Digital Publishing Institute
9. Mrugalska B, Ahmed J (2021) Organizational agility in Industry 4.0: a systematic literature review. Sustainability 13(15):8272. ISSN 2071-1050. https://doi.org/10.3390/su13158272. https://www.mdpi.com/2071-1050/13/15/8272. Number: 15 Publisher: Multidisciplinary Digital Publishing Institute
10. Mubarik MS, Khan SA (2024) Disruptive digital technologies and contemporary supply chains. In The theory, methods and application of managing digital supply chains. Emerald Publishing Limited, pp 15–39. ISBN 978-1-80455-968-0 978-1-80455-969-7. https://doi.org/10.1108/978-1-80455-968-020241002
11. Muzam J (2023) The challenges of modern economy on the competencies of knowledge workers. J Knowl Econ 14(2):1635–1671. ISSN 1868-7873. https://doi.org/10.1007/s13132-022-00979-y
12. Ocasio W, Joseph J (2008) Rise and fall - or transformation?: the evolution of strategic planning at the general electric company, 1940–2006. Long Range Plann 41(3):248–272. ISSN 0024-6301. https://doi.org/10.1016/j.lrp.2008.02.010. https://www.sciencedirect.com/science/article/pii/S0024630108000265
13. Rangel-Sánchez, M.Á., Urbina-González, J.D.J., Carrera-Escobedo, J.L., Guirette-Barbosa, O.A., Murillo-Rodríguez, V.A., Celaya-Padilla, J.M., Durán-Muñoz, H.A. and Cruz-Domínguez, O (2024) Enhancing scrap reduction in electric motor manufacturing for the automotive industry: a case study using the PDCA (Plan–Do–Check–Act) Approach. Appl Sci 14(7):2999. ISSN 2076-3417. https://doi.org/10.3390/app14072999. https://www.mdpi.com/2076-3417/14/7/2999. Number: 7 Publisher: Multidisciplinary Digital Publishing Institute
14. Shu X, Tian K, Ciambrone A, Yao D (2017) Breaking the target: an analysis of target data breach and lessons learned. arXiv:1701.04940 [cs]

Chapter 11
Conclusion and Future Scope of S-ERP in Industry 4.0

11.1 Conclusion

The central theme of this book focuses on Sustainable Enterprise Resource Planning (S-ERP) for Industry 4.0 and its role in fostering a more sustainable and resilient business landscape. Let's have a recap of the core ideas and arguments presented, highlighting their significance. The book introduces S-ERP as an evolution of traditional Enterprise Resource Planning (ERP) systems, designed to integrate sustainability within business operations. It addresses how conventional ERP systems fall short of promoting sustainable development and argues that S-ERP is essential to meet the economic, environmental, and social challenges faced by modern organizations. By leveraging technologies such as IoT, AI, and big data analytics, S-ERP can drive sustainability and resilience in business models, aligning with the triple bottom line (economic, environmental, and social).

The book explores the fundamental relationship between S-ERP and Industry 4.0, emphasizing how the fourth industrial revolution's technological advancements—automation, AI, IoT—transform the business environment. It explains that Industry 4.0 technologies can enhance productivity and innovation but must also address sustainability challenges. S-ERP provides the foundation for this integration by helping organizations monitor their environmental footprint and operational efficiency, ensuring a more holistic business approach.

Implementing S-ERP systems presents unique challenges, including data accuracy, integration with existing systems, and managing sustainability metrics. The book argues for a holistic, step-by-step approach to S-ERP implementation, from employee training to strategic module deployment, to align business processes with sustainability goals. It also provides a roadmap for overcoming these challenges, emphasizing the need for businesses to adapt quickly to the rapidly changing Industry 4.0 environment.

One of the significant challenges in implementing S-ERP is change management. The book stresses the need for organizational flexibility, proper employee training,

K. Anjaria, *Sustainable Enterprise Resource Planning (S-ERP) for Industry 4.0*, Contributions to Management Science, https://doi.org/10.1007/978-981-96-7734-4_11

and integrating sustainable practices into core business functions. Additionally, open innovation is highlighted as essential for fostering collaboration across departments, partners, and stakeholders. S-ERP facilitates this through collaborative platforms, allowing businesses to innovate continuously and adapt to emerging sustainability trends. Ethics plays a pivotal role in the book's discussion, particularly regarding the use of AI and automation within S-ERP systems. Ethical concerns such as job displacement, data privacy, and bias in AI algorithms are addressed. Additionally, the book underscores the importance of robust cybersecurity measures to protect data in the interconnected systems of Industry 4.0. It presents best practices for ensuring the ethical and secure use of technology in S-ERP, including transparency in decision-making processes and the responsible management of digital information.

The book argues that S-ERP can be a catalyst for the circular economy, promoting sustainability by reducing waste and improving resource management. Industry 4.0 technologies, when integrated with S-ERP, enable organizations to better track their environmental impact and implement more sustainable production cycles. The book advocates for using formal models to ensure that organizations transition toward circular economic practices. Analytics, driven by IoT data, plays a crucial role in improving decision-making within S-ERP. The book explores how real-time data from IoT devices can enhance operational efficiency, reduce costs, and improve customer satisfaction. The integration of 5G and 6G technologies further accelerates the potential for S-ERP systems to handle vast amounts of data, contributing to more intelligent, agile, and sustainable business operations.

The digital transformation journey for organizations is seen as incomplete without integrating S-ERP systems, which not only automate business processes but also help businesses meet their long-term sustainability objectives. The book discusses the need for businesses to identify digital blind spots and make strategic investments in S-ERP to enhance customer-centricity and operational efficiency. Finally, the book emphasizes that S-ERP is crucial for enhancing both organizational resilience and agility in the face of evolving technological and sustainability challenges. It explores strategies such as the Resilient-Agility Matrix and Multi-Criteria Decision Analysis (MCDA) for helping organizations adapt to Industry 4.0 while maintaining a focus on sustainability.

In summary, the book outlines the importance of S-ERP as a tool for achieving sustainability in Industry 4.0, offering practical insights into its implementation, challenges, and role in promoting ethical, secure, and efficient business operations. The book argues that Sustainable Enterprise Resource Planning (S-ERP) is essential for integrating sustainability into the core of business operations in the context of Industry 4.0. It posits that traditional ERP systems are inadequate for addressing the modern challenges of economic, environmental, and social sustainability. By leveraging advanced technologies such as IoT, AI, and big data, S-ERP systems can drive both digital transformation and sustainability, ensuring long-term resilience and competitiveness. The book's purpose is to guide organizations in successfully implementing S-ERP, addressing ethical and security concerns, and navigating the complexities of the fourth industrial revolution, thereby making a significant impact on how businesses operate and contribute to a sustainable future.

As we stand at the cusp of the fourth industrial revolution, the need for a sustainable approach to business has never been more urgent. The journey toward integrating Sustainable Enterprise Resource Planning (S-ERP) within Industry 4.0 is not just about adopting new technologies but about rethinking the very fabric of how businesses operate in a rapidly changing world. This book has laid out a roadmap for navigating this complex landscape, offering practical insights and strategies to overcome the challenges of implementation, ethics, and security.

By embracing S-ERP, organizations can not only enhance their operational efficiency but also contribute meaningfully to global sustainability efforts, ensuring that the benefits of Industry 4.0 are realized without compromising the needs of future generations. The convergence of sustainability and technology represents a powerful force for change, one that has the potential to reshape industries, societies, and our collective future. As we move forward, let the principles and practices outlined in this book serve as a guide to building a resilient, sustainable, and ethically grounded enterprise. The challenges are significant, but so too are the opportunities. In embracing S-ERP, you are not just adopting a system; you are committing to a vision of a more sustainable and equitable world—one that balances the demands of the present with the promise of the future. With this vision in mind, the path ahead is clear, and the possibilities are endless.

The themes and messages of this book, centered on S-ERP within Industry 4.0, resonate far beyond the confines of business and technology. They tap into broader societal issues and historical trends, reflecting the ongoing evolution of our global economy, the pressing need for environmental stewardship, and the ethical considerations that accompany technological advancement. The call for sustainability in business practices aligns with a historical trend that began with the industrial revolution, where rapid economic growth often came at the expense of environmental degradation. The consequences of unchecked industrialization—climate change, resource depletion, and pollution—have led to a global movement toward more responsible and sustainable practices. This book's focus on S-ERP represents a continuation of this trend, offering businesses a framework to integrate sustainability into their core operations, thereby contributing to the global effort to mitigate environmental harm and preserve resources for future generations.

The book's discussion on Industry 4.0 and the ethical implications of advanced technologies echoes the broader societal debate about the role of technology in our lives. As with previous industrial revolutions, the advent of automation, AI, and IoT brings both opportunities and challenges. Historically, technological advancements have driven economic growth but also sparked concerns about job displacement, privacy, and inequality. The book's emphasis on ethical considerations and data security within S-ERP reflects a growing awareness that technology must be harnessed responsibly to ensure that its benefits are widely shared and its risks are managed.

In a world increasingly interconnected by globalization, businesses are no longer isolated entities but part of a complex global network. The concept of the triple bottom line—economic, environmental, and social sustainability—championed in this book, mirrors the broader societal shift toward recognizing the interconnectedness of these three pillars. This approach aligns with global movements such as the United Nations' Sustainable Development Goals (SDGs), which seek to balance economic growth with social equity and environmental protection. By advocating for S-ERP, the book underscores the importance of businesses taking a holistic view of their impact on society and the planet, contributing to a more sustainable and equitable global economy. The focus on organizational resilience and agility in the book reflects a broader societal trend of preparing for and adapting to uncertainty, whether from economic fluctuations, environmental crises, or technological disruptions. The lessons from recent global events, such as the COVID-19 pandemic, have highlighted the need for businesses to be flexible and resilient. S-ERP, as discussed in the book, offers a way for organizations to build this resilience, ensuring they can navigate future challenges while remaining committed to sustainable practices. The book's exploration of ethics within S-ERP systems touches on larger historical discussions about corporate social responsibility (CSR) and the role of businesses in society. As corporations have grown in power and influence, there has been increasing pressure for them to act as responsible global citizens. This book aligns with this ongoing discourse, advocating for businesses to not only pursue profit but also consider their ethical obligations to employees, customers, and the broader society.

11.2 Implications and Future Research Directions

The exploration of Sustainable Enterprise Resource Planning (S-ERP) in Industry 4.0 opens a gateway to a future where the convergence of technology and sustainability could redefine the very nature of business and society. As we look ahead, several potential consequences and developments arise from the themes discussed in the book, each carrying profound implications and fueling a sense of curiosity about what lies beyond the horizon.

As more businesses adopt S-ERP systems, we may witness the emergence of truly smart, sustainable enterprises that operate with unprecedented efficiency and minimal environmental impact. These companies will leverage real-time data from IoT devices, advanced AI algorithms, and robust analytics to optimize their resource use, reduce waste, and create value in ways that were previously unimaginable. This could lead to a new standard in corporate responsibility, where sustainability is not just an add-on but a fundamental aspect of business strategy.

The integration of S-ERP within Industry 4.0 will inevitably transform the workforce. Automation and AI may displace certain jobs, but they will also create new opportunities in fields like sustainability management, data science, and ethical AI. The workforce of the future will need to be more adaptable, skilled in both technology and sustainability, and capable of navigating the complexities of a rapidly changing business environment. This shift could spur a renaissance in education and training, with an emphasis on interdisciplinary skills and lifelong learning.

As the adoption of S-ERP grows, there will likely be increased pressure for global standards and regulations to ensure consistency, fairness, and transparency across industries. Governments and international organizations may develop new frameworks to govern the ethical use of AI, data privacy, and sustainability metrics within these systems. This regulatory landscape will shape the future of S-ERP, influencing how businesses operate and compete on a global scale.

The ethical challenges associated with S-ERP systems—such as bias in AI, data privacy, and the social implications of automation—will continue to evolve. Future developments might include more sophisticated tools for auditing and ensuring the ethical behavior of algorithms, as well as broader societal discussions on the limits of automation and the rights of workers in an increasingly automated world. These debates will likely shape the moral compass of future technological advancements, guiding how businesses balance profit with social responsibility.

The book highlights the potential for S-ERP to facilitate the circular economy, but this is just the beginning. Future developments could see the integration of even more advanced technologies—such as blockchain for transparent supply chains or quantum computing for complex sustainability modeling—into S-ERP systems. These innovations could push the boundaries of what is possible in sustainable business practices, leading to a future where zero waste, fully sustainable enterprises become the norm rather than the exception.

As the world faces increasing uncertainty—whether from climate change, economic instability, or geopolitical tensions—S-ERP systems will play a crucial role in helping organizations build resilience. Future developments may see these systems evolving to predict and mitigate risks more effectively, enabling businesses to not only survive but thrive in the face of global challenges. This resilience could extend beyond individual companies to entire industries, fostering a more stable and sustainable global economy. The widespread adoption of S-ERP could lead to a fundamental shift in corporate culture. Companies that embrace these systems may develop a stronger focus on long-term sustainability over short-term profits, fostering a culture of innovation, ethical responsibility, and environmental stewardship. This shift could influence how businesses are perceived by consumers, investors, and society at large, leading to a new era of corporate accountability and transparency.

As we close the pages on the exploration of Sustainable Enterprise Resource Planning (S-ERP) in Industry 4.0, it's clear that the future of business and our planet depends on the actions we take today. The insights and strategies outlined in this book are not just theoretical concepts—they are a blueprint for real, transformative change. Whether we are business leaders, managers, or employees, we all have the power to drive sustainability within our organizations. We can start by advocating for the integration of S-ERP systems that prioritize the triple bottom line—economic, environmental, and social sustainability. Assessing our current operations and identifying areas where sustainable practices can be implemented or enhanced will allow us to improve efficiency and profitability and contribute to a more sustainable future.

The landscape of Industry 4.0 is rapidly evolving, and staying informed is crucial. Committing to ongoing education and exploration of new technologies, ethical considerations, and sustainability practices will help us stay ahead. We can engage with industry forums, attend conferences, and participate in workshops that focus on S-ERP and sustainable business practices. Through continuous learning, we can become catalysts for innovation and ethical leadership in our fields. As we integrate more advanced technologies into our businesses, the ethical implications become increasingly important. We can take active roles in advocating for the ethical use of AI, data privacy, and transparency in decision-making processes within our organizations. Ensuring that technological advancements align with ethical standards and contribute positively to society will help us build a future where technology serves the greater good.

Open innovation and collaboration are key to navigating the complexities of Industry 4.0. We can seek out partnerships with other organizations, universities, and research institutions to share knowledge, resources, and ideas. Encouraging a culture of innovation within our organizations by creating platforms for employees to contribute their insights and suggestions will drive the development of innovative solutions that address both business challenges and global sustainability goals.

Taking proactive steps to incorporate circular economy principles into our organizations' operations will have a significant impact. We can explore how S-ERP systems can help reduce waste, manage resources more efficiently, and create sustainable products and services. By rethinking supply chains, product design, and lifecycle management through the lens of sustainability, we position our businesses as leaders in sustainable innovation. Finally, we all can lead in our organizations and communities by embodying the principles of sustainability and ethical responsibility. Leading by example will show others that it is possible to achieve business success while prioritizing the well-being of our planet and society. By inspiring our colleagues, peers, and industry partners to take action and make a difference, we contribute to widespread change and a more sustainable and equitable world.

Thank you for taking the time to engage with the ideas and concepts presented in this book. Your attention and thoughtful consideration of these topics are invaluable as we collectively work toward a more sustainable, ethical, and innovative future. By exploring these ideas, you have taken an important step toward understanding the critical role that S-ERP in Industry 4.0 can play in shaping the future of business and society. Your commitment to these principles is what will drive meaningful change in our world. We are grateful for your dedication, and we look forward to the impact that you will undoubtedly make.

> Progress is impossible without change, and those who cannot change their minds cannot change anything. — George Bernard Shaw.

As we stand on the brink of a new era in business and technology, let us reflect on this truth. How will we shape the future with the knowledge and tools at our disposal? The choices we make today will define the world of tomorrow. Let us choose wisely, embrace innovation with responsibility, and commit to a path that leads to sustainability, resilience, and shared prosperity. The journey ahead is ours to navigate—let's make it one that future generations will thank us for.

The manufacturer's authorised representative in the EU is Springer Nature Customer Service Centre GmbH, Europaplatz 3, 69115 Heidelberg, Germany. If you have any concerns regarding our products, please contact ProductSafety@springernature.com

Printed and bound by CPI Group (UK) Ltd, Croydon, CR0 4YY
30/09/2026
02225303-0013